QE
770
U6
pt. V
1970

TREATISE ON INVERTEBRATE PALEONTOLOGY

Prepared under Sponsorship of
The Geological Society of America, Inc.

The Paleontological Society   The Society of Economic Paleontologists and Mineralogists
The Palaeontographical Society   The Palaeontological Association

Second Editions Directed and Edited by
CURT TEICHERT

First Editions Directed and Edited by
RAYMOND C. MOORE

Editorial Assistants, LAVON MCCORMICK, ROGER B. WILLIAMS

Part V

Second Edition (Revised and Enlarged)

GRAPTOLITHINA

with sections on ENTEROPNEUSTA and PTEROBRANCHIA

By O. M. B. BULMAN

THE GEOLOGICAL SOCIETY OF AMERICA, INC.
and
THE UNIVERSITY OF KANSAS
BOULDER, COLORADO, and LAWRENCE, KANSAS

1970

© 1970 by The University of Kansas
and
The Geological Society of America, Inc.

All Rights Reserved

First Edition 1955

Second Edition 1970

Library of Congress Catalogue Card
Number: 53-12913

S.B.N. 8137-3123-2

Text Composed by
THE UNIVERSITY OF KANSAS PRINTING SERVICE
Lawrence, Kansas

Illustrations and Offset Lithography
THE MERIDEN GRAVURE COMPANY
Meriden, Connecticut

Binding
RUSSELL RUTTER COMPANY
New York City

Published 1970

Distributed by The Geological Society of America, Inc., P.O. Box 1719, Boulder, Colo., 80302, to which all communications should be addressed.

The *Treatise on Invertebrate Paleontology* has been made possible by (1) grants of funds from The Geological Society of America through the bequest of Richard Alexander Fullerton Penrose, Jr., for initial preparation of illustrations and partial defrayment of organizational expense in 1948, and the United States National Science Foundation, awarded in 1959, and annually since 1966, for continuation of the *Treatise* project; (2) contribution of the knowledge and labor of specialists throughout the world, working in cooperation under sponsorship of The Geological Society of America, The Paleontological Society, The Society of Economic Paleontologists and Mineralogists, The Palaeontographical Society, and The Palaeontological Association; and (3) acceptance by The University of Kansas of publication without any financial gain to the University.

TREATISE ON INVERTEBRATE PALEONTOLOGY

Directed and Edited by RAYMOND C. MOORE
and (for supplements and revisions) CURT TEICHERT

Assistants: LAVON MCCORMICK, ROGER B. WILLIAMS

Advisers: EDWIN B. ECKEL, D. L. CLARK (The Geological Society of America), BERNHARD KUMMEL, J. WYATT DURHAM (The Paleontological Society), N. D. NEWELL, D. M. RAUP (The Society of Economic Paleontologists and Mineralogists), R. V. MELVILLE, W. T. DEAN (The Palaeontographical Society), M. R. HOUSE, J. M. HANCOCK (The Palaeontological Association).

PARTS

Parts of the *Treatise* are distinguished by assigned letters with a view to indicating their systematic sequence while allowing publication of units in whatever order each may be made ready for the press. The volumes are cloth-bound with title in gold on the cover. Copies are available on orders sent to the Publication Sales Department, The Geological Society of America, P.O. Box 1719, Boulder, Colorado 80302. The prices quoted very incompletely cover costs of producing and distributing the several volumes, but on receipt of payment the Society will ship copies without additional charge to any address in the world. Special discounts are available to members of sponsoring societies under arrangements made by appropriate officers of these societies, to whom inquiries should be addressed.

VOLUMES ALREADY PUBLISHED
(Previous to 1970)

Part C. PROTISTA 2 (Sarcodina, chiefly "Thecamoebians" and Foraminiferida), xxxi+900 p., 5311 fig., 1964.

Part D. PROTISTA 3 (chiefly Radiolaria, Tintinnina), xii+195 p., 1050 fig., 1954.

Part E. ARCHAEOCYATHA, PORIFERA, xviii+122 p., 728 fig., 1955.

Part F. COELENTERATA, xvii+498 p., 2700 fig., 1956.

Part G. BRYOZOA, xii+253 p., 2000 fig., 1953.

Part H. BRACHIOPODA, xxxii +927 p., 5198 fig., 1965.

Part I. MOLLUSCA 1 (Mollusca General Features, Scaphopoda, Amphineura, Monoplacophora, Gastropoda General Features, Archaeogastropoda, mainly Paleozoic Caenogastropoda and Opisthobranchia), xxiii+351 p., 1732 fig., 1960.

Part K. MOLLUSCA 3 (Cephalopoda General Features, Endoceratoidea, Actinoceratoidea, Nautiloidea, Bactritoidea), xxviii+519 p., 2382 fig., 1964.

Part L. MOLLUSCA 4 (Ammonoidea), xxii+490 p., 3800 fig., 1957.

Part N. MOLLUSCA 6 (Bivalvia), Volumes 1 and 2 (of 3), xxxviii+952 p., 6198 fig., 1969.

Part O. ARTHROPODA 1 (Arthropoda General Features, Protarthropoda, Euarthropoda General Features, Trilobitomorpha), xix+560 p., 2880 fig., 1959.

Part P. ARTHROPODA 2 (Chelicerata, Pycnogonida, Palaeoisopus), xvii+181 p., 565 fig., 1955.

Part Q. ARTHROPODA 3 (Crustacea, Ostracoda), xxiii+442 p., 3476 fig., 1961.

Part R. ARTHROPODA 4 (Crustacea exclusive of Ostracoda, Myriapoda, Hexapoda), Volumes 1 and 2 (of 3), xxxvi+651 p., 1762 fig., 1969.

Part S. ECHINODERMATA 1 (Echinodermata General Features, Homalozoa, Crinozoa, exclusive of Crinoidea), xxx+650 p., 2868 fig., 1967 [1968].

Part U. ECHINODERMATA 3 (Asterozoans, Echinozoans), xxx+695 p., 3485 fig., 1966.

Part V. GRAPTOLITHINA, xvii+101 p., 358 fig., 1955.

Part W. MISCELLANEA (Conodonts, Conoidal Shells of Uncertain Affinities, Worms, Trace Fossils, Problematica), xxv+259 p., 1058 fig., 1962.

THIS VOLUME

Part V, Second edition (revised and enlarged). GRAPTOLITHINA, xxxii+163 p., 507 fig., 1970.

VOLUMES IN PREPARATION (1970)

Part A. INTRODUCTION.
Part B. PROTISTA 1 (Chrysomonadida, Coccolithophorida, Charophyta, Diatomacea, etc.).
Part J. MOLLUSCA 2 (Gastropoda, Streptoneura exclusive of Archaeogastropoda, Euthyneura).
Part M. MOLLUSCA 5 (Coleoidea).
Part T. ECHINODERMATA 2 (Crinoidea).
Part X. ADDENDA, INDEX.
Part E (revision). Archaeocyatha, Porifera.
Part F (supplement). Coelenterata (Anthozoa Rugosa and Tabulata).
Part G (revision). Bryozoa.
Part L (revision). Ammonoidea.
Part W (supplement). MISCELLANEA (Trace Fossils).

CONTRIBUTING AUTHORS

[Arranged by countries and institutions; an alphabetical list follows. An asterisk preceding name indicates author working on revision of or supplement to a published *Treatise* volume.]

AUSTRALIA

Macquarie University (North Ryde)
J. A. Talent
South Australia Geological Survey (Adelaide)
N. H. Ludbrook
University of Adelaide
M. F. Glaessner
University of Queensland (Brisbane)
Dorothy Hill

AUSTRIA

Universität Wien (Paläontologisches Institut)
Adolph Papp

BELGIUM

Université de Liège
Georges Ubaghs
Université de Louvain
Marius Lecompte

CANADA

Geological Survey of Canada (Ottawa)
J. A. Jeletzky, A. W. Norris, G. W. Sinclair
Institute of Sedimentary & Petroleum Geology (Geological Survey of Canada, Calgary)
D. J. McLaren
National Museum (Ottawa)
A. H. Clarke, Jr.
University of British Columbia (Vancouver)
V. J. Okulitch

DENMARK

Universitet København
Chr. Poulsen

FRANCE

Unattached
André Chavan (Seyssel, Ain)
Université de Paris
Colette Dechaseaux

GERMANY

Bergakademie Freiberg (Geologisches Institut)
A. H. Müller
Freie Universität Berlin
Gerhard Hahn
Hamburg Staatsinstitut
Walter Häntzschel
Natur-Museum und Forschungs-Institut Senckenberg (Frankfurt)
Herta Schmidt, Wolfgang Struve
Unattached
Hertha Sieverts-Doreck (Stuttgart-Möhringen)
Universität Bonn
H. K. Erben, K. J. Müller
Universität Münster
Helmut Hölder
Universität Tübingen
*Jürgen Kullmann, O. H. Schindewolf
Universität Würzburg
Klaus Sdzuy

ITALY
Unattached
 Franco Rasetti (Rome)
Università Modena
 Eugenia Montanaro Gallitelli

JAPAN
Tohoku University (Sendai)
 Kotora Hatai
University of Tokyo
 Tetsuro Hanai

LIBYA
Esso Standard Libya Inc. (Tripoli)
 L. A. Smith

NETHERLANDS
Rijksmuseum van Natuurlijke Historie (Leiden)
 H. Boschma, L. B. Holthuis
Vrije Universiteit Amsterdam
 A. Breimer

NEW ZEALAND
Auckland Institute and Museum
 A. W. B. Powell
Dominion Museum (Wellington)
 R. K. Dell
New Zealand Geological Survey (Lower Hutt)
 C. A. Fleming, J. Marwick

NORWAY
Universitet Oslo
 Gunnar Henningsmoen, T. Soot-Ryen, Leif Størmer

POLAND
Pánstwowe Wydawnictwo Naukowe (Warszawa)
 Gertruda Biernat, Adolf Riedel

SWEDEN
Naturhistoriska Museet Göteborg
 Bengt Hubendick
Rijksmuseum Stockholm
 Valdar Jaanusson
Universitet Lund
 Gerhard Regnéll
Universitet Stockholm
 Ivar Hessland
Universitet Uppsala
 R. A. Reyment

SWITZERLAND
Universität Basel
 Manfred Reichel

UNITED KINGDOM
British Museum (Natural History) (London)
 Leslie Bairstow, *P. L. Cook, Isabella Gordon, *M. K. Howarth, S. M. Manton, N. J. Morris, C. P. Nuttall
British Petroleum Company (Middlesex)
 F. E. Eames
Geological Survey of Great Britain (London)
 Raymond Casey, R. V. Melville
Iraq Petroleum Company (London)
 G. F. Elliott
Queen's University of Belfast
 Margaret Jope, *R. E. H. Reid, *Ronald Tavener-Smith, Alwyn Williams, A. D. Wright
Royal Scottish Museum (Edinburgh)
 J. C. Brower
Unattached
 Dennis Curry (Middlesex), R. P. Tripp (Sevenoaks, Kent), C. J. Stubblefield (London), C. W. Wright (London)
University of Birmingham
 L. J. Wills
University of Cambridge
 O. M. B. Bulman, M. J. S. Rudwick, H. B. Whittington
University College London
 *J. H. Callomon, *D. T. Donovan
University College of Swansea
 D. V. Ager
University of Durham
 *G. P. Larwood
University of Glasgow
 W. D. I. Rolfe, John Weir, C. M. Yonge
University of Hull
 *M. R. House, E. R. Trueman
University of Leicester
 P. C. Sylvester-Bradley

UNITED STATES OF AMERICA
Academy of Natural Sciences of Philadelphia (Pennsylvania)
 A. A. Olsson, Robert Robertson
American Museum of Natural History (New York)
 R. L. Batten, W. K. Emerson, N. D. Newell
Brown University (Providence, R.I.)
 R. D. Staton

California Academy of Sciences
 (San Francisco)
 Eugene Coan, G. D. Hanna, L. G. Hertlein, Barry Roth, A. G. Smith, V. A. Zullo
California Institute of Technology
 (Pasadena)
 H. A. Lowenstam
Carnegie Museum
 (Pittsburgh, Pennsylvania)
 Juan Parodiz
Chevron Oil Field Research Company
 (La Habra, California)
 A. R. Loeblich, Jr.
Continental Oil Company
 (Ponca City, Oklahoma)
 J. A. Eyer
Cornell University (Ithaca, New York)
 W. S. Cole, J. W. Wells
Esso Production Research Company
 (Houston, Texas)
 H. H. Beaver, R. M. Jeffords, S. A. Levinson, L. A. Smith, Joan Stough, J. F. Van Sant
Field Museum of Natural History
 (Chicago)
 Fritz Haas, G. A. Solem
Florida Geological Survey (Tallahassee)
 H. S. Puri
Florida State University (Tallahassee)
 W. H. Heard
Harvard University
 (Cambridge, Massachusetts)
 Kenneth Boss, F. M. Carpenter, W. J. Clench, H. B. Fell, Bernhard Kummel, R. D. Staton, Ruth Turner
Illinois State Geological Survey (Urbana)
 M. L. Thompson
Indiana Geological Survey (Bloomington)
 R. H. Shaver
Johns Hopkins University
 (Baltimore, Maryland)
 Franco Rasetti
Kent State University (Kent, Ohio)
 A. H. Coogan
Louisiana State University (Baton Rouge)
 W. A. van den Bold, H. V. Howe, B. F. Perkins, H. B. Stenzel
New Mexico Institute Mining & Geology
 (Socorro)
 Christina Lochman-Balk
New York State Museum (Albany)
 D. W. Fisher
Ohio State University (Columbus)
 Aurèle La Rocque, W. C. Sweet

Oklahoma Geological Survey (Norman)
 T. W. Amsden, R. O. Fay
Oregon State University (Corvallis)
 A. J. Boucot, Joel Hedgpeth, J. G. Johnson
Paleontological Research Institute
 (Ithaca, New York)
 K. V. W. Palmer
Pennsylvania State University
 (University Park, Pa.)
 *R. J. Cuffey
Princeton University
 (Princeton, New Jersey)
 A. G. Fischer, B. F. Howell
Queens College of the City of New York
 (Flushing)
 *R. M. Finks
Radford College (Radford, Virginia)
 R. L. Hoffman
St. Mary's College of California (St. Mary's)
 A. S. Campbell
San Diego Natural History Museum
 (San Diego, California)
 George Radwin
San Francisco State College
 (San Francisco, California)
 Y. T. Mandra
Shell Development Company
 (Houston, Texas)
 John Wainwright
Sinclair Oil & Gas Company
 (Tulsa, Oklahoma)
 A. L. Bowsher
Smithsonian Institution (Washington, D.C.)
 R. H. Benson, *R. S. Boardman, *A. H. Cheetham, G. A. Cooper, T. G. Gibson, E. G. Kauffman, P. M. Kier, R. B. Manning, David Pawson, John Pojeta, H. A. Rehder
Southern Illinois University (Carbondale)
 *John Utgaard
Stanford University (Stanford, California)
 A. Myra Keen
State Geological Survey of Kansas
 (Lawrence)
 D. E. Nodine Zeller
State University of New York
 (Stony Brook)
 A. R. Palmer
Tulane University
 (New Orleans, Louisiana)
 Emily Vokes, H. E. Vokes
Unattached
 R. Wright Barker (Bellaire, Texas), H. J. Harrington (Houston, Texas)
United States Geological Survey
 (Washington, D.C.)

J. M. Berdan, R. C. Douglass, Mackenzie Gordon, Jr., R. E. Grant, *O. L. Karklins, K. E. Lohman, N. F. Sohl, I. G. Sohn, E. L. Yochelson; Dwight Taylor (Menlo Park, Calif.)

University of Alaska (College)
C. D. Wagner

University of California (Berkeley)
J. W. Durham, C. D. Wagner

University of California (Los Angeles)
N. G. Lane, W. P. Popenoe, Helen Tappan

University of California (San Diego, La Jolla)
M. N. Bramlette, R. R. Hessler, A. R. Loeblich III, W. A. Newman

University of Chicago (Illinois)
J. M. Weller

University of Cincinnati (Ohio)
K. E. Caster, *O. B. Nye

University of Florida (Gainesville)
H. K. Brooks, F. G. Thompson

University of Illinois (Urbana)
*D. B. Blake, H. W. Scott

University of Iowa (Iowa City)
W. M. Furnish, B. F. Glenister, H. L. Strimple

University of Kansas (Lawrence)
A. B. Leonard, Lavon McCormick, R. C. Moore, A. J. Rowell, Curt Teichert, R. H. Thompson

University of Massachusetts (Amherst)
C. W. Pitrat

University of Miami (Florida)
F. M. Bayer, Donald Moore

University of Michigan (Ann Arbor)
J. B. Burch, R. V. Kesling, D. B. Macurda, C. P. Morgan, F. H. T. Rhodes

University of Minnesota (Minneapolis)
F. M. Swain

University of Missouri (Columbia)
R. E. Peck

University of Missouri (Rolla)
Harriet Exline, D. L. Frizzell

University of Oklahoma (Norman)
C. C. Branson

University of Pennsylvania (Philadelphia)
A. J. Boucot, J. G. Johnson

University of Wyoming (Laramie)
D. W. Boyd

Western Reserve University (Cleveland, Ohio)
F. G. Stehli

Wichita State University (Kansas)
Paul Tasch

Yale University (New Haven, Connecticut)
A. L. McAlester

DECEASED

W. J. Arkell, R. S. Bassler, L. R. Cox, L. M. Davies, Julia Gardner, W. H. Hass, H. L. Hawkins, L. H. Hyman, J. B. Knight, M. W. de Laubenfels, A. K. Miller, H. M. Muir-Wood, Alexander Petrunkevitch, Emma Richter, Rudolf Richter, W. K. Spencer, M. A. Stainbrook, L. W. Stephenson, E. C. Stumm, O. W. Tiegs, Johannes Wanner, T. H. Withers, Arthur Wrigley

Alphabetical List

Ager, D. V., Swansea, Wales (Univ. College)
Amsden, T. W., Norman, Okla. (Oklahoma Geol. Survey)
Arkell, W. J., (deceased)
Bairstow, Leslie, London (British Museum Nat. History)
Barker, R. W., Bellaire, Tex. (unattached)
Bassler, R. S. (deceased)
Batten, R. L., New York (American Museum Nat. History)
Bayer, F. M., Miami, Fla. (Inst. Marine Sci., Univ. Miami)
Beaver, H. H., Houston, Tex. (Esso Prod. Research Co.)
Benson, R. H., Washington, D.C. (Smithsonian Inst.)
Berdan, J. M., Washington, D.C. (U.S. Geol. Survey)
Biernat, Gertruda, Warszawa, Pologne (Pánstwowe Wydawnictwo Naukowe)

*Blake, D. B., Urbana, Ill. (Univ. Illinois)
*Boardman, R. S., Washington, D.C. (Smithsonian Inst.)
Bold, W. A. van den, Baton Rouge, La. (Louisiana State Univ.)
Boschma, H., Leiden, Netherlands (Rijksmuseum van Natuurlijke Historie)
Boss, Kenneth, Cambridge, Mass. (Harvard Univ.)
Boucot, A. J., Corvallis, Ore. (Oregon State Univ.)
Bowsher, A. L., Tulsa, Okla. (Sinclair Oil & Gas Co.)
Boyd, D. W., Laramie, Wyo. (Univ. Wyoming)
Bramlette, M. N., La Jolla, Calif. (Scripps Inst. Oceanography)
Branson, C. C., Norman, Okla. (Univ. Oklahoma)
Breimer, A., Amsterdam, Netherlands (Inst. Aardwetensch. Vrije Univ.)
Brooks, H. K., Gainesville, Fla. (Univ. Florida)
Brower, J. C., Edinburgh, Scot. (Royal Scottish Museum)

Bulman, O. M. B., Cambridge, Eng. (Univ. Cambridge)
Burch, J. B., Ann Arbor, Mich. (Univ. Michigan)
*Callomon, J. H., London (Univ. College)
Campbell, A. S., St. Mary's, Calif. (St. Mary's College)
Carpenter, F. M., Cambridge, Mass. (Harvard Univ.)
Casey, Raymond, London (Geol. Survey Great Britain)
Caster, K. E., Cincinnati, Ohio (Univ. Cincinnati)
Chavan, André, Seyssel (Ain), France (unattached)
*Cheetham, A. H., Washington, D.C. (Smithsonian Inst.)
Clarke, A. H., Jr., Ottawa, Ontario, Canada (Natl. Museum)
Clench, W. J., Cambridge, Mass. (Harvard Univ.)
Coan, Eugene, San Francisco, Calif. (Calif. Acad. Sci.)
Cole, W. S., Ithaca, N.Y. (Cornell Univ.)
Coogan, H. A., Kent, Ohio (Kent State Univ.)
*Cook, P. L., London (British Museum Nat. History)
Cooper, G. A., Washington, D.C. (Smithsonian Inst.)
Cox, L. R. (deceased)
*Cuffey, R. J., University Park, Pa. (Pennsylvania State Univ.)
Curry, Dennis, Middlesex, Eng. (unattached)
Davies, L. M. (deceased)
Dechaseaux, Colette, Paris (Laboratoire de Paléont. des Vertébrés et de Paléont. Humaine)
Dell, R. K., Wellington, N. Z. (Dominion Museum)
*Donovan, D. T., London, Eng. (Univ. College)
Douglass, R. C., Washington, D.C. (U.S. Geol. Survey)
Durham, J. W., Berkeley, Calif. (Univ. California)
Eames, F. E., Middlesex, Eng. (British Petroleum Co.)
Elliott, G. F., London (Iraq Petroleum Co.)
Emerson, W. K., New York (American Museum Nat. History)
Erben, H. K., Bonn, W. Germany (Univ. Bonn)
Exline, Harriet, Rolla, Mo. (Univ. Missouri)
Eyer, J. A., Ponca City, Okla. (Continental Oil Co.)
Fay, R. O., Norman, Okla. (Oklahoma Geol. Survey)
Fell, H. B., Cambridge, Mass. (Harvard Univ.)
*Finks, R. M., Flushing, N.Y. (Queens College)
Fischer, A. G., Princeton, N. J., (Princeton Univ.)
Fisher, D. W., Albany, N. Y. (New York State Museum)
Fleming, C. A., Lower Hutt, N. Z. (New Zealand Geol. Survey)
Frizzell, D. L., Rolla, Mo. (Univ. Missouri)
Furnish, W. M., Iowa City, Iowa (Univ. Iowa)
Gardner, Julia (deceased)
Gibson, T. G., Washington, D.C. (Smithsonian Inst.)

Glaessner, M. F., Adelaide, S. Australia (Univ. Adelaide)
Glenister, B. F., Iowa City, Iowa (Univ. Iowa)
Gordon, Isabella, London (British Museum Nat. History)
Gordon, Mackenzie, Jr., Washignton, D.C. (U.S. Geol. Survey)
Grant, R. E., Washington, D.C. (U.S. Geol. Survey)
Haas, Fritz, Chicago, Ill. (Field Museum Nat. History)
Hahn, Gerhard, Berlin (Freie Univ.)
Hanai, Tetsuro, Tokyo (Univ. Tokyo)
Hanna, G D., San Francisco, Calif. (California Acad. Sci.)
Häntzschel, Walter, Hamburg, Germany (Geol. Staatsinst.)
Harrington, H. J., Houston, Tex. (unattached)
Hass, W. H. (deceased)
Hatai, Kotora, Sendai, Japan (Tohoku Univ.)
Hawkins, H. L. (deceased)
Heard, W. H., Tallahassee, Fla. (Florida State Univ.)
Hedgpeth, J. W., Newport, Ore. (Oregon State Univ.)
Henningsmoen, Gunnar, Oslo, Norway (Univ. Oslo)
Hertlein, L. G., San Francisco, Calif. (California Acad. Sci.)
Hessland, Ivar, Stockholm, Sweden (Univ. Stockholm)
Hessler, R. R., La Jolla, Calif. (Scripps Inst. Oceanography)
Hill, Dorothy, Brisbane, Australia (Univ. Queensland)
Hölder, Helmut, Münster, Germany (Univ. Münster)
Hoffman, R. L., Radford, Va. (Radford College)
Holthuis, L. B., Leiden, Netherlands (Rijksmuseum van Natuurlijke Historie)
*House, M. R., Hull, Eng. (Univ. Hull)
*Howarth, M. K., London (British Museum Nat. History)
Howe, H. V., Baton Rouge, La. (Louisiana State Univ.)
Howell, B. F., Princeton, N. J. (Princeton Univ.)
Hubendick, Bengt, Göteborg, Sweden (Naturhistoriska Museet)
Hyman, L. H. (deceased)
Jaanusson, Valdar, Stockholm (Rijksmuseum)
Jeffords, R. M., Houston, Tex. (Esso Prod. Research Co.)
Jeletzky, J. A., Ottawa, Ontario, Canada (Geol. Survey Canada)
Johnson, J. G., Corvallis, Ore. (Oregon State Univ.)
Jope, Margaret, Belfast, N. Ireland (Queen's Univ. of Belfast)
*Karklins, O. L., Washington, D.C. (U.S. Geol. Survey)
Kauffman, E. G., Washington, D.C. (Smithsonian Inst.)

Keen, A. Myra, Stanford, Calif. (Stanford Univ.)
Kesling, R. V., Ann Arbor, Mich. (Univ. Michigan)
Kier, P. M., Washington, D.C. (Smithsonian Inst.)
Knight, J. B. (deceased)
*Kullmann, Jürgen, Tübingen, W.Germany (Univ. Tübingen)
Kummel, Bernhard, Cambridge, Mass. (Harvard Univ.)
Lane, N. G., Los Angeles, Calif. (Univ. California)
La Rocque, Aurèle, Columbus, Ohio (Ohio State Univ.)
*Larwood, G. P., Durham, Eng. (Univ. Durham)
Laubenfels, M. W. de (deceased)
Lecompte, Marius, Louvain, Belgium (Univ. Louvain)
Leonard, A. B., Lawrence, Kans. (Univ. Kansas)
Levinson, S. A., Houston, Tex. (Esso Prod. Research Co.)
Lochman-Balk, Christina, Socorro, N. Mex. (New Mexico Inst. Mining & Technology)
Loeblich, A. R., Jr., La Habra, Calif. (Chevron Oil Field Research Co.)
Loeblich, A. R., III, La Jolla, Calif. (Scripps Inst. Oceanography)
Lohman, K. E., Washington, D.C. (U.S. Geol. Survey)
Lowenstam, H. A., Pasadena, Calif. (California Inst. Technology)
Ludbrook, N. H., Adelaide, S. Australia (South Australia Geol. Survey)
McAlester, A. L., New Haven, Conn. (Yale Univ.)
McCormick, Lavon, Lawrence, Kans. (Univ. Kansas)
McLaren, D. J., Calgary, Alberta, Canada (Inst. Sed. & Petrol. Geology)
Macurda, D. B., Ann Arbor, Mich. (Univ. Michigan)
Mandra, Y. T., San Francisco, Calif. (San Francisco State College)
Manning, R. B., Washington, D.C. (Smithsonian Inst.)
Manton, S. M., London (British Museum Nat. History)
Marwick, J., Lower Hutt, N. Z. (New Zealand Geol. Survey)
Melville, R. V., London (Geol. Survey Great Britain)
Miller, A. K. (deceased)
Montanaro Gallitelli, Eugenia, Modena, Italy (Univ. Modena)
Moore, Donald, Miami, Fla. (Univ. Miami, Inst. Marine Sci.)
Moore, R. C., Lawrence, Kans. (Univ. Kansas)
Morgan, C. P., Ann Arbor, Mich. (Univ. Michigan)
Morris, N. J., London (British Museum Nat. History)
Müller, A. H., Freiberg, Germany (Geol. Inst. Bergakad.)
Müller, K. J., Bonn, Germany (Univ. Bonn)
Muir-Wood, H. M. (deceased)
Newell, N. D., New York (American Museum Nat. History)

Newman, W. A., La Jolla, Calif. (Scripps Inst. Oceanography)
Norris, A. W., Ottawa, Ontario, Canada (Geol. Survey of Canada)
Nuttall, C. P., London (British Museum Nat. History)
*Nye, O. B., Cincinnati, Ohio (Univ. Cincinnati)
Okulitch, V. J., Vancouver, Canada (Univ. British Columbia)
Olsson, A. A., Coral Gables, Fla. (Acad. Nat. Sci. Philadelphia)
Palmer, A. R., Stony Brook, Long Island, N. Y. (State Univ. New York)
Palmer, K. V. W., Ithaca, N. Y. (Paleont. Research Inst.)
Papp, Adolph, Wien, Austria (Univ. Wien)
Parodiz, Juan, Pittsburgh, Pa. (Carnegie Museum)
Pawson, David, Washington, D.C. (Smithsonian Inst.)
Peck, R. E., Columbia, Mo. (Univ. Missouri)
Perkins, B. F., Baton Rouge, La. (Louisiana State Univ.)
Petrunkevitch, Alexander (deceased)
Pitrat, C. W., Amherst, Mass. (Univ. Massachusetts)
Pojeta, John, Washington, D.C. (Smithsonian Inst.)
Popenoe, W. P., Los Angeles, Calif. (Univ. Calif.)
Poulsen, Chr., København (Univ. København)
Powell, A. W. B., Auckland, N. Z. (Auckland Inst. & Museum)
Puri, H. S., Tallahassee, Fla. (Florida Geol. Survey)
Radwin, George, San Diego, Calif. (San Diego Nat. History Museum)
Rasetti, Franco, Rome, Italy (Unattached)
Regnéll, Gerhard, Lund, Sweden (Univ. Lund)
Rehder, H. A., Washington, D.C. (Smithsonian Inst.)
Reichel, Manfred, Basel, Switzerland (Univ. Basel)
*Reid, R. E. H., Belfast, N. Ireland (Queen's Univ. of Belfast)
Reyment, R. A., Uppsala, Sweden (Univ. Uppsala)
Rhodes, F. H. T., Ann Arbor, Mich. (Univ. Michigan)
Richter, Emma (deceased)
Richter, Rudolf (deceased)
Riedel, Adolf, Warszawa, Pologne (Pánstwowe Wydawnictwo Naukowe)
Robertson, Robert, Philadelphia, Pa. (Acad. Nat. Sci.)
Rolfe, W. D. I., Glasgow, Scotland (Univ. Glasgow)
Roth, Barry, San Francisco, Calif. (Calif. Acad. Sci.)
Rowell, A. J., Lawrence, Kans. (Univ. Kansas)
Rudwick, M. J. S., Cambridge, Eng. (Univ. Cambridge)
Schindewolf, O. H., Tübingen, Germany (Univ. Tübingen)
Schmidt, Herta, Frankfurt, Germany (Natur Museum u. Forsch.-Inst. Senckenberg)

Scott, H. W., Urbana, Ill. (Univ. Illinois)
Sdzuy, Klaus, Würzburg, Germany (Univ. Würzburg)
Shaver, R. H., Bloomington, Ind. (Indiana Geol. Survey & Univ. Indiana)
Sieverts-Doreck, Hertha, Stuttgart-Möhringen, Germany (unattached)
Sinclair, G. W., Ottawa, Ontario, Canada (Geol. Survey Canada)
Smith, A. G., San Francisco, Calif. (California Acad. Sci.)
Smith, L. A., Houston, Tex. (Esso Prod. Research Co.)
Sohl, N. F., Washington, D.C. (U.S. Geol. Survey)
Sohn, I. G., Washington, D.C. (U.S. Geol. Survey)
Solem, G. A., Chicago, Ill. (Field Museum Nat. History)
Soot-Ryen, T., Oslo (Univ. Oslo)
Spencer, W. K. (deceased)
Stainbrook, M. A. (deceased)
Staton, R. D., Providence, R.I. (Brown Univ.)
Stehli, F. G., Cleveland, Ohio (Western Reserve Univ.)
Stenzel, H. B., Baton Rouge, La. (Louisiana State Univ.)
Stephenson, L. W. (deceased)
Størmer, Leif, Oslo (Univ. Oslo)
Stough, Joan, Houston, Tex. (Esso Prod. Research Co.)
Strimple, H. L., Iowa City, Iowa (Univ. Iowa)
Struve, Wolfgang, Frankfurt, Germany (Natur-Museum u. Forsch.-Inst. Senckenberg)
Stubblefield, C. J., London (unattached)
Stumm, E. C. (deceased)
Swain, F. M., Minneapolis, Minn. (Univ. Minnesota)
Sweet, W. C., Columbus, Ohio (Ohio State Univ.)
Sylvester-Bradley, P. C., Leicester, Eng. (Univ. Leicester)
Talent, J. A., North Ryde, N.S.W. (Macquarie Univ.)
Tappan, Helen, Los Angeles, Calif. (Univ. California)
Tasch, Paul, Wichita, Kans. (Wichita State Univ.)
*Tavener-Smith, Ronald, Belfast, N. Ireland (Queen's Univ. of Belfast)
Taylor, Dwight, Menlo Park, Calif. (U.S. Geol. Survey)
Teichert, Curt, Lawrence, Kans. (Univ. Kansas)
Thompson, F. G., Gainesville, Fla. (Univ. Florida)
Thompson, M. L., Urbana, Ill. (Illinois State Geol. Survey)
Thompson, R. H., Lawrence, Kans. (Univ. Kansas)
Tiegs, O. W. (deceased)
Tripp, R. P., Sevenoaks, Kent, Eng. (unattached)
Trueman, E. R., Hull, Eng. (Univ. Hull)
Turner, Ruth, Cambridge, Mass. (Harvard Univ.)
Ubaghs, Georges, Liège (Univ. Liège)
*Utgaard, John, Carbondale, Ill. (Southern Illinois Univ.)
Van Sant, J. F., Houston, Tex. (Esso Prod. Research Co.)
Vokes, Emily, New Orleans, La. (Tulane Univ.)
Vokes, H. E., New Orleans, La. (Tulane Univ.)
Wagner, C. D., College, Alaska (Univ. Alaska)
Wainwright, John, Houston, Tex. (Shell Development Co.)
Wanner, Johannes (deceased)
Weir, John, Tayport, Fife, Scotland (Univ. Glasgow)
Weller, J. M., Chicago, Ill. (Univ. Chicago)
Wells, J. W., Ithaca, N. Y. (Cornell Univ.)
Whittington, H. B., Cambridge, Eng. (Univ. Cambridge)
Williams, Alwyn, Belfast, N. Ireland (Queen's Univ. of Belfast)
Wills, L. J., Birmingham, Eng. (Univ. Birmingham)
Withers, T. H (deceased)
Wright, A. D., Belfast, N. Ireland (Queen's Univ. of Belfast)
Wright, C. W., London (unattached)
Wrigley, Arthur (deceased)
Yochelson, E. L., Washington, D.C. (U.S. Geol. Survey)
Yonge, C. M., Glasgow, Scotland (Univ. Glasgow)
Zeller, D. E. Nodine, Lawrence, Kans. (State Geol. Survey Kansas)
Zullo, V. A., San Francisco, Calif. (California Acad. Sci.)

EDITORIAL PREFACE

The present volume is the first of a number of revised editions of parts of the *Treatise on Invertebrate Paleontology* planned for publication during the next several years, directed and edited by CURT TEICHERT. The first volume of the *Treatise,* Part G, was published in 1953. Seven more parts followed in the 1950's. Experience has shown that after a period of about 10 years, many volumes, or at least sections of volumes, begin to decline in usefulness, except in fields where progress in research is slow. The first edition of Part V was published in 1955 and research on graptolites has proceeded rapidly in succeeding years as demonstrated by the increased size of this second edition.

In general, style and format of the revised editions and supplements will conform with those of earlier *Treatise* volumes. Much of the contents will represent the results of new research by many paleontologists.

The general Editorial Preface prepared by R. C. MOORE which has been reprinted, with modifications, in all parts of the *Treatise* is included here because it also applies to revised editions.

The aim of the *Treatise on Invertebrate Paleontology,* as originally conceived and consistently pursued, is to present the most comprehensive and authoritative, yet compact statement of knowledge concerning invertebrate fossil groups that can be formulated by collaboration of competent specialists in seeking to organize what has been learned of this subject up to the mid-point of the present century. Such work has value in providing a most useful summary of the collective results of multitudinous investigations and thus should constitute an indispensable text and reference book for all persons who wish to know about remains of invertebrate organisms preserved in rocks of the earth's crust. This applies to neozoologists as well as paleozoologists and to beginners in study of fossils as well as to thoroughly trained, long-experienced professional workers, including teachers, stratigraphical geologists, and individuals engaged in research on fossil invertebrates. The making of a reasonably complete inventory of present knowledge of invertebrate paleontology may be expected to yield needed foundation for future research and it is hoped that the *Treatise* will serve this end.

The *Treatise* is divided into parts which bear index letters, each except the initial and concluding ones being defined to include designated groups of invertebrates. The chief purpose of this arrangement is to provide for independence of the several parts as regards date of publication, because it is judged desirable to print and distribute each segment as soon as possible after it is ready for press. Pages in each part bear the assigned index letter joined with numbers beginning with 1 and running consecutively to the end of the part.

The outline of subjects to be treated in connection with each large group of invertebrates includes (1) description of morphological features, with special reference to hard parts, (2) ontogeny, (3) classification, (4) geological distribution, (5) evolutionary trends and phylogeny, (6) paleoecology, and (7) systematic description of genera, subgenera, and higher taxonomic units. A selected list of references is furnished in each part of the *Treatise.*

Features of style in the taxonomic portions of this work have been fixed by the Editor with aid furnished by advice from representatives of the societies which have undertaken to sponsor the *Treatise.* It is the Editor's responsibility to consult with authors and coordinate their work, seeing that manuscript properly incorporates features of adopted style. Especially he has been called on to formulate policies in respect to many questions of nomenclature and procedure. The subject of family and subfamily names is reviewed briefly in a following section of this preface, and features of *Treatise* style in generic descriptions are explained.

A generous grant of $35,000 was made in 1948 by the Geological Society of America for initial work in preparing *Treatise* illustrations. Administration of expenditures has been in charge of the Editor and most of the work by photographers and artists has been done under his direction at the University of Kansas, but sizable parts of this program have also been carried forward in Washington, London, and many other places.

In December, 1959, the National Science Foundation of the United States, through its Division of Biological and Medical Sciences and the Program Director for Systematic Biology, made a grant in the amount of $210,000 for the purpose of aiding the completion of yet-unpublished volumes of the *Treatise*. Payment of this sum was provided to be made in installments distributed over a five-year period, with administration of disbursements handled by the University of Kansas. An additional grant (No. GB 4544) of $102,800 was made by the National Science Foundation in January, 1966, for the two-year period 1966-67, and this was extended for the calendar year 1968 by payment of $25,700 in October, 1967. This grant was extended further by payments of $57,800 in 1968 and $66,600 in 1969 for calendar years 1969-70. Expenditures planned are primarily for needed assistance to authors and may be arranged through approved institutions located anywhere. Important help for the Director-Editor of the *Treatise* has been made available from the grant. Grateful acknowledgment to the Foundation is expressed on behalf of the societies sponsoring the *Treatise*, the University of Kansas, and innumerable individuals benefited by the *Treatise* project.

ZOOLOGICAL NAMES

Many questions arise in connection with zoological names, especially including those that relate to their acceptability and to alterations of some which may be allowed or demanded. Procedure in obtaining answers to these questions is guided and to a large extent governed by regulations published (1961) in the *International Code of Zoological Nomenclature* (hereinafter cited simply as the *Code*). The prime object of the *Code* is to promote stability and universality in the scientific names of animals, ensuring also that each name is distinct and unique while avoiding restrictions on freedom of taxonomic thought or action. Priority is a basic principle, but under specified conditions its application can be modified. This is all well and good, yet nomenclatural tasks confronting the zoological taxonomist are formidable. They warrant the complaint of some that zoology, including paleozoology, is the study of animals rather than of names applied to them.

Several ensuing pages are devoted to aspects of zoological nomenclature that are judged to have chief importance in relation to procedures adopted in the *Treatise*. Terminology is explained, and examples of style employed in the nomenclatural parts of systematic descriptions are given.

TAXA GROUPS

Each taxonomic unit (taxon, pl., taxa) of the animal and protistan kingdoms belongs to some one or another rank in the adopted hierarchy of classificatory divisions. In part, this hierarchy is defined by the *Code* to include a species-group of taxa, a genus-group, and a family-group. Units of lower rank than subspecies are excluded from zoological nomenclature and those higher than superfamily of the family-group are not regulated by the *Code*. It is natural and convenient to discuss nomenclatural matters in general terms first and then to consider each of the taxa groups separately. Especially important is provision that within each taxa group classificatory units are coordinate (equal in rank), whereas units of different taxa groups are not coordinate.

FORMS OF NAMES

All zoological names are divisible into groups based on their form (spelling). The first-published form (or forms) of a name is defined as original spelling (*Code*, Art. 32) and any later-published form (or forms) of the same name is designated as subsequent spelling (Art. 33). Obviously, original and subsequent spellings of a given name may or may not be identical and this affects consideration of their correctness. Further, examination of original spellings of names shows that by no means all can be distinguished as correct. Some are incorrect, and the same is true of subsequent spellings.

Original Spellings

If the first-published form of a name is consistent and unambiguous, being identical wherever it appears, the original spelling is defined as correct unless it contravenes some stipulation of the *Code* (Arts. 26-31), unless the original publication contains clear evidence of an inadvertent error, in the sense of the *Code,* or among names belonging to the family-group, unless correction of the termination or the stem of the type-genus is required. An unambiguous original spelling that fails to meet these requirements is defined as incorrect.

If a name is spelled in more than one way in the original publication, the form adopted by the first reviser is accepted as the correct original spelling, provided that it complies with mandatory stipulations of the *Code* (Arts. 26-31), including its provision for automatic emendations of minor sort.

Incorrect original spellings are any that fail to satisfy requirements of the *Code,* or that represent an inadvertent error, or that are one of multiple original spellings not adopted by a first reviser. These have no separate status in zoological nomenclature and therefore cannot enter into homonymy or be used as replacement names. They call for correction wherever found. For example, a name originally published with a diacritic mark, apostrophe, diaeresis, or hyphen requires correction by deleting such features and uniting parts of the name originally separated by them, except that deletion of an umlaut from a vowel is accompanied by inserting "e" after the vowel.

Subsequent Spellings

If a name classed as a subsequent spelling is identical with an original spelling, it is distinguishable as correct or incorrect on the same criteria that apply to the original spelling. This means that a subsequent spelling identical with a correct original spelling is also correct, and one identical with an incorrect original spelling is also incorrect. In the latter case, both original and subsequent spellings require correction wherever found (authorship and date of the original incorrect spelling being retained).

If a subsequent spelling differs from an original spelling in any way, even by the omission, addition, or alteration of a single letter, the subsequent spelling must be defined as a different name (except that such changes as altered terminations of adjectival specific names to obtain agreement in gender with associated generic names, of family-group names to denote assigned taxonomic rank, and corrections for originally used diacritic marks, hyphens, and the like are excluded from spelling changes conceived to produce a different name).

Altered subsequent spellings other than the exceptions noted may be either intentional or unintentional. If demonstrably intentional, the change is designated as an emendation. Emendations are divisible into those classed as justifiable and those comprising all others classed as unjustifiable. Justifiable emendations are corrections of incorrect original spellings, and these take the authorship and date of the original spellings. Unjustifiable emendations are names having their own status in nomenclature, with author and date of their publication; they are junior objective synonyms of the name in its original form.

Subsequent spellings that differ in any way from original spellings, other than previously noted exceptions, and that are not classifiable as emendations are defined as incorrect subsequent spellings. They have no status in nomenclature, do not enter into homonymy, and cannot be used as replacement names.

AVAILABLE AND UNAVAILABLE NAMES

Available Names

An available zoological name is any that conforms to all mandatory provisions of the *Code.* Such names are classifiable in groups which are usefully recognized in the *Treatise,* though not explicitly differentiated in the *Code.* They are as follows:

1) So-called *"inviolate names"* include all available names that are not subject to any sort of alteration from their originally published form. They comprise correct original spellings and commonly include correct subsequent spellings, but include no names classed as emendations. Here belong most genus-group names (including those for collective groups), some of which differ in spelling from others by only a single letter.

2) Names may be termed *"perfect names"* if, as originally published (with or without duplication by subsequent authors), they meet all mandatory requirements, needing no correction of any kind, but nevertheless are legally alterable in such

ways as changing the termination (e.g., many species-group names, family-group names, suprafamilial names). This group does not include emended incorrect original spellings (e.g., *Oepikina,* replacement of *Öpikina*).

3) *"Imperfect names"* are available names that as originally published (with or without duplication by subsequent authors) contain mandatorily emendable defects. Incorrect original spellings are imperfect names. Examples of emended imperfect names are: among species-group names, *guerini* (not *Guérini*), *obrienae* (not *O'Brienae*), *terranovae* (not *terra-novae*), *nunezi* (not *Nuñezi*), *Spironema rectum* (not *Spironema recta,* because generic name is neuter, not feminine); among genus-group names, *Broeggeria* (not *Bröggeria*), *Obrienia* (not *O'Brienia*), *Maccookites* (not *McCookites*; among family-group names, Oepikidae (not Öpikidae), Spironematidae (not Spironemidae, incorrect stem), Athyrididae (not Athyridae, incorrect stem). The use of "variety" for named divisions of fossil species, according to common practice of some paleontologists, gives rise to imperfect names, which generally are emendable (*Code,* Art. 45e) by omitting this term so as to indicate the status of this taxon as a subspecies.

4) *"Vain names"* are available names consisting of unjustified intentional emendations of previously published names. The emendations are unjustified because they are not demonstrable as corrections of incorrect original spellings as defined by the *Code* (Art. 32,c). Vain names have status in nomenclature under their own authorship and date. They constitute junior objective synonyms of names in their original form. Examples are: among species-group names, *geneae* (published as replacement of original unexplained masculine, *geni,* which now is not alterable), *ohioae* (invalid change from original *ohioensis*); among genus-group names, *Graphiodactylus* (invalid change from original *Graphiadactyllis*); among family-group names, Graphiodactylidae (based on junior objective synonym having invalid vain name).

5) An important group of available zoological names can be distinguished as *"transferred names."* These comprise authorized sorts of altered names in which the change depends on transfer from one taxonomic rank to another, or possibly on transfers in taxonomic assignment of subgenera, species, or subspecies. Most commonly the transfer calls for a change in termination of the name so as to comply with stipulations of the *Code* on endings of family-group taxa and agreement in gender of specific names with associated generic names. Transferred names may be derived from any of the preceding groups except the first. Examples are: among species-group names, *Spirifer ambiguus* (masc.) to *Composita ambigua* (fem.), *Neochonetes transversalis* to *N. granulifer transversalis* or vice versa; among genus-group names, *Schizoculina* to *Oculina (Schizoculina)* or vice versa; among family-group names, Orthidae to Orthinae or vice versa, or superfamily Orthacea derived from Orthidae or Orthinae; among suprafamilial taxa (not governed by the *Code*), order Orthida to suborder Orthina or vice versa. The authorship and date of transferred names are not affected by the transfers, but the author responsible for the transfer and the date of his action may appropriately be recorded in such works as the *Treatise.*

6) Improved or *"corrected names"* include both mandatory and allowable emendations of imperfect names and of suprafamilial names, which are not subject to regulation as to name form. Examples of corrected imperfect names are given with the discussion of group 3. Change from the originally published ordinal name Endoceroidea (TEICHERT, 1933) to the presently recognized Endocerida illustrates a "corrected" suprafamilial name. Group 6 names differ from those in group 5 in not being dependent on transfers in taxonomic rank or assignment, but some names are classifiable in both groups.

7) *"Substitute names"* are available names expressly proposed as replacements for invalid zoological names, such as junior homonyms. These may be classifiable also as belonging in groups 1, 2, or 3. The glossary appended to the *Code* refers to these as "new names" *(nomina nova)* but they are better designated as substitute names, since their newness is temporary and relative. The first-published substitute name that complies with the definition here given

takes precedence over any other. An example is *Marieita* LOEBLICH & TAPPAN, 1964, as substitute for *Reichelina* MARIE, 1955 (*non* ERK, 1942).

8) *"Conserved names"* include a relatively small number of species-group, genus-group, and family-group names which have come to be classed as available and valid by action of the International Commission on Zoological Nomenclature exercising its plenary powers to this end or ruling to conserve a junior synonym in place of a rejected "forgotten" name *(nomen oblitum)* (Art. 23,b). Currently, such names are entered on appropriate "Official Lists," which are published from time to time.

It is useful for convenience and brevity of distinction in recording these groups of available zoological names to employ Latin designations in the pattern of *nomen nudum* (abbr., *nom. nud.*) and others. Thus we may recognize the preceding numbered groups as follows: 1) *nomina inviolata* (sing., *nomen inviolatum*, abbr., *nom. inviol.*), 2) *nomina perfecta (nomen perfectum, nom. perf.)*, 3) *nomina imperfecta (nomen imperfectum, nom. imperf.)*, 4) *nomina vana (nomen vanum, nom. van.)*, 5) *nomina translata (nomen translatum, nom. transl.)*, 6) *nomina correcta (nomen correctum, nom. correct.)*, 7) *nomina substituta (nomen substitutum, nom. subst.)*, 8) *nomina conservata (nomen conservatum, nom. conserv.)*.

Additional to the groups differentiated above, the *Code* (Art. 17) specifies that a zoological name is not prevented from availability a) by becoming a junior synonym, for under various conditions this may be re-employed, b) for a species-group name by finding that original description of the taxon relates to more than a single taxonomic entity or to parts of animals belonging to two or more such entities, c) for species-group names by determining that it first was combined with an invalid or unavailable genus-group name, d) by being based only on part of an animal, sex of a species, ontogenetic stage, or one form of a polymorphic species, e) by being originally proposed for an organism not considered to be an animal but now so regarded, f) by incorrect original spelling which is correctable under the *Code*, g) by anonymous publication before 1951, h) by conditional proposal before 1961, i) by designation as a variety or form before 1961, j) by concluding that a name is inappropriate (Art. 18), or k) for a specific name by observing that it is tautonymous (Art. 18).

It is worthy of mention that names published for collective groups (see later discussion under "Genus-Group Names") are authorized by the *Code* (Art. 42c) for use in zoological nomenclature and therefore may be construed to be available names which are treated for convenience exactly as if they were generic names.

Unavailable Names

All zoological names which fail to comply with mandatory provisions of the *Code* are unavailable names and have no status in zoological nomenclature. None can be used under authorship and date of their original publication as a replacement name *(nom. subst.)* and none preoccupies for purposes of the Law of Homonymy. Names identical in spelling with some, but not all, unavailable names can be classed as available if and when they are published in conformance to stipulations of the *Code* and they are then assigned authorship and take date of the accepted publication. Different groups of unavailable names can be discriminated, as follows.

9) *"Naked names"* include all those that fail to satisfy provisions stipulated in Article 11 of the *Code*, which states general requirements of availability, and in addition, if published before 1931, that were unaccompanied by a description, definition, or indication (Arts. 12, 16), and if published after 1930, that lacked accompanying statement of characters purporting to serve for differentiation of the taxon, or definite bibliographic reference to such a statement, or that were not proposed expressly as replacement *(nom. subst.)* of a pre-existing available name (Art. 13,a). Examples of "naked names" are: among species-group taxa, *Valvulina mixta* PARKER & JONES, 1865 (=*Cribrobulimina mixta* CUSHMAN, 1927, available and valid); among genus-group taxa, *Orbitolinopsis* SILVESTRI, 1932 (=*Orbitolinopsis* HENSON, 1948, available but classed as invalid junior synonym of *Orbitolina* D'ORBIGNY, 1850); among family-group taxa, Aequilateralidae D'ORBIGNY, 1846 (lacking type-genus), Hélicostègues

D'ORBIGNY, 1826 (vernacular not latinized by later authors, Art. 11,e,iii), Poteriocrinidae AUSTIN & AUSTIN, 1843 (=fam. Poteriocrinoidea AUSTIN & AUSTIN, 1842) (neither 1843 or 1842 names complying with Art. 11,e, which states that "a family-group name must, when first published, be based on the name then valid for a contained genus," such valid name in the case of this family being *Poteriocrinites* MILLER, 1821).

10) *"Denied names"* include all those that are defined by the *Code* (Art. 32,c) as incorrect original spellings. Examples are: Specific names, *nova-zelandica, mülleri, 10-brachiatus;* generic names, *M'Coyia, Størmerella, Römerina, Westgårdia;* family name, Růžičkinidae. Uncorrected "imperfect names" are "denied names" and unavailable, whereas corrected "imperfect names" are available.

11) *"Impermissible names"* include all those employed for alleged genus-group taxa other than genus and subgenus (Art. 42,a) (e.g., supraspecific divisions of subgenera), and all those published after 1930 that are unaccompanied by definite fixation of a type species (Art. 13,b). Examples of impermissible names are: *Martellispirifer* GATINAUD, 1949, and *Mirtellispirifer* GATINAUD, 1949, indicated respectively as a section and subsection of the subgenus *Cyrtospirifer; Fusarchaias* REICHEL, 1949, without definitely fixed type species (=*Fusarchaias* REICHEL, 1952, with *F. bermudezi* designated as type species).

12) *"Null names"* include all those that are defined by the *Code* (Art. 33,b) as incorrect subsequent spellings, which are any changes of original spelling not demonstrably intentional. Such names are found in all ranks of taxa.

13) *"Forgotten names"* are defined (Art. 23,b) as senior synonyms that have remained unused in primary zoological literature for more than 50 years. Such names are not to be used unless so directed by ICZN.

Latin designations for the discussed groups of unavailable zoological names are as follows: 9) *nomina nuda* (sing., *nomen nudum,* abbr., *nom. nud.*), 10) *nomina negata (nomen negatum, nom. neg.),* 11) *nomina vetita (nomen vetitum, nom. vet.),* 12) *nomina nulla (nomen nullum, nom. null.),* 13) *nomina oblita (nomen oblitum, nom. oblit.).*

VALID AND INVALID NAMES

Important distinctions relate to valid and available names, on one hand, and to invalid and unavailable names, on the other. Whereas determination of availability is based entirely on objective considerations guided by Articles of the *Code,* conclusions as to validity of zoological names partly may be subjective. A valid name is the correct one for a given taxon, which may have two or more available names but only a single correct name, generally the oldest. Obviously, no valid name can also be an unavailable name, but invalid names may include both available and unavailable names. Any name for a given taxon other than the valid name is an invalid name.

A sort of nomenclatorial no-man's-land is encountered in considering the status of some zoological names, such as *"doubtful names," "names under inquiry,"* and *"forgotten names."* Latin designations of these are *nomina dubia, nomina inquirenda,* and *nomina oblita,* respectively. Each of these groups may include both available and unavailable names, but the latter can well be ignored. Names considered to possess availability conduce to uncertainty and instability, which ordinarily can be removed only by appealed action of ICZN. Because few zoologists care to bother in seeking such remedy, the "wastebasket" names persist.

SUMMARY OF NAME GROUPS

Partly because only in such publications as the *Treatise* is special attention to groups of zoological names called for and partly because new designations are now introduced as means of recording distinctions explicitly as well as compactly, a summary may be useful. In the following tabulation valid groups of names are indicated in boldface type, whereas invalid ones are printed in italics.

DEFINITIONS OF NAME GROUPS

nomen conservatum (nom. conserv.). Name unacceptable under regulations of the *Code* which is made valid, either with original or altered spelling, through procedures specified by the *Code* or by action of ICZN exercising its plenary powers.

nomen correctum (nom. correct.). Name with intentionally altered spelling of sort required or allowable by the *Code* but not dependent on transfer from one taxonomic rank to another ("improved name"). (*See Code,* Arts. 26-b, 27, 29, 30-a-3, 31, 32-c-i, 33-a; in addition change of endings for suprafamilial taxa not regulated by the *Code.*)

nomen imperfectum (nom. imperf.). Name that as originally published (with or without subsequent identical spelling) meets all mandatory requirements of the *Code* but contains defect needing correction ("imperfect name"). (*See Code,* Arts. 26-b, 27, 29, 32-c, 33-a.)

nomen inviolatum (nom. inviol.). Name that as originally published meets all mandatory requirements of the *Code* and also is not correctable or alterable in any way ("inviolate name").

nomen negatum (nom. neg.). Name that as originally published (with or without subsequent identical spelling) constitutes invalid original spelling, and although possibly meeting all other mandatory requirements of the *Code,* cannot be used and has no separate status in nomenclature ("denied name"). It is to be corrected wherever found.

nomen nudum (nom. nud.). Name that as originally published (with or without subsequent identical spelling) fails to meet mandatory requirements of the *Code* and having no status in nomenclature, is not correctable to establish original authorship and date ("naked name").

nomen nullum (nom. null.). Name consisting of an unintentional alteration in form (spelling) of a previously published name (either available name, as *nom. inviol., nom. perf., nom imperf., nom. transl.*; or unavailable name, as *nom. neg., nom. nud., nom. van.,* or another *nom. null.*) ("null name").

nomen oblitum (nom. oblit.). Name of senior synonym unused in primary zoological literature in more than 50 years, not to be used unless so directed by ICZN ("forgotten name").

nomen perfectum (nom. perf.). Name that as originally published meets all mandatory requirements of the *Code* and needs no correction of any kind but which nevertheless is validly alterable by change of ending ("perfect name").

nomen substitutum (nom. subst.). Replacement name published as substitute for an invalid name, such as a junior homonym (equivalent to "new name").

nomen translatum (nom. transl.). Name that is derived by valid emendation of a previously published name as result of transfer from one taxonomic rank to another within the group to which it belongs ("transferred name").

nomen vanum (nom. van.). Name consisting of an invalid intentional change in form (spelling) from a previously published name, such invalid emendation having status in nomenclature as a junior objective synonym ("vain name").

nomen vetitum (nom. vet.). Name of genus-group taxon not authorized by the *Code* or, if first published after 1930, without definitely fixed type species ("impermissible name").

Except as specified otherwise, zoological names accepted in the *Treatise* may be understood to be classifiable either as *nomina inviolata* or *nomina perfecta* (omitting from notice *nomina correcta* among specific names) and these are not discriminated. Names which are not accepted for one reason or another include junior homonyms, senior synonyms classifiable as *nomina negata* or *nomina nuda,* and numerous junior synonyms which include both objective *(nomina vana)* and subjective types; rejected names are classified as completely as possible.

NAME CHANGES IN RELATION TO TAXA GROUPS

SPECIES-GROUP NAMES

Detailed consideration of valid emendation of specific and subspecific names is unnecessary here because it is well understood and relatively inconsequential. When the form of adjectival specific names is changed to obtain agreement with the gender of a generic name in transferring a species from one genus to another, it is never needful to label the changed name as a *nom. transl.* Likewise, transliteration of a letter accompanied by a diacritical mark in manner now called for by the *Code* (as in changing originally published *bröggeri* to *broeggeri*) or elimination of a hyphen (as in changing originally published *cornuoryx* to *cornuoryx*) does not require *"nom. correct."* with it.

GENUS-GROUP NAMES

So rare are conditions warranting change of the originally published valid form of generic and subgeneric names that lengthy discussion may be omitted. Only elimination of diacritical marks of some names in this category seems to furnish basis for valid emendation. It is true that many changes of generic and subgeneric names have been published, but virtually all of these are either *nomina vana* or *nomina nulla.* Various names which formerly were classed as homonyms are not now, for two names that differ only by a single letter (or in original publication by presence or absence of a diacritical mark) are construed to be entirely distinct.

A category of genus-group taxa and names for them calls for special notice. This comprises assemblages of identifiable species which cannot with any certainty be placed in a known genus. Such assemblages are recognized by the *Code* as valid zoological entities called **collective groups**, with names for them "treated as generic names in the meaning of the *Code*" (Art. 42c). They differ from genera in that collective groups require no type species. Particularly for dealing with fossil assemblages of dissociated skeletal remains of echinoderms (chiefly crinoids) procedures based on definition of collective groups must find place in the *Treatise*. Names for these will uniformly be labeled as applied to collective groups with accompanying abbreviation "*coll. coll.*" (for Latin *collectio collectiva*, collective group or assemblage), thus distinguishing them from names for genera. An example is *Pentagonopentagonalis* YELTYSHEVA, 1955 *(coll. coll.)*, no type species. The species *P. bilobatus* YELTYSHEVA, 1960, is available as the type species of *Obuticrinus* YELTYSHEVA in YELTYSHEVA & STUKALINA, 1963, in accordance with its original designation as such by these authors.

Examples in use of classificatory designations for genus-group names as previously given are the following, which also illustrate designation of type species as explained later.

Kurnatiophyllum THOMPSON, 1875 [**K. concentricum;* SD GREGORY, 1917] [=*Kumatiophyllum* THOMPSON, 1876 *(nom. null.);* Cymatophyllum THOMPSON, 1901 *(nom. van.);* Cymatiophyllum LANG, SMITH & THOMAS, 1940 *(nom. van.)*].

Stichophyma POMEL, 1872 [**Manon turbinatum* RÖMER, 1841; SD RAUFF, 1893] [=*Stychophyma* VOSMAER, 1885 *(nom. null.);* Sticophyma MORET, 1924 *(nom. null.)*].

Stratophyllum SMYTH, 1933 [**S. tenue*] [=*Ethmoplax* SMYTH, 1939 *(nom. van. pro Stratophyllum);* Stratiphyllum LANG, SMITH & THOMAS, 1940 *(nom. van. pro Stratophyllum* SMYTH) (non *Stratiphyllum* SCHEFFEN, 1933)].

Placotelia OPPLIGER, 1907 [**Porostoma marconi* FROMENTEL, 1859; SD DELAUBENFELS, herein] [=*Plakotelia* OPPLIGER, 1907 *(nom. neg.)*].

Walcottella DELAUBENFELS, 1955 [*nom. subst., pro Rhopalicus* SCHRAMM, 1936 (*non* FÖRSTER, 1856)].

Cyrtograptus CARRUTHERS, 1867 [*nom. correct.* LAPWORTH, 1873 (*pro Cyrtograpsus* CARRUTHERS, (1867), *nom. conserv.* proposed BULMAN, 1955 (ICZN 1963, p. 105, Opinion 650)].

Pentagonopentagonalis YELTYSHEVA, 1955 *(coll. coll.),* for species based on crinoid-stem parts [no type species] (ICZN pend.).

FAMILY-GROUP NAMES; USE OF "NOM. TRANSL."

The *Code* specifies the endings only for subfamily (-inae) and family (-idae) but all family-group taxa are defined as coordinate, signifying that for purposes of priority a name published for a taxon in any category and based on a particular type genus shall date from its original publication for a taxon in any category, retaining this priority (and authorship) when the taxon is treated as belonging to a lower or higher category. By exclusion of -inae and -idae, respectively reserved for subfamily and family, the endings of names used for tribes and superfamilies must be unspecified different letter combinations. These, if introduced subsequent to designation of a subfamily or family based on the same nominate genus, are *nomina translata*, as is also a subfamily that is elevated to family rank or a family reduced to subfamily rank. In the *Treatise* it is desirable to distinguish the valid alteration comprised in the changed ending of each transferred family-group name by the abbreviation *"nom. transl."* and record of the author and date belonging to this alteration. This is particularly important in the case of superfamilies, for it is the author who introduced this taxon that one wishes to know about rather than the author of the superfamily as defined by the *Code,* for the latter is merely the individual who first defined some lower-rank family-group taxon that contains the nominate genus of the superfamily. The publication of the author containing introduction of the superfamily *nomen translatum* is likely to furnish the information on taxonomic considerations that support definition of the unit.

Examples of the use of *"nom. transl."* are the following.

Subfamily STYLININAE d'Orbigny, 1851
[*nom. transl.* EDWARDS & HAIME, 1857 (*ex* Stylinidae D'ORBIGNY, 1851)]

Superfamily ARCHAEOCTONOIDEA Petrunkevitch, 1949
[*nom. transl.* PETRUNKEVITCH, 1955 (*ex* Archaeoctonidae PETRUNKEVITCH, 1949)]

Superfamily CRIOCERATITACEAE Hyatt, 1900
[*nom. transl.* WRIGHT, 1952 (*ex* Crioceratitidae HYATT, 1900)]

FAMILY-GROUP NAMES; USE OF "NOM. CORRECT."

Valid name changes classed as *nomina correcta* do not depend on transfer from one category of family-group units to another but most commonly involve correction of the stem of the nominate genus; in addition, they include somewhat arbitrarily chosen modification of ending for names of tribe or superfamily. Examples of the use of "*nom. correct.*" are the following.

Family STREPTELASMATIDAE Nicholson, 1889
[*nom. correct.* WEDEKIND, 1927 (*pro* Streptelasmidae NICHOLSON, 1889, *nom. imperf.*)]

Family PALAEOSCORPIIDAE Lehmann, 1944
[*nom. correct.* PETRUNKEVITCH, 1955 (*pro* Palaeoscorpionidae LEHMANN, 1944, *nom. imperf.*)]

Family AGLASPIDIDAE Miller, 1877
[*nom. correct.* STØRMER, 1959 (*pro* Aglaspidae MILLER, 1877, *nom. imperf.*)]

Superfamily AGARICIICAE Gray, 1847
[*nom. correct.* WELLS, 1956 (*pro* Agaricioidae VAUGHAN & WELLS, 1943, *nom. transl. ex* Agariciidae GRAY, 1847)]

FAMILY-GROUP NAMES; USE OF "*NOM. CONSERV.*"

It may happen that long-used family-group names are invalid under strict application of the *Code*. In order to retain the otherwise invalid name, appeal to ICZN is needful. Examples of use of *nom. conserv.* in this connection, as cited in the *Treatise,* are the following.

Family ARIETITIDAE Hyatt, 1874
[*nom. correct.* HAUG, 1885 (*pro* Arietidae HYATT, 1875) *nom. conserv.* proposed ARKELL, 1955 (ICZN pend.)]

Family STEPHANOCERATIDAE Neumayr, 1875
[*nom. correct.* FISCHER, 1882 (*pro* Stephanoceratinen NEUMAYR, 1875, invalid vernacular name), *nom. conserv.* proposed ARKELL, 1955 (ICZN pend.)]

FAMILY-GROUP NAMES; REPLACEMENTS

Family-group names are formed by adding letter combinations (prescribed for family and subfamily but not now for others) to the stem of the name belonging to genus (nominate genus) first chosen as type of the assemblage. The type genus need not be the oldest in terms of receiving its name and definition, but it must be the first-published as name-giver to a family-group taxon among all those included. Once fixed, the family-group name remains tied to the nominate genus even if its name is changed by reason of status as a junior homonym or junior synonym, either objective or subjective. Seemingly, the *Code* (Art. 39) requires replacement of a family-group name only in the event that the nominate genus is found to be a junior homonym, and then a substitute family-group name is accepted if it is formed from the oldest available substitute name for the nominate genus. Authorship and date attributed to the replacement family-group name are determined by first publication of the changed family group-name, but for purposes of the Law of Priority, they take the date of the replaced name. Numerous long-used family-group names are incorrect in being *nomina nuda,* since they fail to satisfy criteria of availability (Art. 11,e). These also demand replacement by valid names.

The aim of family-group nomenclature is greatest possible stability and uniformity, just as in case of other zoological names. Experience indicates the wisdom of sustaining family-group names based on junior subjective synonyms if they have priority of publication, for opinions of different workers as to the synonymy of generic names founded on different type species may not agree and opinions of the same worker may alter from time to time. The retention similarly of first-published family-group names which are found to be based on junior objective synonyms is less clearly desirable, especially if a replacement name derived from the senior objective synonym has been recognized very long and widely. To displace a much-used family-group name based on the senior objective synonym by disinterring a forgotten and virtually unused family-group name based on a junior objective synonym because the latter happens to have priority of publication is unsettling.

Replacement of a family-group name may be needed if the former nominate genus is transferred to another family-group. Then the first-published name-giver of a family-group assemblage in the remnant taxon is to be recognized in forming a replacement name.

FAMILY-GROUP NAMES; AUTHORSHIP AND DATE

All family-group taxa having names based on the same type genus are attributed to the author who first published the name for any of these assemblages, whether tribe, subfamily, or family (superfamily being almost inevitably a later-conceived taxon). Accordingly, if a family is divided into subfamilies or a subfamily into tribes, the name of no such subfamily or tribe can antedate the family name. Also, every family containing differentiated subfamilies must have a nominate *(sensu stricto)* subfamily, which is based on the same type genus as that for the family, and the author and date set down for the nominate subfamily invariably are identical with those of the family, without reference to whether the author of the family or some subsequent author introduced subdivisions.

Changes in the form of family-group names of the sort constituting *nomina correcta,* as previously discussed, do not affect authorship and date of the taxon concerned, but in publications such as the *Treatise* it is desirable to record the authorship and date of the correction.

SUPRAFAMILIAL TAXA

International rules of zoological nomenclature as given in the *Code* (1961) are limited to stipulations affecting lower-rank categories (infrasubspecies to superfamily). Suprafamilial categories (suborder to phylum) are either unmentioned or explicitly placed outside of the application of zoological rules. The *Copenhagen Decisions on Zoological Nomenclature* (1953, Arts. 59-69) proposed to adopt rules for naming suborders and higher taxonomic divisions up to and including phylum, with provision for designating a type genus for each, hopefully in such manner as not to interfere with the taxonomic freedom of workers. Procedures for applying the Law of Priority and Law of Homonymy to suprafamilial taxa were outlined and for dealing with the names for such units and their authorship, with assigned dates, when they should be transferred on taxonomic grounds from one rank to another. The adoption of terminations of names, different for each category but uniform within each, was recommended.

The Colloquium on zoological nomenclature which met in London during the week just before the XVth International Congress of Zoology convened in 1958 thoroughly discussed the proposals for regulating suprafamilial nomenclature, as well as many others advocated for inclusion in the new *Code* or recommended for exclusion from it. A decision which was supported by a wide majority of the participants in the Colloquium was against the establishment of rules for naming taxa above family-group rank, mainly because it was judged that such regulation would unwisely tie the hands of taxonomists. For example, if a class or order was defined by some author at a given date, using chosen morphologic characters (e.g., gills of pelecypods), this should not be allowed to freeze nomenclature, taking precedence over another later-proposed class or order distinguished by different characters (e.g., hinge-teeth of pelecypods). Even the fixing of type genera for suprafamilial taxa might have small value, if any, hindering taxonomic work rather than aiding it. At all events, no legal basis for establishing such types and for naming these taxa has yet been provided.

The considerations just stated do not prevent the Editor of the *Treatise* from making "rules" for dealing with suprafamilial groups of animals described and illustrated in this publication. At least a degree of uniform policy is thought to be needed, especially for the guidance of *Treatise*-contributing authors. This policy should accord with recognized general practice among zoologists, but where general practice is indeterminate or nonexistent our own procedure in suprafamilial nomenclature needs to be specified as clearly as possible. This pertains especially to decisions about names themselves, about citation of authors and dates, and about treatment of suprafamilial taxa which on taxonomic grounds are changed from their originally assigned rank. Accordingly, a few "rules" expressing *Treatise* policy are given here, some with examples of their application.

1) The name of any suprafamilial taxon must be a Latin or latinized uninominal noun of plural form, or treated as such, a) with a capital initial letter, b) without diacritical mark, apostrophe, diaeresis, or hyphen, and c) if component consisting of a numeral, numerical adjective, or adverb is used, this must be written in full (e.g., Stethostomata, Trionychi, Septemchitonina, Scorpiones, Subselliflorae). No uniformity in choice of ending for taxa of a given rank is demanded (e.g., orders named Gorgon*acea,* Millepor*ina,* Rugosa, Scleractin*ia,* Stromatopor*oidea,* Phalang*ida*).

2) Names of suprafamilial taxa may be constructed in almost any way, a) intended to indicate morphological attributes (e.g., Lamellibranchiata, Cyclostomata, Toxoglossa), b) based on the stem of an included genus (e.g., Bellerophontina, Nautilida, Fungiina), or c) arbitrary combina-

tions of letters, (e.g., Yuania), but none of these can be allowed to end in -idae or -inae, reserved for family-group taxa. A class or subclass (e.g., Nautiloidea), order (e.g., Nautilida), or suborder (e.g., Nautilina) named from the stem of an included genus may be presumed to have that genus (e.g., *Nautilus*) as its objective type. No suprafamilial name identical in form to that of a genus or to another published suprafamilial name should be employed (e.g., order Decapoda Latreille, 1803, crustaceans, and order Decapoda Leach, 1818, cephalopods; suborder Chonetoidea Muir-Wood, 1955, and genus *Chonetoidea* Jones, 1928). Worthy of notice is the classificatory and nomenclatural distinction between suprafamilial and family-group taxa which respectively are named from the same type genus, since one is not considered to be transferable to the other (e.g., suborder Bellerophontina Ulrich & Scofield, 1897; superfamily Bellerophontacea M'Coy, 1851; family Bellerophontidae M'Coy, 1851). Family-group names and suprafamilial names are not coordinate.

3) The Laws of Priority and Homonymy lack any force of international agreement as applied to suprafamilial names, yet in the interest of nomenclatural stability and the avoidance of confusion these laws are widely applied by zoologists to taxa above the family-group level wherever they do not infringe on taxonomic freedom and long-established usage.

4) Authors who accept priority as a determinant in nomenclature of a suprafamilial taxon may change its assigned rank at will, with or without modifying the terminal letters of the name, but such change(s) cannot rationally be judged to alter the authorship and date of the taxon as published originally. a) A name revised from its previously published rank is a "transferred name" *(nom. transl.),* as illustrated in the following.

Order CORYNEXOCHIDA Kobayashi, 1935

[*nom. transl.* MOORE, 1955 (*ex* suborder Corynexochida KOBAYASHI, 1935)]

b) A name revised from its previously published form merely by adoption of a different termination, without changing taxonomic rank, is an "altered name" *(nom. correct.).* Examples follow.

Order DISPARIDA Moore & Laudon, 1943

[*nom. correct.* MOORE, 1952 (*pro* order Disparata MOORE & LAUDON, 1943)]

Suborder AGNOSTINA Salter, 1864

[*nom. correct.* HARRINGTON & LEANZA, 1957 (*pro* suborder Agnostini SALTER, 1864)]

c) A suprafamilial name revised from its previously published rank with accompanying change of termination (which may or may not be intended to signalize the change of rank) is construed to be primarily a *nom. transl.* (compare change of ending for family-group taxa -idae to -inae, or vice versa, and to superfamily) but if desired it could be recorded as *nom. transl. et correct.*

Order ORTHIDA Schuchert & Cooper, 1931

[*nom. transl.* MOORE, 1952 (*ex* suborder Orthoidea SCHUCHERT & COOPER, 1931)]

5) The authorship and date of nominate subordinate and superordinate taxa among suprafamilial taxa are considered in the *Treatise* to be identical since each actually or potentially has the same type. Examples are given below.

Subclass ENDOCERATOIDEA Teichert, 1933

[*nom. transl.* TEICHERT, 1964 (*ex* superorder Endoceratoidea SHIMANSKIY & ZHURAVLEVA, 1961, *nom. transl. ex* order Endoceroidea TEICHERT, 1933)]

Order ENDOCERIDA Teichert, 1933

[*nom. correct.* TEICHERT, 1964 (*pro* order Endoceroidea TEICHERT, 1933)]

Suborder ENDOCERINA Teichert, 1933

[*nom. correct.* TEICHERT, 1964 (*pro* suborder Endoceracea SCHINDEWOLF, 1935, *nom. transl. ex* order Endoceroidea TEICHERT, 1933)]

6) A suprafamilial taxon may or may not contain a family-group taxon or taxa having the same type genus, and if it does, the respective suprafamilial and family-group taxa may or may not be nominate (having names with the same stem). The zoological *Code* (Art. 61) affirms that "each taxon [of any rank] has, actually or potentially, its type." Taxa above the family-group level which may be designated as having the same type genus (such designations not being stipulated or recognized by any articles of the zoological *Code*) are considered to have identical authorship and date if the stem of names employed is the same (illustrated in preceding paragraph), but otherwise their authorship and date are accepted as various. Examples showing both suprafamilial and familial taxa in a group of spiders follow.

Class ARACHNIDA Lamarck, 1801

[*nom. correct.* NEWPORT, 1830 (*pro* class—not family—Arachnidae LAMARCK, 1801) (type, *Araneus* CLERCK, 1757, validated ICZN, 1948)]

Subclass CAULOGASTRA Pocock, 1893

[type, *Araneus* CLERCK, 1757]

Superorder LABELLATA Petrunkevitch, 1949

[type, *Araneus* CLERCK, 1757]

Order ARANEIDA Clerck, 1757

[*nom. correct.* DALLAS, 1864 (*pro* Araneidea BLACKWALL, 1861, *pro* Araneides LATREILLE, 1801, *pro* Aranei CLERCK, 1757, validated ICZN, 1948) (type, *Araneus* CLERCK, 1757)]

Suborder DIPNEUMONINA Latreille, 1817

[*nom. correct.* PETRUNKEVITCH, 1955 (*pro* Dipneumones LATREILLE, 1817) (type, *Araneus* CLERCK, 1757)]

Division TRIONYCHI Petrunkevitch, 1933

[type, *Araneus* CLERCK, 1757]

Superfamily ARANEOIDEA Leach, 1815

[*nom. transl.* PETRUNKEVITCH, 1955 (*ex* Araneides LEACH, 1815) (type, *Araneus* CLERCK, 1757)]

Family ARANEIDAE Leach, 1815

[*nom. correct.* PETRUNKEVITCH, 1955 (*pro* Araneadae LEACH, 1819, *pro* Araneides LEACH, 1815) (type, *Araneus* CLERCK, 1757)]

Subfamily ARANEINAE Leach, 1815

[*nom. transl.* SIMON, 1892 (*ex* Araneidae LEACH, 1815) (type, *Araneus* CLERCK, 1757)]

TAXONOMIC EMENDATION

Emendation has two measurably distinct aspects as regards zoological nomenclature. These embrace 1) alteration of a name itself in various ways for various reasons, as has been reviewed, and 2) alteration or taxonomic scope or concept in application of a given zoological name, whatever its hierarchical rank. The latter type of emendation primarily concerns classification and inherently is not associated with change of name, whereas the other type introduces change of name without necessary expansion, restriction, or other modification in applying the name. Little attention generally has been paid to this distinction in spite of its significance.

Most zoologists, including paleozoologists, who have signified emendation of zoological names refer to what they consider a material change in application of the name such as may be expressed by an importantly altered diagnosis of the assemblage covered by the name. The abbreviation *"emend."* then may accompany the name, with statement of the author and date of the emendation. On the other hand, a multitude of workers concerned with systematic zoology think that publication of *"emend."* with a zoological name is valueless, because more or less alteration of taxonomic sort is introduced whenever a subspecies, species, genus, or other assemblage of animals is incorporated under or removed from the coverage of a given zoological name. Inevitably associated with such classificatory expansions and restrictions is some degree of emendation affecting diagnosis. Granting this, still it is true that now and then somewhat radical revisions are put forward, generally with published statement of reasons for changing the application of a name. To erect a signpost at such points of most significant change is worthwhile, both as aid to subsequent workers in taking account of the altered nomenclatural usage and as indication that not to-be-overlooked discussion may be found at a particular place in the literature. Authors of contributions to the *Treatise* are encouraged to include records of all specially noteworthy emendations of this nature, using the abbreviation *"emend."* with the name to which it refers and citing the author and date of the emendation.

In Part G (Bryozoa) and Part D (Protista 3) of the *Treatise,* the abbreviation *"emend."* is employed to record various sorts of name emendations, thus conflicting with usage of *"emend."* for change in taxonomic application of a name without alteration of the name itself. This is objectionable. In Part E (Archaeocyatha, Porifera) and later-issued divisions of the *Treatise,* use of *"emend."* is restricted to its customary sense, that is, significant alteration in taxonomic scope of a name such as calls for noteworthy modifications of a diagnosis. Other means of designating emendations that relate to form of a name are introduced.

STYLE IN GENERIC DESCRIPTIONS

CITATION OF TYPE SPECIES

The name of the type species of each genus and subgenus is given next following the generic name with its accompanying author and date, or after entries needed for definition of the name if it is involved in homonymy. The originally published combination of generic and trivial names for this species is cited, accompanied by an asterisk (*), with notation of the author and date of original publication. An exception in this procedure is made, however, if the species was first published in the same paper and by the same author as that containing definition of the genus which it serves as type; in such case, the initial letter of the generic name followed by the trivial name is given without repeating the name of the author and date, for this saves needed space. Examples of these two sorts of citations are as follows:

Diplotrypa NICHOLSON, 1879 [*Favosites petropolitanus* PANDER, 1830].

Chainodictyon FOERSTE, 1887 [*C. laxum*].

If the cited type species is a junior synonym of some other species, the name of this latter also is given, as follows:

Acervularia SCHWEIGGER, 1819 [*A. baltica* (=*Madrepora ananas* LINNÉ, 1758)].

It is judged desirable to record the manner of establishing the type species, whether by original designation or by subsequent designation.

Fixation of type species originally. The type species of a genus or subgenus, according to provisions of the *Code*, may be fixed in various ways originally (that is, in the publication containing first proposal of the generic name) or it may be fixed in specified ways subsequent to the original publication. Fixation of the type species of a genus or subgenus in an original publication is stipulated by the *Code* (Art. 68) in order of precedence as 1) *original designation* (in the *Treatise* indicated as OD) when the type species is explicitly stated or (before 1931) indicated by "n. gen., n. sp." (or its equivalent) applied to a single species included in a new genus, 2) defined by use of *typus* or *typicus* for one of the species included in a new genus (adequately indicated in the *Treatise* by the specific name), 3) established by *monotypy* if a new genus or subgenus includes only one originally included species which is neither OD nor TYP (in the *Treatise* indicated as M), and 4) fixed by *tautonymy* if the genus-group name is identical to an included species name not indicated as type belonging to one of the three preceding categories (indicated in the *Treatise* as T).

Fixation of type species subsequently. The type species of many genera are not determinable from the publication in which the generic name was introduced and therefore such genera can acquire a type species only by some manner of subsequent designation.

Most commonly this is established by publishing a statement naming as type species one of the species originally included in the genus, and in the *Treatise* fixation of the type species in this manner is indicated by the letters "SD" accompanied by the name of the subsequent author (who may be the same person as the original author) and the date of publishing the subsequent designation. Some genera, as first described and named, included no mentioned species and these necessarily lack a type species until a date subsequent to that of the original publication when one or more species are assigned to such a genus. If only a single species is thus assigned, it automatically becomes the type species and in the *Treatise* this subsequent monotypy is indicated by the letters "SM." Of course, the first publication containing assignment of species to the genus which originally lacked any included species is the one concerned in fixation of the type species, and if this named two or more species as belonging to the genus but did not designate a type species, then a later "SD" designation is necessary. Examples of the use of "SD" and "SM" as employed in the *Treatise* follow.

Hexagonaria GÜRICH, 1896 [*Cyathophyllum hexagonum* GOLDFUSS, 1826; SD LANG, SMITH & THOMAS, 1940].

Muriceides STUDER, 1887 [*M. fragilis* WRIGHT & STUDER, 1889; SM WRIGHT & STUDER, 1889].

Another mode of fixing the type species of a genus that may be construed as a special sort of subsequent designation is action of the International Commission on Zoological Nomenclature using its plenary powers. Definition in this way may set aside application of the *Code* so as to arrive at a decision considered to be in the best interest of continuity and stability of zoological nomenclature. When made, it is binding and commonly is cited in the *Treatise* by the letters "ICZN," accompanied by the date of announced decision and (generally) reference to the appropriate numbered Opinion.

Worthy of repetition is the lack of requirement of a type species for definition of collective groups *(coll. coll.)*, but when differentiated and named these are treated for convenience as genera in the meaning of the *Code* (Art. 42c).

HOMONYMS

Most generic names are distinct from

all others and are indicated without ambiguity by citing their originally published spelling accompanied by name of the author and date of first publication. If the same generic name has been applied to 2 or more distinct taxonomic units, however, it is necessary to differentiate such homonyms, and this calls for distinction between junior homonyms and senior homonyms. Because a junior homonym is invalid, it must be replaced by some other name. For example, *Callopora* HALL, 1851, introduced for Paleozoic trepostome bryozoans, is invalid because GRAY in 1848 published the same name for Cretaceous-to-Recent cheilostome bryozoans, and BASSLER in 1911 introduced the new name *Hallopora* to replace HALL's homonym. The *Treatise* style of entry is:

Hallopora BASSLER, 1911, *nom. subst.* [*pro Callopora* HALL, 1851 (*non* GRAY, 1848)].

In like manner, a needed replacement generic name may be introduced in the *Treatise* (even though first publication of generic names otherwise in this work is avoided). The requirement that an exact bibliographic reference must be given for the replaced name commonly can be met in the *Treatise* by citing a publication recorded in the list of references, using its assigned index number, as shown in the following example.

Mysterium DE LAUBENFELS, *nom. subst.* [*pro Mystrium* SCHRAMMEN, 1936 (ref. 40, p. 60) (*non* ROGER, 1862)] [*Mystrium porosum* SCHRAMMEN, 1936].

For some replaced homonyms, a footnote reference to the literature is necessary. A senior homonym is valid, and in so far as the *Treatise* is concerned, such names are handled according to whether the junior homonym belongs to the same major taxonomic division (class or phylum) as the senior homonym or to some other; in the former instance, the author and date of the junior homonym are cited as:

Diplophyllum HALL, 1851 [*non* SOSHKINA, 1939] [*D. caespitosum*].

Otherwise, no mention of the existence of a junior homonym generally is made.

Synonymic homonyms. An author sometimes publishes a generic name in two or more papers of different date, each of which indicates that the name is new. This is a bothersome source of errors for later workers who are unaware that a supposed first publication which they have in hand is not actually the original one. Although the names were separately published, they are identical and therefore definable as homonyms; at the same time they are absolute synonyms. For the guidance of all concerned, it seems desirable to record such names as synonymic homonyms and in the *Treatise* the junior one of these is indicated by the abbreviation "jr. syn. hom."

Identical family-group names not infrequently are published as new names by different authors, the author of the later-introduced name being ignorant of previous publication(s) by one or more other workers. In spite of differences in taxonomic concepts as indicated by diagnoses and grouping of genera and possibly in assigned rank, these family-group taxa are nomenclatural homonyms, based on the same type genus, and they are also synonyms. Wherever encountered, such synonymic homonyms are distinguished in the *Treatise* as in dealing with generic names.

SYNONYMS

Citation of synonyms is given next following record of the type species and if two or more synonyms of differing date are recognized, these are arranged in chronological order. Objective synonyms are indicated by accompanying designation "(obj.)," others being understood to constitute subjective synonyms. Examples showing *Treatise* style in listing synonyms follow.

Calapoecia BILLINGS, 1865 [*C. anticostiensis;* SD LINDSTRÖM, 1883] [=*Columnopora* NICHOLSON, 1874; *Houghtonia* ROMINGER, 1876].

Staurocyclia HAECKEL, 1882 [*S. cruciata* HAECKEL, 1887] [=*Coccostaurus* HAECKEL, 1882 (obj.); *Phacostaurus* HAECKEL, 1887 (obj.)].

A synonym which also constitutes a homonym is recorded as follows:

Lyopora NICHOLSON & ETHERIDGE, 1878 [*Palaeopora? favosa* M'COY, 1850] [=*Liopora* LANG, SMITH & THOMAS, 1940 (*non* GIRTY, 1915)].

Some junior synonyms of either objective or subjective sort may take precedence desirably over senior synonyms wherever uniformity and continuity of nomenclature are served by retaining a widely used but technically rejectable name for a generic assemblage. This requires action of ICZN using its plenary powers to set aside the unwanted name and validate the wanted one, with placement of the concerned names on appropriate official lists. In the *Treatise* citation of such a conserved generic name is given in the manner shown by the following example.

Tetragraptus SALTER, 1863 [*nom. correct.* HALL, 1865 (*pro Tetragrapsus* SALTER, 1863), *nom. conserv.* proposed BULMAN, 1955, ICZN pend.] [**Fucoides serra* BRONGNIART, 1828 (=*Graptolithus bryonoides* HALL, 1858)].

ABBREVIATIONS

Abbreviations used in this division of the *Treatise* are explained in the following alphabetically arranged list.

Abhandl., *Abhandlung(en)*
aff., *affinis* (related to)
Afhandl., *Afhandlingar*
Afr., Africa, -an
Am., America, -n
Ann., *Annaes, Année, Annalen, Annales,* Annual, *Annuaire*
approx., approximately
Arg., Argentina
Atl., Atlantic
auctt., *auctorum* (of authors)

B.C., British Columbia
Bendigon., Bendigonian
Boh., Bohemia
Brit., Britain, British
Bull., Bulletin

C., Central
ca., *circa*
Cam., Cambrian
Can., Canada
Canad., Canadian
Carb., Carboniferous
Castlemain., Castlemanian
cf., *confer* (compare)
Chewton., Chewtonian
Coll., Collection(s)
Contrib., Contribution(s)
Cret., Cretaceous
Czech., Czechoslovakia

Darriwil., Darriwilian
dec., decade
Denkschr., *Denkschrift(en)*
Dept., Department
Dev., Devonian
diagram., diagrammatic

E., East
ed., edited, editor
edit., edition
e.g., *exempli gratia* (for example)
emend., *emendatus(-a)*
Eng., England
enl., enlarged
Eoc., Eocene

err., *errore* (by error)
Est., Estonia
et al., *et alii* (and others, persons)
etc., *et cetera* (and others, objects)
Eu., Europe
excl., excluding

F., Formation
fam., family
fig., figure(s)
Förhandl., *Förhandlingar*

Gedinn., Gedinnian
Geol., Geological, *Geologique, Geologische,* Geology
Ger., German, Germany
Gr., Great., Group

Handl., *Handlingar*
hom., homonym

ICZN, International Commission on Zoological Nomenclature
i.e., *id est* (that is)
illus., illustrated, -ions
incl., inclined, including
Ind., Indiana
indet., indeterminate
IndoPac., Indo-Pacific
Internatl., International
Ire., Ireland
Is., Island(s)

Jahrb., *Jahrbuch*
Jahresber., *Jahresbericht*
Jahrg., *Jahrgang*
Jour., Journal

Kans., Kansas
Ky., Kentucky

L., low., Lower
lat., lateral
Lief., *Lieferung*
Llandov., Llandovery
long., longitudinal

Ls., Limestone

M., mid., Middle
M, monotypy
Medd., *Meddelelser*
Mem., Memoir(s), *Memoria, Memorie*
Mém., *Mémoire*(s)
Minn., Minnesota
Misc., Miscellaneous
mm., millimeter(s)
Mon., Monograph, *Monographia, Monographie*
Monatsber., *Monatsberichte*

n, n, new
N., North
N.Am., North America(n)
Namur., Namurian
Nat., Natural
Nev., Nevada
Newf., Newfoundland
Niag., Niagaran
no., number
nom. conserv., *nomen conservatum* (conserved name)
nom. correct., *nomen correctum* (corrected or intentionally altered name)
nom. dub., *nomen dubium* (doubtful name)
nom. imperf., *nomen imperfectum* (imperfect name)
nom. neg., *nomen negatum* (denied name)
nom. nov., *nomen novum* (new name)
nom. nud., *nomen nudum* (naked name)
nom. null., *nomen nullum* (null, void name)
nom. oblit., *nomen oblitum* (forgotten name)
nom. subst., *nomen substitutum* (substitute name)
nom. transl., *nomen translatum* (transferred name)

nom. van., *nomen vanum* (vain, void name)
nom. vet., *nomen vetitum* (impermissible name)
Nor., Norway
NW., Northwest
N.Y., New York
N.Z., New Zealand

obj., objective
OD, original designation
Okla., Oklahoma
Ont., Ontario
Opin., Opinion
Ord., Ordovician

p., page(s)
Pac., Pacific
Paleont., Paleontological
pend., pending
Philos., Philosophical
pl., plate(s), plural
Pol., Poland
Proc., Proceedings
Prof., Professional
pt., part(s)
publ., publication(s), published

Quart., Quarterly
Que., Quebec

Rec., Recent
reconstr., reconstructed, -ion
Rept., Report(s)
restor., restoration

S., Sea, South
S.Am., South America
schem., schematic
Sci., Science
Scot., Scotland
SD, subsequent designation
sec., section(s)
secc., *seccion*
ser., serial, series, *seriia*
Sh., Shale
Sib., Siberia
Siegen., Siegenian
Sil., Silurian
Sitzungsber., *Sitzungsberichte*
s.lat., *sensu lato* (in the wide sense, broadly defined)
Sl., Slate
SM, subsequent monotypy
Soc., *Société*, Society
sp., species (spp., plural)
Spec., Special

s.str., *sensu stricto* (in the strict sense, narrowly defined)
suppl., supplement(s)

Tenn., Tennessee
Tert., Tertiary
Trans., Transactions
Trempeal., Trempealeauan

U., up., Upper
U.S., United States
USA, United States of America
USSR, Union of Soviet Socialist Republics

v., volumes(s)
var., variety
Vict., Victoria

W., West
Wis., Wisconsin

Yapeen, Yapeenian

Z., Zone
Zeitschr., *Zeitschrift*
Zool., Zoological, *Zoologici, Zoologisch*

REFERENCES TO LITERATURE

Each part of the *Treatise* is accompanied by a selected list of references to paleontological literature consisting primarily of recent and comprehensive monographs available but also including some older works recognized as outstanding in importance. The purpose of giving these references is to aid users of the *Treatise* in finding detailed descriptions and illustrations of morphological features of fossil groups, discussions of classifications and distribution, and especially citations of more or less voluminous literature. Generally speaking, publications listed in the *Treatise* are not original sources of information concerning taxonomic units of various rank but they tell the student where he may find them; otherwise it is necessary to turn to such aids as the *Zoological Record* or NEAVE's *Nomenclator Zoologicus*. References given in the *Treatise* are arranged alphabetically by authors and accompanied by index numbers which serve the purpose of permitting citation most concisely in various parts of the text; these citations of listed papers are enclosed invariably in parentheses and, except in Parts C and N, are distinguishable from dates because the index numbers comprise no more than 3 digits. The systematic descriptions given in part C are accompanied by a reference list containing more than 2,000 entries with the index numbers marked by an asterisk, and in Part N (containing over 1,000 entries), they are italicized.

The following is a statement of the full names of serial publications which are cited in abbreviated form in the lists of references in the present volume. The information thus provided should be useful in library research work. The list is alphabetized according to the serial titles which were employed at the time of original publication. Those following in brackets are those under which the publication may be found currently in the *Union List of Serials,* the United States Library of Congress listing, and most library card catalogues. The names of serials published in Cyrillic are transliterated; in the reference lists these titles, which may be abbreviated, are accompanied by transliterated authors' names and titles, with English translation of the title. The place of publication is added (if not included in the serial title).

xxvii

The method of transliterating Cyrillic letters that is adopted as "official" in the *Treatise* is that suggested by the Geographical Society of London and the U.S. Board on Geographic Names. It follows that names of some Russian authors in transliterated form derived in this way differ from other forms, possibly including one used by the author himself. In *Treatise* reference lists the alternative (unaccepted) form is given enclosed by square brackets (e.g., Chernyshev [Tschernyschew], T.N.).

List of Serial Publications

Académie Royale de Belgique, Classe des Sciences, Bulletin; Mémoires. Bruxelles.
Académie des Sciences de l'URSS, Comptes Rendus; Institut Paléontologique, Travaux; Institut Paléozoologique, Travaux [Akademiya Nauk SSSR, Doklady]. Leningrad.
Académie Tchèque des Sciences, Bulletin International, Classe des Sciences Mathématiques. Praha.
Acta Geologica Polonica. Warszawa.
Acta Palaeontologia Sinica. Peking.
Acta Palaeontologica Polonica [Polska Akademia Nauk, Komitet Geologiczny]. Warszawa.
Acta Universitatis Asiae Mediae. Tashkent.
[K.] Akademie der Wissenschaften zu Wien, mathematische-naturwissenschaftliche Klasse, Denkschriften; Sitzungsberichte.
Akademiya Nauk SSSR. Kirgizskii Filial. Institut Geologii.
Akademiya Nauk SSSR. Sibirskoe Otdelenie, Institut Geologii i Geofiziki, Trudy. Novosibirsk.
Akademiya Nauk Uzbekskoy SSR, Doklady. Tashkent.
American Association for the Advancement of Science, Proceedings; Publications. Washington, D.C.
American Journal of Science. New Haven, Conn.
American Museum of Natural History, Bulletins; Memoirs; Novitates. New York.
Annales Musei Zoologici Polonici. Warszawa.
Annales de la Société Géologique du Nord. Lille.
Annals and Magazine of Natural History. London.
Arkiv för Mineralogi och Geologi (*see* K. Svenska Vetenskaps Akademien). Stockholm.
Aufschluss, Der. Zeitschrift für die Freunde der Mineralogie und Geologie. Göttingen.
Australia Bureau of Mineral Resources, Geology and Geophysics, Bulletins; Explanatory Notes; Paleontological Papers; Reports. Canberra.
Belfast Naturalists' Field Club, Proceedings. Belfast.
British Association for the Advancement of Science, Reports. London.
British Museum (Natural History), Geology, Bulletins. London.
Bulletin of American Paleontology. Ithaca, N.Y.
Canada, Geological Survey of, Department of Mines and Resources, Mines and Geology Branch, Bulletins; Memoirs; Museum Bulletins; Canadian Organic Remains. Ottawa, Montreal.
Canadian Field Naturalist. Ottawa.
Canadian Journal of Earth Sciences. National Research Council, Canada. Ottawa.
Canadian Naturalist and Geologist. Montreal.

Cincinnati, Quarterly Journals of Science.
Dansk Geologisk Forening, Meddelelser. København.
Denison University, Scientific Laboratories, Bulletins; Journals. Granville, Ohio.
Deutsch Akademie der Wissenschaften zu Berlin, Abhandlungen; Monatsberichte, Geologie und Mineralogie.
Deutsche Geologische Gesellschaft, Zeitschrift. Berlin, Hannover.
Eesti NSV Teaduste Akadeemia, Geoloogia Instituudi, Uurimused. [Akademiya Nauk Estonskoi SSR, Instituta Geologii, Trudy.] Tallinn.
Faculdade de Ciencias, Universidade de Lisboa, Revista.
[K.] Fysiografiska Sällskapet i Lund, Förhandlingar; Handlingar.
Geologica Balkanica. Sofia.
Geological Magazine. London, Hertford.
Geological Society of America, Bulletins; Memoirs; Special Papers. Boulder, Colo.
Geological Society of China, Bulletins. Peking.
Geological Society of London; Memoirs; Proceedings; Quarterly Journals; Transactions.
Geologie von Thüringen, Beiträge. Jena.
Geologiska Föreningen, Stockholm, Föhhandlingar.
Geologists' Association, Proceedings. London.
Gesellschaft von Freunden der Naturwissenschaften in Gera, Jahresbericht.
Indiana, Department of Conservation Geological Survey, Bulletins; Reports. Bloomington, Ind.
Journal of Geology. Chicago.
Journal of Paleontology, Tulsa, Okla.
Karlova, Universita. Práce, Geologicko-paleontologický ústav. Praha.
Královské České Společnost Nauk Prague, Třída Matematicko-Přírodovědecká; Rozjravy; Věstník.
Linnean Society of London (Zoology), Journals; Proceedings; Transactions.
Lund Universitet, Årsskrift.
Meddelelser om Grønland (Kommissionen Videnskabelige Undersøgelser i Grønland). København.
Missouri University Museum, Bulletins. Columbia.
Musée Royal d'Histoire Naturelle de Belgique, Annales; Bulletins; Mémoires (continued as Institut Royal des Sciences Naturelles de Belgique). Bruxelles.
Národního Musea Časopis, Oddíl Přírodovědny Ročnik. Praha.
Natur. Deutsche und Osterreichische Naturwissenschaftliche Gesellschaft. Leipzig.
Nauchno-issledovatelskii Institut Geologii Arktiki,

Ministerstvo Geologii i Okhrany Nedr SSSR, Trudy. Leningrad.
Neues Jahrbuch für Geologie und Paläontologie (Before 1950, Neues Jahrbuch für Mineralogie, Geologie, und Paläontologie), Abhandlungen; Beilage-Bande; Monatshefte. Stuttgart.
New York State Cabinet of Natural History, Annual Reports. Albany, N.Y.
New York State Geological Survey, Annual Reports; Natural History of New York; Palaeontology of New York. Albany.
New York State Museum of Natural History, Annual Reports; Bulletins. Albany.
Norsk Geologisk Tidsskrift (Norsk Geologisk Forening). Oslo.
Norske Videnskaps-Akademi i Oslo, Skrifter.
Osnovy Paleontologii Spravochnik, dlya Paleontologiv i Geologov SSSR. Yu. A. Orlov, ed. Akademiya Nauk SSSR. Moskva.
Ottawa Naturalist (see Canadian Field Naturalist).
Palaeontographica. Stuttgart, Kassel.
Palaeontographical Society, Memoirs; Monographs. London.
Palaeontologia Polonica. Warszawa.
Paläontologische Zeitschrift. Berlin, Stuttgart.
Palaeontology (Palaeontological Association). London.
Paleontologicheskiy Zhurnal. Akademiya Nauk SSSR. Moskva.
Public Museum, Bulletins. Milwaukee, Wis.
Quekett Microscopical Club, Journals. London.
Revista de la Asociación Geológica Argentina. Buenos Aires.
Royal Physical Society of Edinburgh, Proceedings.
Royal Society of Canada, Proceedings and Transactions. Ottawa, Canada.
Royal Society of London, Philosophical Transactions; Proceedings.
Royal Society of New South Wales, Journals; Proceedings. Sydney.
Royal Society of New Zealand, Proceedings; Transactions. Wellington.
Royal Society of Victoria, Proceedings. Melbourne.
Scientia Sinica. Academia Sinica. Peking.
Société Géologique de France, Comptes Rendus des Séances; Bulletin; Mémoires. Paris.
Société Géologique et Minéralogique de Bretagne, Mémoires. Rennes.
Société Impériale des Naturalistes de Moscou, Bulletin.
Státního Geologického Ústavu Československe Republiky, Věstník. Praha.
Studia Geologica Polonica. Polska Akademia Nauk Zaklad Nauk Geologicznych. Warszawa.
[K.] Svenska Vetenskapsakademien, Arkiv för Mineralogi och Geologi; Arkiv för Zoologi; Handlingar. Stockholm.
Sveriges Geologiska Undersökning, Afhandlingar; Årsbok. Stockholm.
Tartu Ülikooli Geoloogia-Instituudi Toimetused. Acta de Commentationes Universitatis Tartuensis (Dorpatensis).
Texas, University of, Bulletins; Publications. Austin.
United Kingdom, Geological Survey of, Memoirs. London.
United States Geological Survey, Bulletins; Monographs; Professional Papers. Washington, D.C.
United States National Museum, Bulletins; Proceedings. Washington, D.C.
Uppsala, University of, Geological Institution Bulletins.
Uspekhi Sovremennoi Biologii. Moskva.
Ústředního Ústavu Geologického, Rozpravy; Sborník; Věstník. Praha.
Victoria Department of Mines, Mining and Geological Journal. Melbourne.
Voprosy Paleontologii (Leningradskii Gosudarstvennyi Universitet, Nauchno-Issledovatelskii Institut Zemnoi Kory, Paleontologischeskaya Laboratoriya).
Vsesoyuznyi Paleontologicheskoe Obshchestvo, Ezhegodnik. Moskva.
Wisconsin, Geological and Natural History Survey, Bulletins. Madison.
Zeitschrift für Geschiebeforschung. Berlin.
Zeitschrift für Naturwissenschaften. Halle.

SOURCES OF ILLUSTRATIONS

At the end of figure captions an index number is given to supply record of the author of illustrations used in the *Treatise,* reference being made either (1) to publications cited in reference lists or (2) to the names of authors with or without indication of individual publications concerned. Previously unpublished illustrations are marked by the letter "n" (signifying "new") with the name of the author.

STRATIGRAPHIC DIVISIONS

Classification of rocks forming the geologic column as commonly cited in the *Treatise* in terms of units defined by concepts of time is reasonably uniform and firm throughout most of the world as regards major divisions (e.g., series, systems, and rocks representing eras) but it is variable and unfirm as regards smaller divisions (e.g., substages, stages, and subseries), which are provincial in application. Users of the *Treatise* have suggested the desirability of publishing reference lists showing the stratigraphic arrangement of at least the most commonly cited divisions. Accordingly, a tabulation of European and North American units, which broadly is applicable also to other continents, is given here.

Generally Recognized Divisions of Geologic Column

EUROPE

ROCKS OF CENOZOIC ERA
NEOGENE SYSTEM[1]
  Pleistocene Series (including Recent)
  Pliocene Series
  Miocene Series
PALEOGENE SYSTEM
  Oligocene Series
  Eocene Series
  Paleocene Series

ROCKS OF MESOZOIC ERA
CRETACEOUS SYSTEM
  Upper Cretaceous Series
    Maastrichtian Stage[2]
    Campanian Stage[2]
    Santonian Stage[2]
    Coniacian Stage[2]
    Turonian Stage
    Cenomanian Stage

  Lower Cretaceous Series
    Albian Stage
    Aptian Stage
    Barremian Stage[3]
    Hauterivian Stage[3]
    Valanginian Stage[3]
    Berriasian Stage[3]

JURASSIC SYSTEM
  Upper Jurassic Series
    Portlandian Stage[4]
    Kimmeridgian Stage
    Oxfordian Stage

  Middle Jurassic Series
    Callovian Stage (or Upper Jurassic)
    Bathonian Stage
    Bajocian Stage

  Lower Jurassic Series (Liassic)
    Toarcian Stage
    Pliensbachian Stage
    Sinemurian Stage
    Hettangian Stage

TRIASSIC SYSTEM
  Upper Triassic Series
    Rhaetian Stage[5]
    Norian Stage
    Carnian Stage
  Middle Triassic Series
    Ladinian Stage
    Anisian Stage (Virglorian)
  Lower Triassic Series
    Scythian Stage (Werfenian)

NORTH AMERICA

ROCKS OF CENOZOIC ERA
NEOGENE SYSTEM[1]
  Pleistocene Series (including Recent)
  Pliocene Series
  Miocene Series
PALEOGENE SYSTEM
  Oligocene Series
  Eocene Series
  Paleocene Series

ROCKS OF MESOZOIC ERA
CRETACEOUS SYSTEM
  Gulfian Series (Upper Cretaceous)
    Navarroan Stage
    Tayloran Stage
    Austinian Stage

    Woodbinian (Tuscaloosan) Stage
  Comanchean Series (Lower Cretaceous)
    Washitan Stage

    Fredericksburgian Stage
    Trinitian Stage

  Coahuilan Series (Lower Cretaceous)
    Nuevoleonian Stage

    Durangoan Stage

JURASSIC SYSTEM
  Upper Jurassic Series
    Portlandian Stage
    Kimmeridgian Stage
    Oxfordian Stage

  Middle Jurassic Series
    Callovian Stage (or Upper Jurassic)
    Bathonian Stage
    Bajocian Stage

  Lower Jurassic Series (Liassic)
    Toarcian Stage
    Pliensbachian Stage
    Sinemurian Stage
    Hettangian Stage

TRIASSIC SYSTEM
  Upper Triassic Series
    (Not recognized)
    Norian Stage
    Carnian Stage
  Middle Triassic Series
    Ladinian Stage
    Anisian Stage
  Lower Triassic Series
    Scythian Stage

ROCKS OF PALEOZOIC ERA

PERMIAN SYSTEM

Upper Permian Series
Tatarian Stage[6]
Kazanian Stage[7]
Kungurian Stage

Lower Permian Series
Artinskian Stage[8]
Sakmarian Stage
Asselian Stage

CARBONIFEROUS SYSTEM

Upper Carboniferous Series
Stephanian Stage

Westphalian Stage

Namurian Stage

Lower Carboniferous Series
Visean Stage

Tournaisian Stage
Strunian Stage

DEVONIAN SYSTEM

Upper Devonian Series

Famennian Stage

Frasnian Stage

Middle Devonian Series
Givetian Stage

Couvinian Stage

Lower Devonian Series
Emsian Stage
Siegenian Stage
Gedinnian Stage

SILURIAN SYSTEM

Ludlow Series

Wenlock Series

ROCKS OF PALEOZOIC ERA

PERMIAN SYSTEM

Upper Permian Series
Ochoan Stage
Guadalupian Stage

Lower Permian Series
Leonardian Stage
Wolfcampian Stage

PENNSYLVANIAN SYSTEM

Kawvian Series (Upper Pennsylvanian)
Virgilian Stage
Missourian Stage
Oklan Series (Middle Pennsylvanian)
Desmoinesian Stage
Bendian Stage
Ardian Series (Lower Pennsylvanian)
Morrowan Stage

MISSISSIPPIAN SYSTEM

Tennesseean Series (Upper Mississippian)
Chesteran Stage

Meramecian Stage
Waverlyan Series (Lower Mississippian)
Osagian Stage
Kinderhookian Stage

DEVONIAN SYSTEM

Chautauquan Series (Upper Devonian)

Conewangoan Stage
Cassadagan Stage

Senecan Series (Upper Devonian)
Chemungian Stage
Fingerlakesian Stage

Erian Series (Middle Devonian)
Taghanican Stage
Tioughniogan Stage
Cazenovian Stage

Ulsterian Series (Lower Devonian)
Onesquethawan Stage
Deerparkian Stage
Helderbergian Stage

SILURIAN SYSTEM

Cayugan Series
Includes age equivalents of middle and upper Ludlow (in New York)

Niagaran Series
Includes age equivalents of upper Llandovery, Wenlock, and lower Ludlow (in New York)

Landovery Series	**Medinan Series** Includes age equivalents of lower and middle Llandovery (in New York)
ORDOVICIAN SYSTEM	**ORDOVICIAN SYSTEM**

	Cincinnatian Series (Upper Ordovician)
	    Richmondian Stage
	    Maysvillian Stage
	    Edenian Stage

Ashgill Series	

Caradoc Series	Champlainian Series (Middle Ordovician)
	    Mohawkian Stage
	        Trentonian Substage
	        Blackriveran Substage
	    Chazyan Stage

Llandeilo Series Llanvirn Series	
Arenig Series Tremadoc Series[9]	Canadian Series (Lower Ordovician)
CAMBRIAN SYSTEM	**CAMBRIAN SYSTEM**

Upper Cambrian Series	Croixian Series (Upper Cambrian)
	    Trempealeauan Stage
	    Franconian Stage
	    Dresbachian Stage

Middle Cambrian Series Lower Cambrian Series	Albertan Series (Middle Cambrian) Waucoban Series (Lower Cambrian)
EOCAMBRIAN SYSTEM (?Upper Proterozoic) **ROCKS OF PRECAMBRIAN AGE**	**EOCAMBRIAN SYSTEM (?Upper Proterozoic)** **ROCKS OF PRECAMBRIAN AGE**

CURT TEICHERT AND RAYMOND C. MOORE

---

[1] Considered by some to exclude post-Pliocene deposits.
[2] Classed as division of Senonian Subseries.
[3] Classed as division of Neocomian Subseries.
[4] Includes Purbeckian deposits.
[5] Interpreted as lowermost Jurassic in some areas.
[6] Includes some Lower Triassic and equivalent to upper Thuringian (Zechstein) deposits.
[7] Equivalent to lower Thuringian (Zechstein) deposits.
[8] Equivalent to upper Autunian and part of Rotliegend deposits.
[9] Classed as uppermost Cambrian by some geologists.

PART V

GRAPTOLITHINA
WITH SECTIONS ON ENTEROPNEUSTA AND PTEROBRANCHIA

By O. M. B. BULMAN
[Cambridge University]

CONTENTS

	PAGE
INTRODUCTION	*V*5
General features	*V*5
Outline of classification	*V*6
Historical notes on classification of Graptolithina	*V*6
Classification of Hemichordata	*V*7
Morphological terms applied to Graptolithina and other Hemichordata	*V*8
Glossary of morphological terms	*V*8
Stratigraphical note	*V*12
HEMICHORDATA	*V*12
Phylum Hemichordata Bateson, 1885, *emend*. Fowler, 1892	*V*12
ENTEROPNEUSTA	*V*13
Class Enteropneusta Gegenbaur, 1870	*V*13
PTEROBRANCHIA	*V*13
Class Pterobranchia Lankester, 1877	*V*13
Morphology	*V*13
Order Rhabdopleurida Fowler, 1892	*V*14
Order Cephalodiscida Fowler, 1892	*V*16
PLANCTOSPHAEROIDEA	*V*17
Class Planctosphaeroidea van der Horst, 1936	*V*17
GRAPTOLITHINA	*V*17
Diagnosis and general features	*V*17
Class Graptolithina Bronn, 1846	*V*17
Pioneer work on graptolites	*V*17
Techniques	*V*18
Preparation of specimens	*V*18
Illustration	*V*20
Structure and composition of periderm	*V*21
Graptolite affinities	*V*22

	PAGE
Nature of graptolite zooid	V22
Dendroidea	V25
Order Dendroidea Nicholson, 1872	V25
Morphology	V26
Thecae	V26
Stolothecae	V26
Autothecae	V27
Bithecae	V28
Thecal grouping	V28
Mode of branching	V31
Dissepiments and anastomosis	V32
Development	V32
Paleoecology	V34
Geographic distribution	V35
Stratigraphic distribution	V35
Classification	V36
Systematic descriptions	V36
Family Dendrograptidae Roemer in Frech, 1897	V36
Family Anisograptidae Bulman, 1950	V39
Family Ptilograptidae Hopkinson, 1875, in Hopkinson & Lapworth, 1875	V41
Family Acanthograptidae Bulman, 1938	V41
Tuboidea	V44
Order Tuboidea Kozłowski, 1938	V44
Morphology	V44
Thecae	V44
Stolothecae	V44
Autothecae	V44
Bithecae	V45
Conothecae	V46
Form of rhabdosome and thecal grouping	V46
Development	V46
Systematic descriptions	V47
Family Tubidendridae Kozłowski, 1949	V47
Family Idiotubidae Kozłowski, 1949	V47
Camaroidea	V49
Order Camaroidea Kozłowski, 1938	V49
Morphology	V49
Autothecae	V49
Bithecae	V50
Stolon system	V50
Systematic descriptions	V50
Family Bithecocamaridae Bulman, 1955	V50
Family Cysticamaridae Bulman, 1955	V50
Crustoidea	V51
Order Crustoidea Kozłowski, 1962	V51

	PAGE
Morphology	*V*51
Autothecae	*V*51
Bithecae	*V*51
Stolothecae	*V*51
Graptoblasts and cysts	*V*51
Systematic descriptions	*V*52
Family Wimanicrustidae Bulman, n. fam.	*V*52
Family Hormograptidae Bulman, n. fam.	*V*52
Stolonoidea	*V*53
Order Stolonoidea Kozłowski, 1938	*V*53
Morphology	*V*53
Systematic descriptions	*V*53
Family Stolonodendridae Bulman, 1955	*V*53
Dendroidea, Tuboidea, Camaroidea, Crustoidea, Stolonoidea, taxonomic position uncertain	*V*54
Graptoloidea	*V*57
Order Graptoloidea Lapworth, 1875	*V*57
Morphology	*V*57
General features	*V*57
Sicula	*V*59
Thecae	*V*60
General relations	*V*60
Interthecal septum	*V*62
Median septum	*V*62
Principal types	*V*62
Monograptid trends and their significance	*V*66
Apertural processes, spines and localized thickening of periderm	*V*66
Nema	*V*69
Regeneration	*V*70
Abnormalities in development	*V*71
Development	*V*71
General discussion	*V*71
Position and formation of porus	*V*72
Initial bud	*V*73
Later development	*V*73
Dichograptid type	*V*75
Isograptid type	*V*75
Leptograptid type	*V*75
Diplograptid type	*V*77
Monograptid type	*V*79
Pericalycal type	*V*79
Branching of rhabdosome	*V*82
Cladia	*V*85
Synrhabdosomes	*V*89
Paleoecology	*V*91

	PAGE
Significance of graptolite facies	V91
Buoyancy mechanism	V93
Geographic distribution	V95
Stratigraphic distribution	V96
Anisograptid fauna	V97
Dichograptid fauna	V97
Diplograptid fauna	V99
Monograptid fauna	V99
British and Australian graptolite zones	V100
Principles of classification	V100
Phylogeny	V103
Didymograptina	V103
Glossograptina	V106
Diplograptina	V107
Monograptina	V109
Systematic descriptions	V109
Suborder Didymograptina Lapworth, 1880, *emend.* Bulman, herein	V109
Family Dichograptidae Lapworth, 1873	V109
Multiramous forms	V111
Pauciramous forms	V115
Family Sinograptidae Mu, 1957	V117
Family Abrograptidae Mu, 1958	V118
Family Corynoididae Bulman, 1944	V119
Family Nemagraptidae Lapworth (*ex* Hopkinson MS), 1873	V119
Family Dicranograptidae Lapworth, 1873	V121
Suborder Glossograptina Jaanusson, 1960	V122
Family Glossograptidae Lapworth, 1873	V122
Family Cryptograptidae Hadding, 1915, *emend.* Bulman, herein	V123
Suborder Diplograptina Lapworth, 1880, *emend.* Bulman, herein	V123
Family Diplograptidae Lapworth, 1873	V124
Family Lasiograptidae Lapworth, 1879	V126
Family Dicaulograptidae Bulman, n. fam.	V128
Family Peiragraptidae Jaanusson, 1960	V128
Family Retiolitidae Lapworth, 1873	V128
Subfamily Retiolitinae Lapworth, 1873	V128
Subfamily Archiretiolitinae Bulman, 1955	V130
Subfamily Plectograptinae Bouček & Münch, 1952	V130
Family Dimorphograptidae Elles & Wood, 1908	V131
Suborder Monograptina Lapworth, 1880	V132
Family Monograptidae Lapworth, 1873	V132
Family Cyrtograptidae Bouček, 1933	V134
Subfamily Cyrtograptinae Bouček, 1933	V134
Subfamily Linograptinae Obut, 1957	V135
Graptolithina incertae sedis	V136
Group Graptoblasti Kozłowski, 1949	V136

	PAGE
Group Acanthastida Kozłowski, 1949	V138
Group Graptovermida Kozłowski, 1949	V138
Unrecognizable genera	V139
REFERENCES	V139
ADDENDUM	V149
Classification of the graptolite family Monograptidae Lapworth, 1873 (O. M. B. Bulman and R. B. Rickards)	V149
Introduction	V149
Acceptable genera	V150
Rastrites	V150
Monoclimacis	V151
Pristiograptus	V151
Saetograptus	V151
Cucullograptus and Lobograptus	V152
Genera of dubious value	V152
Genera based on rhabdosome shape	V152
Genera based mainly on thecal form	V154
Indeterminate genera	V156
Selected references	V156
INDEX	V158

INTRODUCTION

This section of the *Treatise* is concerned essentially with the Graptolithina, but it is prefaced with a short account of living and fossil Hemichordata *(Balanoglossus, Cephalodiscus,* and *Rhabdopleura)*. The reason for this is that although the graptolites are an extinct group, confined to the Paleozoic, morphological discoveries during the past 20 years have made it seem probable that they may be allied more nearly to some of the Hemichordata than to any other living group, and accordingly they are here provisionally regarded as a separate class of that phylum. An account of morphology of the soft parts, particularly of *Rhabdopleura,* may therefore help the student to visualize the kind of zooid which probably inhabited the graptolite rhabdosome, though it must be borne in mind that the analogy is tentative. The evidence concerning graptolite affinities is discussed here on p. *V*22 and the broader classification of these living protochordates adopted here accords with that used by BARRINGTON (1965), where also will be found a concise discussion of the relations of the protochordates to other Deuterostomia.

It is a pleasure to acknowledge the patient editorial assistance of Professors CURT TEICHERT and RAYMOND C. MOORE and the help of numerous of my students and co-workers in Britain and overseas in the preparation of this second edition of *Treatise* Part V. Likewise, I express special thanks to LAVON MCCORMICK and ROGER B. WILLIAMS, of the *Treatise* editorial staff at the University of Kansas, for painstaking work by them on typescripts and illustrations.

All figures have been specially drawn and are "after" rather than "from" the sources indicated.

GENERAL FEATURES

While the tunicates (Urochordata) and Acrania (Cephalochordata) are currently accepted as protochordate members of the phylum Chordata, the Hemichordata are now regarded by most authorities as constituting an independent phylum. It is extremely difficult, however, to give a concise, comprehensive survey of the general features of the varied and specialized groups comprised in this phylum. The Enteropneusta and Pterobranchia are virtually the only two living classes, for the Planctosphaeroidea scarcely merit consideration here, and the Graptolithina are an extinct class provisionally assigned to the phylum; the affinities of the Graptolithina and the nature of the graptolite zooid are discussed on p. *V*22.

The Enteropneusta lack any coenoecium (external skeleton) common to the other two classes, but show significant resemblances to the Pterobranchia in their soft-part morphology and ontogeny. The Graptolithina show significant coenoecial resemblances to the Pterobranchia, though their zooids are known only by inference. Thus the features linking the enteropneusts to the pterobranchs are inapplicable to the Graptolithina, and those connecting the Graptolithina and the Pterobranchia are irrelevant to the Enteropneusta.

The Enteropneusta and the Pterobranchia belong to the Deuterostomia because the anus develops from the blastopore and the mouth represents a new opening. Also they possess an enterocoelic coelom divided into anterior, median, and posterior chambers. As Hemichordata, they possess in addition pharyngeal openings (absent in *Rhabdopleura*), but they are distinct from the Chordata because they lack a notochord and an endostyle (an organ homologous with the thyroid gland of the Chordata). The body shows a division into proboscis or cephalic shield, collar, and trunk, and following metamorphosis of the tornaria larva, the larval pterobranch bears a strong resemblance to the wormlike enteropneusts, with a tripartite body and terminal anus.

The graptolithine and pterobranch coenoecium or rhabdosome consists of tubes, or thecae, which may comprise both fusellar and cortical tissue of scleroproteic composition and in at least some orders of both classes a comparable stolon system occurs. The external layer of living tissue postulated in the Graptolithina has no counterpart in any known pterobranch (or indeed in any other hemichordate).

OUTLINE OF CLASSIFICATION

HISTORICAL NOTES ON CLASSIFICATION OF GRAPTOLITHINA

The name *Graptolithus* was applied by LINNÉ in 1735 (*Systema Naturae*, edit. 1) to inorganic markings (such as dendritic incrustations) simulating fossils, and when in his 12th edition of *Systema Naturae* (1768) he included *G. sagittarius* and *G. scalaris*, these, too, were considered to be inorganic. The former nominal species is possibly a fossil plant, and the latter probably a graptolite. In his *Skånska Resa* (1751) he had described and figured a "Fossil or graptolite of a strange kind," now believed to represent *Climacograptus scalaris* and *Monograptus triangulatus;* and in 1821 the name *Graptolithus* was used by WAHLENBERG for definite graptolite remains. HISINGER, MURCHISON, and others described many more, and BRONN *(Index Palaeontologicus)* listed species known to him up to 1846, placing them in a subdivision of the Anthozoa. Numerous genera and subgenera came to be described during the second half of the century (see p. *V*100) and following LAPWORTH, 1873, the name *Graptolithus* was abandoned; it was formally suppressed in 1954 (ICZN, Opinion 197) and placed on the list of rejected generic names in view of the doubtful nature of the genolectotype and the originally-expressed intention to denote inorganic objects.

Several older writers had used some kind of key arrangement of genera in systematic sections of their work, and NICHOLSON (1872) in his uncompleted *Monograph* divided his family Graptolitidae into four "sections": Monoprionidae and Diprionidae for uniserial and biserial rhabdosomes (based on BARRANDE's "subgenera" *Monoprion* and *Diprion*), Tetraprionidae, and Dendroidea. This seems to be the first positive separation of the Dendroidea from the true graptolites and NICHOLSON wrote:

"These forms are very doubtfully referable to the Graptolitidae." The first formal classification into families was that of LAPWORTH, 1873, prefaced by a short but penetrating analysis of the structure and development of graptolite rhabdosome. To a remarkable extent this still constitutes the basis of current classification and it is quoted below:

Lapworth's (1873) Arrangement of Graptolite Rhabdosomes

RHABDOPHORA (Allman)
Section I. GRAPTOLITIDAE
Family I. Monograptidae
Family II. Nemagraptidae (HOPKINSON MS)
Family III. Dichograptidae
Family IV. Dicranograptidae
Family V. Diplograptidae
Family VI. Phyllograptidae
Section II. RETIOLOIDEA
Family VII. Glossograptidae (provisional Family)
Family VIII. Retiolitidae

With renaming of the Nemagraptidae as Leptograptidae (LAPWORTH, 1879), merging of the Phyllograptidae in Dichograptidae, and addition of the Dimorphograptidae (ELLES & WOOD, 1908), together with the suppression of Section II, Retioloidea, we find virtually the classification used in the *Monograph of British Graptolites* (ELLES & WOOD, 1901-19). ALLMAN's term Rhabdophora, erected by him as a suborder of the Hydroida, became redundant when the Retioloidea were merged with the Graptoloidea, and HOPKINSON's comparable term Cladophora (HOPKINSON & LAPWORTH, 1875) is synonymous with Dendroidea (NICHOLSON, 1872).

Although WIMAN (1895), retained the "group" Retioloidea, he based the Dendroidea not only on the dendroid rhabdosome habit but, more important, on thecal polymorphism. In spite of this, the dendroids were again reduced to family rank by FRECH (1897), who included them with the Dichograptidi in his Axonolipa, distinguished from all other graptolites which constituted his Axonophora. This emphasis on the presence or absence of a virgula was, of course, quite excessive, for FRECH did not appreciate the identity of the virgula with the nema. Moreover, it is a feature difficult to apply systematically, and RUEDEMANN (1904, 1908) adopted the FRECH grouping but with a different familial composition. The terms Axonolipa and Axonophora are often useful adjectively, but are not now accepted as having any taxonomic value.

Other families have been added to the Graptoloidea and the Dendroidea in recent years, and several new orders have been added to the Graptolithina by KOZŁOWSKI.

CLASSIFICATION OF HEMICHORDATA

The following tabulation records numbers of genera in suprageneric divisions of Enteropneusta, Pterobranchia, and Graptolithina, accompanied by statements of stratigraphic ranges. Family-group taxa which contain subgenera are accompanied by two figures, the number of included genera being indicated by the first and subgenera additional to nominotypical subgenera by the second. Thus, the figures 16;1 indicate 16 genera and 1 subgenus in addition to the nominotypical one.

Main Divisions of Enteropneusta, Pterobranchia, and Graptolithina

Hemichordata *(phylum)*. (202;10). *M.Cam.-Rec.*
Enteropneusta *(class)*. (12). *Rec.*
Pterobranchia *(class)*. (7;3). *L.Ord.(Tremadoc)-Rec.*
Rhabdopleurida *(order)*. (3). *L.Ord.-Rec.*
Rhabdopleuridae (3). *L.Ord.-Rec.*
Cephalodiscida *(order)* (4;3). *L.Ord.(Tremadoc)-Rec.*
Eocephalodiscidae (1). *L.Ord.(Tremadoc).*
Cephalodiscidae (3;3). *Ord., ?Tert., Rec.*
Planctosphaeroidea *(class)*. *Rec.*
Graptolithina *(class)*. (183;7). *M.Cam.-Carb.*
Dendroidea *(order)* (21;3). *?M.Cam., U.Cam.-Carb.(Namur.)*
Dendrograptidae (9;3). *?M.Cam., U.Cam.-Carb.*
Anisograptidae (11). *L.Ord.(Tremadoc), ?U.Ord.*
Ptilograptidae (1). *L.Ord.-U.Sil.*
Acanthograptidae (5). *?U.Cam., L.Ord.-M.Dev.*
Tuboidea *(order)* (12). *?U.Cam., L.Ord.-Sil.*
Tubidendridae (2). *L.Ord.(Tremadoc)-Sil.*
Idiotubidae (10). *?U.Cam.(USSR), L.Ord.-Sil.*
Camaroidea *(order)* (5). *Ord.*
Bithecocamaridae (1). *L.Ord.*

Cysticamaridae (4). *L.Ord.*
Crustoidea *(order)* (7). *L.Ord.-U.Ord.*
  Wimanicrustidae (6). *L.Ord.-U.Ord.*
  Hormograptidae (1). *U.Ord.*
Stolonoidea *(order)* (2). *Ord.*
  Stolonodendridae (2). *Ord.*
Taxonomic Position Uncertain (24). *M.Cam.-Dev.*
Graptoloidea *(order)* (108;4). *L.Ord.(Arenig)-L.Dev.(?Emsian).*
  Didymograptina *(suborder)* (54). *Ord.*
    Dichograptidae (36). *L.Ord.*
      Multiramous forms (21). *L.Ord.*
        Goniograpti (13). *L.Ord.*
        Temnograpti (2). *L.Ord.*
        Schizograpti (5). *L.Ord.*
        Dichograpti (1). *L.Ord.*
      Pauciramous forms (15). *L.Ord.-U.Ord.*
        Tetragrapti (3). *L.Ord.*
        Didymograpti (12). *L.Ord.-U.Ord.*
    Sinograptidae (6). *L.Ord.(U.Arenig-L.Llanvirn).*
    Abrograptidae (3). *Ord.(?U.Arenig-Nemagraptus gracilis Z.)*
    Corynoididae (2). *U.Ord.*
    Nemagraptidae (5). *?L.Ord., U.Ord.*
    Dicranograptidae (2). *L.Ord.-U.Ord.*
  Glossograptina *(suborder)* (5). *Ord.*
    Glossograptidae (4). *Ord.*
    Cryptograptidae (1). *L.Ord.-U.Ord.*
  Diplograptina *(suborder)* (35;3). *L.Ord.-U.Sil.*
    Diplograptidae (9;3). *L.Ord.-L.Sil.*
    Lasiograptidae (5). *Ord.*
    Dicaulograptidae (1). *L.Ord.*
    Peiragraptidae (1). *U.Ord.*
    Retiolitidae (16). *U.Ord.-U.Sil.*
      Retiolitinae (5). *U.Ord.-M.Sil.*
      Archiretiolitinae (6). *U.Ord.*
      Plectograptinae (5). *?L.Sil., M.Sil.-U.Sil.*
    Dimorphograptidae (3). *L.Sil.*
  Monograptina *(suborder)* (14;1). *L.Sil.-L.Dev.*
    Monograptidae (6;1). *L.Sil.-L.Dev.*
    Cyrtograptidae (8). *L.Sil.-L.Dev.*
      Cyrtograptinae (3). *M.Sil.(Wenlock).*
      Linograptinae (5). *L.Sil.-L.Dev.*
Graptolithina Incertae Sedis (4). *Ord.*
  Group Graptoblasti (2). *L.Ord.(Tremadoc-Llandeilo).*
  Group Acanthastida (1). *L.Ord.(Tremadoc).*
  Group Graptovermida (1). *L.Ord.(Tremadoc).*
Unrecognizable genera (38).

MORPHOLOGICAL TERMS APPLIED TO GRAPTOLITHINA AND OTHER HEMICHORDATA

Description of the morphology of various-rank divisions of the Hemichordata recognized in this book is given in several places under appropriate headings. Accordingly, it has seemed very desirable to organize a single alphabetically arranged glossary of morphological terms containing concise definitions and indicating typographically the relative importance attached to the different terms. Thus, most commonly used terms are printed in boldface capital letters (as **AUTOTHECA**), useful but less important terms in boldface small letters (as **clathria**), and least important (in part obsolete) terms in italic letters (as *solid axis*).

Glossary of Morphological Terms

adapertural plate. Portion of apertural lobe in cucullograptids fused to ventral or dorsal wall of theca.

amplexograptid theca. Strongly geniculate theca with deep and long, rounded apertural excavations, generally with infragenicular selvage and typically with low rounded apertural lappets.

anastomosis. Temporary fusion, as of adjacent branches to form an ovoid mesh.

ancora (ancora stage). Anchor-shaped initial growth stage of retiolitids, apparently formed of virgella with two distal bifurcations.

angular fuselli. Extremely thin growth-bands of fusellar tissue filling angle between apertural margin and wall of succeeding theca (see Fig. 93,5a).

annulus (*pl.,* **annuli**). Internal ring on sicula and (rarely) early thecae of some monograptids, composed of fine irregularly laminated tissue.

apertural spine. Projection originating on margin of aperture; commonly single, less commonly paired.

appendix. Reticulate tubular structure at distal end of rhabdosome in Plectograptinae.

aseptate. Biserial rhabdosome lacking median septum.

auriculate. Expanded, earlike lateral lobes in highly modified thecae; e.g., cucullograptids, Crustoidea.

AUTOTHECA. Larger type of regularly-developed graptolite thecae, possibly containing female zooid (e.g., in Dendroidea) or hermaphrodite zooid (in Graptoloidea). (*See* also stolotheca.)

axil. Base of V-shaped bifurcation of dichotomously branched rhabdosomes, and especially bifurcation of dicranograptids.

axonolipous. Graptoloid rhabdosomes which are not scandent and therefore do not enclose nema.

axonophorous. Scandent biserial and uniserial graptoloids in which nema (virgula) is enclosed within rhabdosome or embedded in dorsal wall.

basal disc. Discoidal plate developed from apex of sicula for attachment of sessile graptolites, as in Dendroidea.

biform. Graptoloid rhabdosome (especially monograptids) with proximal and distal thecae of conspicuously different form.

bilateral. Graptoloid rhabdosome disposed more or less symmetrically to right and left of sicula.

bipolar. Bilateral monograptid rhabdosome with sicular cladium or pseudocladium.

BISERIAL. Scandent graptoloid rhabdosome with two series of thecae enclosing nema (virgula). (See also dipleural, monopleural.)

BITHECA. Smaller type of regularly-developed graptolite thecae, absent in Graptoloidea, possibly containing male zooid.

branch. See stipe.

branching, dichotomous. Division of stipe in which two branches diverge symmetrically from parent stipe.

branching, lateral. Division of stipe in which branch diverges at angle to parent stipe, which continues its original direction of growth.

budding individual. Term formerly used for stolotheca (now obsolete).

camara. Inflated proximal portion of autotheca in Camaroidea.

central disc. Web of sclerotized tissue uniting proximal ends of stipes in certain horizontal graptoloid rhabdosomes.

cladium. Rhabdosome developed from sicular or thecal aperture in Cyrtograptidae. (See also metacladium, procladium, pseudocladium.)

clathria. Skeletal framework of rods (lists) composing rhabdosome, in some supporting reticulum or attenuated periderm.

climacograptid theca. Strongly geniculate theca with straight or slightly convex supragenicular wall parallel to axis of rhabdosome and relatively short (narrow) apertural excavation.

COENOECIUM. Tubular exoskeleton of colonies or associations of Pterobranchia.

collum. Erect tubular (distal) portion of autotheca in Camaroidea.

colony. See rhabdosome.

common canal. Term sometimes used for continuous tubular cavity collectively formed by prothecae of graptoloid; rarely involving some portion of metathecae.

complete septum. See median septum.

CONOTHECA. Relatively large, conical theca with small circular aperture, irregularly developed on rhabdosomes of certain Tuboidea.

corona (corona stage). Inflated reticulate proximal end of retiolitids succeeding ancora stage in development.

CORTICAL TISSUE. Outer layer of finely, rather irregularly laminated tissue composing graptolite periderm *(q.v.).*

CROSSING CANAL. Proximal (prothecal) portion of graptoloid theca which grows across axis of sicula to develop on side opposite that of its origin.

cryptoseptate. Biserial rhabdosomes in which median septum is composed of peridermal rods arranged as in septate forms, but lacking peridermal septal membrane.

cysts. Vesicles of varying size and shape occurring in autothecal cavities of Crustoidea.

declined. Graptoloid rhabdosome with branches hanging below the sicula, subtending an angle less than 180° between their ventral sides (see Fig. 38).

deflexed. Similar to declined but with distal extremities of stipes tending to horizontal (see Fig. 38).

dendroid (habit of growth). Bushy colony formed by irregular branching.

denticulate. Sharply pointed thecal apertures provided with short spine or mucro.

diad budding. Mode of budding in Tuboidea resulting in two zooids at each nodal division, lacking regularity of thecal succession.

DICALYCAL THECA. Graptoloid theca giving rise to two buds (c.f. normal asexual reproduction in which single bud is produced by each zooid).

dicellograptid theca. Geniculate theca characterized by introversion, usually accompanied by some degree of isolation of apertural region.

dichograptid theca. Straight, almost parallel-sided tubular theca.

dichotomous. See branching.

DIPLEURAL. Biserial graptoloid rhabdosome in which two stipes are in back-to-back contact so that each stipe has two external walls.

dissepiment. Strand of cortical periderm serving to connect adjacent branches in dendroid rhabdosome (especially *Dictyonema*).

DISTAL. Last-formed part (of stipe, theca, etc.) farthest away from point of origin.

DORSAL. Side of stipe opposite thecal apertures, or comparable side of thecal aperture; not necessarily related to position of growth, but presumably related to dorsal side of zooid.

everted. Plane of aperture facing outward (cf. introverted, retroverted); may be associated with angular fuselli *(q.v.).*

extensiform. Didymograptid with horizontal stipes.

flabellate (habit of growth). Rhabdosome fan-shaped, with stipes spread out in single plane.

FUSELLAR TISSUE. Inner layer of periderm, generally composed of alternating L and R growth bands or fuselli.

genicular spine. Sharp projection originating on geniculum; commonly single, rarely paired.

GENICULUM. Angular bend in direction of growth of graptoloid theca, especially climacograptid or lasiograptid; hence supragenicular, infragenicular.

glyptograptid theca. Sinuous theca with smooth curve in place of angular geniculum and convex supragenicular wall.

gonangium. Term formerly used for bitheca (now obsolete).

gymnocaulus. Unsclerotized stolon situated behind terminal bud in *Rhabdopleura,* from which zooids are proliferated.

gymnograptid theca. Sharply geniculate theca with extremely short supragenicular wall directed inward distally and generally with deep and long, rounded apertural excavation.

horizontal. Graptoloid rhabdosome with stipes disposed in plane at right angles to axis of sicula.

hydrosome. Obsolete term for rhabdosome.

hydrotheca. Obsolete term for autotheca of dendroids and other groups and for theca of graptoloids.

incomplete septum. *See* median septum.

INITIAL BUD. Outgrowth through foramen in sicular wall producing first theca of rhabdosome; stolotheca of dendroids, etc., prothecal portion of first theca of graptoloids.

interthecal septum. Peridermal membrane separating overlapping thecal cavities in Graptoloidea, comprising dorsal wall of one theca and part of ventral wall of succeeding theca.

introverted. Plane of aperture facing inward (dorsally), resulting from excessive growth of ventral wall of theca usually accompanied by sigmoidal curvature of thecal axis.

isolation. Separation of distal (metathecal) portions of thecae from stipe, as in *Rastrites,* or distal portions of autothecae of Dendroidea, etc.

lacinia. Delicate skeletal network, extraneous to rhabdosome proper, supported on spines.

lacuna stage. Final period in development of porus in monograptids, where notch or sinus is closed by fusellar growth bands.

languette. Laterally expanded ventral apertural process of theca.

lappet. Broad, rounded, lateral apertural process of theca (or sicula).

lasiograptid theca. Sharply geniculate theca with supragenicular wall directed inwards distally and deep, moderately long, rounded apertural excavation; less extreme than gymnograptid.

leptograptid theca. Theca with rounded geniculum and very long supragenicular wall typically parallel to axis of stipe.

list. Skeletal rod strengthening periderm in Graptoloidea, a unit of clathria.

lophophore. Paired arms or groups of arms, ciliated and bearing tentacles, situated adjacent to mouth of zooid; functionally food-collecting and respiratory.

MEDIAN SEPTUM. Partition in biserial graptoloids separating two series of thecae. Its relationships in monopleural rhabdosomes are imperfectly known. In dipleural forms it appears to be a single membrane and arises between daughter thecae of dicalycal theca; thus a complete **septum** arises between the 4th $(th2^2)$ and 5th $(th3^1)$ thecae, an incomplete septum arises between some later pair, and a partial septum occurs on one side only, thecae appearing to alternate on opposite sides.

mesial. Middle portion of free ventral wall (supragenicular wall) of theca; hence mesial spine.

metacladium. Term proposed for thecal or sicular cladium as opposed to procladium or main stipe.

METASICULA. Distal portion of sicula composed of normal fusellar growth bands. (*See* also prosicula.)

METATHECA. Distal portion of graptoloid theca, morphologically equivalent to autotheca of dendroids, etc. (*see* also protheca).

microfusellar tissue. Fusellar substance composed of extremely fine and somewhat irregular growth bands; genicular flanges are generally composed of microfusellar tissue.

microtheca. Type of autotheca occurring in Tuboidea, with narrow terminal portion and differently oriented apertures.

monofusellar tissue. Type of fusellar substance laid down in single, not alternating, series of growth bands.

MONOPLEURAL. Biserial graptoloid rhabdosome in which two stipes are in contact laterally (Glossograptina) so that each stipe has only one external wall (*see* Fig. 62).

monopodial growth. Type of colonial growth with permanent terminal zooid behind which new zooids arise as stem elongates. (Cf. sympodial.)

multiramous. Branches numerous.

NEMA (*pl.,* **NEMATA**). Threadlike extension of apex of prosicula, extending embryonic *nema prosiculae;* probably solid in adult rhabdosomes. May have served for attachment or may terminate in disc of attachment or vanelike "float" structures.

OBVERSE. Aspect of graptoloid rhabdosome (especially early growth stages or biserial forms) in which sicula is most completely visible. (Cf. reverse).

occlusion. Sealing of thecal aperture by sclerotized film.

orders (of branching). Successive divisions of dichotomous branches, or successive generations of cladia.

orthograptid theca. Straight, parallel-sided, tubular theca of biserial graptoloid.

partial septum. *See* median septum.

pauciramous. Branches comparatively few.

pectocaulus. Sclerotized stolon (or "black stolon") embedded in lower surface of mature parts of coenoecium of *Rhabdopleura.*

pendent. With approximately parallel branches hanging below sicula (*see* Fig. 38).

pericalycal. Mode of development of scandent (monopleural) rhabdosomes associated with dicalycal $th1^1$ and left-handed origin of $th1^2$, sicula becoming largely enclosed on both sides during subsequent development. (Cf. platycalycal.)

PERIDERM. Horny substance of scleroproteic

composition forming skeleton of Graptolithina, comprising inner (fusellar) layer with growth bands and growth lines and outer (cortical) layer of finely laminated tissue.

platycalycal. Mode of development, especially of scandent dipleural rhabdosomes, associated with dicalycal $th2^1$ and concentration of budding on reverse side. (Cf. pericalycal.)

polymorphic. Colony comprising more than one kind of zooid, or rhabdosome with more than one type of theca.

porus. Circular opening in wall of sicula through which initial bud passes to exterior; generally produced by resorption, but in monograptids arises as apertural notch (sinus) during growth of sicula.

preoral lobe. Anterior glandular lobe or disc in pterobranchs, which secretes coenoecium.

procladium. Term proposed for main stipe of cladia-bearing rhabdosome, normal cladia then being distinguished as metacladia.

PROSICULA. Proximal, initially formed part of sicula, apparently secreted as single conical unit with faintly marked spiral thread; at later stage longitudinal fibers are added.

prosoblastic. Type of diplograptid development in which $th2^1$ and ultimately $th1^2$ grow upward (distally) from their origin. (Cf. streptoblastic.)

PROTHECA. Proximal portion of graptoloid theca before differentiation of succeeding theca; morphological equivalent of stolotheca of dendroids and other groups.

prothecal fold. Inverted U-shaped curvature of part of protheca (usually initial portion) giving noded appearance to dorsal margin of stipe in certain axonolipous graptoloids, or similarly placed swellings (rare) in monograptids.

PROXIMAL. First-formed portion (of rhabdosome, stipe, theca, etc.) nearest point of origin.

pseudocladium. Term proposed for regenerated portion of bipolar rhabdosome lacking sicula.

pseudovirgula. Virgula of thecal or sicular cladium, originating as thecal or sicular apertural spine.

quadriserial. Scandent graptoloid rhabdosome composed of four rows of thecae in "back-to-back" contact *(Phyllograptus)*.

reclined. Graptoloid rhabdosome with branches growing upward, subtending an angle less than 180° between their dorsal sides (*see* Fig. 38).

reflexed. Similar to reclined, but with distal extremities of the stipes tending to horizontal (*see* Fig. 38).

reticulum. Delicate irregular network, usually supported on clathria, replacing continuous periderm in retiolitids.

retroverted. Thecal apertures facing proximally in consequence of hooked or reflexed shape of metatheca, following excessive growth of dorsal wall of theca.

REVERSE. Aspect of graptoloid rhabdosome (especially early growth stages or biserial forms) in which sicula is more or less concealed by crossing canal(s).

RHABDOSOME. Sclerotized exoskeleton of entire graptolithine colony; includes compound rhabdosomes with cladia, but not associations of rhabdosomes. (See synrhabdosome.)

root. Irregular branching structure (cortical tissue) developed from apex of sicula serving for attachment of sessile dendroids, etc.

scalariform. Preservational view presenting ventral (thecal) aspect of graptoloid rhabdosome, especially biserial forms.

SCANDENT. Graptoloid rhabdosome with stipes growing erect (distally), enclosing or including nema (virgula) (*see* Fig. 38).

sclerotized. Hardening due to secretion of scleroproteic substances by zooid(s). (It is now known that chitin is completely lacking in graptolite periderm.)

scopulae. Peculiar ramifying fibrous development from edges of median septum (as in lasiograptids) comparable with lacinia.

selvage. Thickened margin, especially of aperture.

septal. Related to septum.

septum. *See* interthecal septum, median septum.

SICULA. Skeleton of initial zooid of colony, comprising conical prosicula and tubular distal metasicula.

sinus stage. Initial phase in development of porus in monograptids, consisting of notch in apertural margin.

solid axis. Obsolete term for virgula of graptoloids.

STIPE. One branch of branched rhabdosome or entire colony of unbranched rhabdosome.

stolon. Thin sclerotized sheath presumably surrounding unsclerotized thread of soft tissue, from which thecae appear to originate in Dendroidea and other groups; comparable to pectocaulus of pterobranchs.

STOLOTHECA. One of three principal types of theca (cf. autotheca and bitheca) enclosing main stolon and proximal portions of daughter stolotheca, autotheca and bitheca; probably secreted by immature autozooid and constituting in effect proximal portion of autotheca; equivalent to protheca of graptoloid.

streptoblastic. Type of diplograptid development in which significant portion of proximal parts of $th1^2$, $th2^1$ and even $th2^2$, grow downward. (Cf. prosoblastic.)

sympodial growth. Type of colonial growth in which each zooid is in turn terminal zooid of its branch. (Cf. monopodial.)

synrhabdosome. Association of several (usually biserial) graptoloids attached distally by their nemata to common center.

THECA. Sclerotized tube or cup enclosing any zooid of rhabdosome (other than sicula); term generally used to denote autotheca of Graptoloidea, which are not polymorphic.

thecal grouping. More or less regular association of groups of autothecae and bithecae forming small branches (twigs), particularly in acanthograptids.

thecorhiza. Encrusting basal disc in Tuboidea, composed principally of stolothecae, from which autothecae and bithecae arise singly, in clusters, or as branches.

triad budding. Mode of budding in Dendroidea and Crustoidea in which three zooids are produced at each division, with regular succession of thecae. *See* Wiman rule.

triangulate theca. Type of isolate monograptid theca, triangular in lateral view, with retroflexed aperture.

twig. *See* thecal grouping.

umbellate theca. Type of autotheca in some Tuboidea, characterized by enlarged, reflexed, umbrella-shaped hood extending back over aperture of preceding autotheca (*see* Fig. 25).

UNISERIAL. Rhabdosome or stipe of graptoloid consisting of single row of thecae only. (Cf. biserial, quadriserial.)

VENTRAL. Side of stipe on which thecal apertures are situated or comparable side of thecal aperture; not necessarily related to position of growth of rhabdosome, but assumed to be related to ventral side of zooid.

vesicular diaphragm. Globular swelling on main stolon at nodes or points of origin of daughter stolons.

VIRGELLA. Spine developed during growth of metasicula, embedded in sicular wall and projecting freely from its apertural margin.

virgellarium. Umbrella-shaped structure developed at tip of virgella in linograptids.

virgula. Term commonly used for nema of scandent graptoloids.

Wiman rule. Process of budding resulting in regularly alternating triads of autotheca, bitheca, and stolotheca, diagnostic of Dendroidea.

ZOOID. Soft-bodied individual inhabiting theca or coenoecial tube (e.g., thecal zooid, siculozooid).

STRATIGRAPHICAL NOTE

Following current practice in the *Treatise,* the Tremadoc Series is classed as the lowermost division of the Ordovician, not, as in most British works, as uppermost Cambrian. A good deal of confusion relates to the Lower-Middle and the Middle-Upper Ordovician boundaries in various parts of the world. I follow here the solution accepted by WHITTINGTON & WILLIAMS (1964) of recognizing only two divisions, Lower and Upper, drawing the boundary at the base of the *Nemagraptus gracilis* Zone. This species is widely distributed and constitutes a horizon that can be recognized confidently in most graptolite sequences. The Silurian is divided into the conventional Lower, Middle, and Upper (Llandovery, Wenlock, and Ludlow of the British succession), but it should be noted that the last is extended up to the base of the *Monograptus uniformis* Zone, which is currently accepted as the boundary between the Silurian and Devonian systems.

HEMICHORDATA

Phylum HEMICHORDATA
Bateson, 1885, emend. Fowler, 1892

[*nom. transl.* HYMAN, 1959, p. 74 (*ex* class Hemichordata BATESON, 1885, p. 111)] [=Stomochorda DAWYDOFF, 1948, p. 367 (subphylum)]

For reasons discussed in the section on "General Features" (p. *V*5), it is not possible to give a collective diagnosis covering living and extinct classes here assigned to the Hemichordata. Where the organism is known, it exhibits the essential embryological features of the Deuterostomia and also possesses pharyngeal openings (except in *Rhabdopleura*), but it lacks the endostyle and notochord distinctive of the phylum Chordata. When present, the coenoecium or rhabdosome consists of fusellar tissue, with or without an external laminated cortical tissue, and the substance of this is scleroproteic in composition.

The Hemichordata comprise the two extant classes, Enteropneusta and Pterobranchia. The former are unknown fossil, but the pterobranchs are known as exceedingly rare fossils dating back to the Tremadoc, their representatives thus being contemporary with the graptolites. The extremely rare living organism *Planctosphaera* is generally assigned to a separate (third) class, and for taxonomic convenience the Graptolithina are accepted here as a fourth and extinct class of the phylum. *M.Cam.-Rec.*

ENTEROPNEUSTA

Class ENTEROPNEUSTA Gegenbaur, 1870

[*nom. correct.* HAECKEL, 1879 (*pro* Enteropneusti GEGENBAUR, 1870, p. 158)] [=Hemichordata BATESON, 1885]

Free, with elongate wormlike body and pronounced division into an acorn-shaped proboscis (protosoma), collar (mesosoma), and trunk (metasoma); branchial apparatus well developed as a long double row of pores strengthened by a cuticular branchial skeleton. *Rec.*

The class is unknown fossil;[1] among numerous living genera, the following may be mentioned:

Balanoglossus DELLE CHIAJE, 1829, p. 141 [*B. clavigerus;* M]. Widely distributed around Atlantic and Pacific coasts.
Ptychodera ESCHSCHOLTZ, 1825. p. 740 [*P. flava;* M]. IndoPac.-W.Indies.
Saccoglossus SCHIMKEWITSCH, 1892, p. 93 [*Balanoglossus mereschkowskii* WAGNER, 1885, p. 46; M]. N.Atl.-White Sea-Japan-N.Z.——FIG. 1. *S. pusillus* RITTER, White Sea; ×0.7 (48).

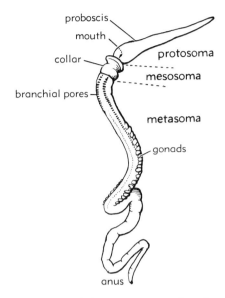

FIG. 1. A typical enteropneustan, *Saccoglossus pusillus,* showing the principal external features of the body (48).

PTEROBRANCHIA

Class PTEROBRANCHIA Lankester, 1877

[Class Pterobranchia LANKESTER, 1877, p. 448]

Fixed colonial or pseudocolonial organisms; body compact and without conspicuous division into three parts; middle segment (mesosoma) small but with one or more pairs of arms furnished with ciliated tentacles (=lophophore); posterior segment (metasoma) with long stalk or peduncle by which the zooid may be attached; branchial apparatus rudimentary; cuticular skeleton external. *L.Ord.(Tremadoc)-Rec.*

---

[1] What appears to be a giant abyssal enteropneust was photographed at the end of a "spiral" fecal cast at a deep-water Pacific station, as described by BOURNE & HEEZEN (*Science*, v. 150, 1965, p. 60), and the form of this cast is reminiscent of some trace fossils such as *Taphrhelminthopsis* that have been described from Alpine flysch. Comparable tracks are known, however, to be made by other types of organism. Burrows attributed to Enteropneusta have been described from the Muschelkalk of the Holy Cross Mountains by KAŹMIERCZAK & PSZCZÓŁKOWSKI (1969) with references to previous records from German Trias by SOERGEL (1923) and MAGDEFRAU (1932).

MORPHOLOGY

The body of pterobranchiates is small (2 to 7 mm. in *Cephalodiscus*) or even microscopic (less than 0.5 mm. in *Rhabdopleura*), and its most conspicuous feature is the **lophophore** structure developed from the mesosomal or collar segment, which gives it a pronounced bilateral symmetry and a superficially polyzoan appearance (Fig. 2,*1*). The lophophore consists of one pair of arms in *Rhabdopleura,* and many pairs in *Cephalodiscus,* and contains an extension of the collar coelom into the arms and tentacles. The preoral lobe (protosoma) forms a glandular cephalic disc which posteriorly overhangs the mouth. Only in embryonic stages is the mesosomal segment clearly differentiated from the trunk segment, and from the ventral side of the sac-like trunk arises the **peduncle** or contractile stalk. In *Rhabdopleura* this is of considerable length and serves to attach the organism at the base of its tube to the

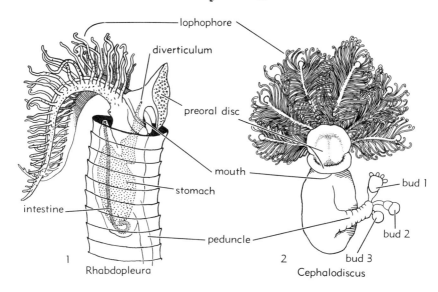

Fig. 2. Enlarged drawings of *(1) Rhabdopleura* and *(2) Cephalodiscus* showing the principal external features of the body in the Pterobranchia (48).

pectocaulus or stolon; it is shorter in *Cephalodiscus* and free, the organisms living not in true colonies but in associations.

The mouth opens into a large pharyngeal region, from the roof of which is given off anteriorly a small diverticulum formerly regarded as the homologue of the notochord. Laterally the pharynx is developed into a branchial region with a single pair of branchial pores in *Cephalodiscus*, whereas in *Rhabdopleura* the whole branchial structure is rudimentary. Posteriorly, the pharynx leads into a capacious stomach, from which a straight intestine doubles back to the anal pore situated on a dorsal prominence in the front part of the trunk segment (Fig. 2). The gonads are paired in *Cephalodiscus*, single in *Rhabdopleura;* the sexes are separate except in certain species of *Cephalodiscus* where hermaphrodite individuals occur. Males and females usually are indistinguishable, but some species are dimorphic and in *C. sibogae* the males are degenerate. Asexual reproduction (budding) is common in both genera.

The blood system comprises few main vessels, centered on a cardiopericardial vesicle situated in the protosoma; this is claimed to be homologous with the madreporic vesicle of larval echinoderms. The nervous system is rudimentary, with a central ganglion near the base of the lophophore.

A cuticular exoskeleton is secreted both by *Cephalodiscus* and *Rhabdopleura*.

Order RHABDOPLEURIDA Fowler, 1892

[Rhabdopleurida Fowler, 1892, p. 297]

Truly colonial animals with zooids attached by a contractile stalk to the stolon or pectocaulus; zooids provided with one pair of arms; gonads unpaired; no branchial pores. The skeleton (coenoecium) consists of an irregularly branching system of sclerotized tubes, attached to the surface of a pebble or shell, from which slender free zooidal tubes rise erect. Creeping and zooidal tubes are alike composed of regular growth bands that are clearly defined by transverse growth lines, and the pectocaulus is embedded in the base of the creeping tube. *L.Ord.-Rec.*

Growth of the colony is by distal extension of the soft stolon (gymnocaulus) bearing at its extremity a permanent terminal bud ("blastozooid *inachevé*"). This terminal bud secretes the adnate or creeping tube as it advances. According to Schepotieff, this tube is a closed, pointed tube; but

LANKESTER describes it as an open-ended tube. Normal zooidal buds develop successively behind the terminal bud on the gymnocaulus, forming a linear series with the youngest always nearest to the terminal bud (Fig. 3). As each develops, it becomes sealed off by a transverse partition across the creeping tube, and at about this stage the gymnocaulus becomes sclerotized to form the black stolon or pectocaulus (some 20 microns in diameter) and becomes embedded in the lower wall of the creeping tube. Each zooid in turn then forms by resorption a circular pore at the distal end of its chamber, and the zooid emerges, secreting as it grows upward the slender, erect, free portion of the zooidal tube. Branching occurs when one of the buds develops into a terminal bud instead of a normal zooid and starts to form its own creeping tube. Initial stages of development of the colony are only very imperfectly known.

Growth bands of the creeping tube are laid down in the form of half segments deposited alternately to left and right, so that the growth lines exhibit a characteristic median zigzag suture; the free zooidal tube consists of complete rings of periderm, each intersected by a single oblique suture marking the beginning and end of its formation. The initial rings at the base of the free zooidal tube are of course laid down discordantly on the growth bands of the creeping tube.

Family RHABDOPLEURIDAE Harmer, 1905

[Rhabdopleuridae HARMER, 1905, p. 5]

Characters of the order. *L.Ord.-Rec.*

Rhabdopleura ALLMAN, 1869, p. 58 [*R. normani;

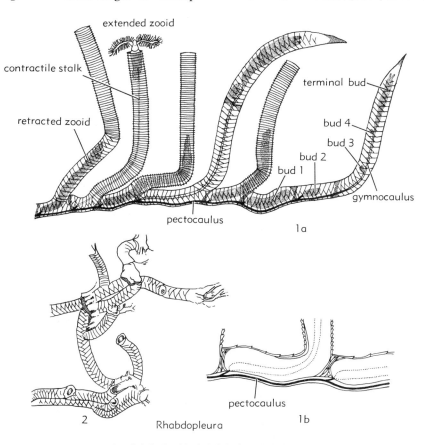

FIG. 3. Rhabdopleurida (Rhabdopleuridae) (p. *V*15).

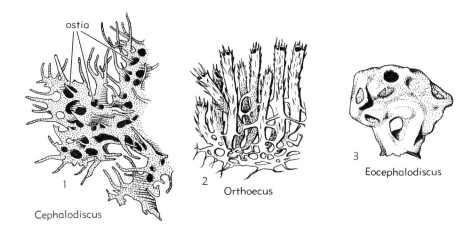

FIG. 4. Cephalodiscida (Cephalodiscidae) (p. V17).

M]. U.Cret.(Pol.)-Eoc.(Eng.)-Rec.(E.Atl.-S.Pac.-Antarctic).——FIG. 3,1. *R. normani, Rec.; 1a, portion of coenoecium of living specimen showing expanded and retracted zooids, developing buds, terminal bud, and characteristic growth lines of creeping and zooidal tubes, ×20 (114); 1b, long sec. of part of coenoecium showing pectocaulus, transv. partitions, and relations of growth bands, ×25 (23).——FIG. 3,2. R. eocenica THOMAS & DAVIS, M.Eoc., S.Eng.; ×12 (235).

Rhabdopleurites KOZŁOWSKI, 1967, p. 126 [*R. primaevus, p. 127; OD]. Similar to Rhabdopleura. Ord.(glacial boulder), Eu.(Pol.).

Rhabdopleuroides KOZŁOWSKI, 1961, p. 4 [*R. exspectatus; M]. Coenoecial tubes attached throughout their length; aperture with languette. L.Ord. (glacial boulders), Eu.(Pol.).

Order CEPHALODISCIDA Fowler, 1892

[Cephalodiscida FOWLER, 1892, p. 297]

Zooids forming free unattached associations, not true colonies; lophophore composed of several pairs of arms; gonads paired; one pair of branchial pores; coenoecium extremely variable and generally irregular in form. *L.Ord.(Tremadoc)-Rec.*

The coenoecium of cephalodiscids is extremely variable in form, encrusting, dendroid or compact, and it may be elaborately spined. In the majority of species, separate zooidal tubes are formed, usually connected by cuticular substance; or somewhat rarely, completely embedded in it; in a few forms, the superficial openings (ostia) lead into a general cavity occupied by all the zooids and their buds. Where distinct zooidal tubes are present, usually they do not communicate with one another, and buds produced from the peduncle or stalk free themselves from the parent before secreting their own tubes. Zooids are able to leave their tubes and creep about the coenoecium, and in this way to secrete the connective cuticular tissue. In the less compact types of coenoecium, the zooidal tubes are seen to be formed of growth bands comparable with those of *Rhabdopleura* but irregular in form and spacing.

Family EOCEPHALODISCIDAE Kozłowski, 1949

[Eocephalodiscidae KOZŁOWSKI, 1949, p. 195]

Chambers relatively few (about 10), forming a compact, minute, unspined coenoecium. *L.Ord.(Tremadoc).*

Eocephalodiscus KOZŁOWSKI, 1949, p. 195 [*E. polonicus; OD]. L.Ord.(Tremadoc), Pol.——FIG. 4,3. *E. polonicus; ×20 (114).

Family CEPHALODISCIDAE Harmer, 1905

[Cephalodiscidae HARMER, 1905, p. 5]

Coenoecium relatively large, variable in form, with or without individual zooidal tubes, or rarely absent altogether. *Ord., ?Tert., Rec.*

Cephalodiscus M'INTOSH, 1882, p. 348 [*C. dodecalophus;* M]. Several subgenera based on form of coenoecium. *Rec.*, S.Hemis.(almost exclusively). [Silicified tubes from *M.Eoc.*, France, provisionally referred to this genus.]

C. (Cephalodiscus) [=*Demiothecia* RIDEWOOD, 1906, p. 191]. Coenocium branching, each ostium leading into cavity which is occupied in common by all zooids and their buds. *Rec.*, Antarctic.——FIG. 4,*1. C. (C.) hodgsoni* (RIDEWOOD); approx. ×1.5 (48).

C. (Idiothecia) LANKESTER in RIDEWOOD, 1906, p. 191 [*Cephalodiscus nigrescens* LANKESTER, 1905, p. 400; SD BULMAN, herein]. Coenoecium branching, composed of individual zooidal tubes embedded in common coenoecial substance. *Rec.*, Antarctic.

C. (Orthoecus) ANDERSSON, 1907, p. 11 [*Cephalodiscus solidus*, p. 11; SD BULMAN, herein]. Zooids with individual tubes embedded in common coenoecial substance to form irregular mass. *Rec.*, Pac.(E.Indies).——FIG. 4,*2. C. (O.)* sp.; approx. ×1.5 (48).

C. (Acoelothecia) JOHN, 1931, p. 259 [*C. (A.) kempi;* M]. Colony in form of branched network of spines and bars without definite coenoecial cavities. *Rec.*, Antarctic (Falkland Is.).

Atubaria SATO, 1936, p. 105 [*A. heterolopha;* M]. Without any coenoecium. *Rec.*, Pac.(Japan).

Pterobranchites KOZŁOWSKI, 1967, p. 123 [*P. antiquus;* OD]. Coenoecium of irregularly aggregated tubes and elongated vesicles. *L.Ord.*(glacial boulder), Eu.(Pol.).

PLANCTOSPHAEROIDEA

Class PLANCTOSPHAEROIDEA van der Horst, 1936

[Planctosphaeroidea VAN DER HORST, 1936, p. 612]

This class is based on two specimens from the Bay of Biscay believed to represent the larval form of an unknown type of Hemichordata. *Rec.*

GRAPTOLITHINA

DIAGNOSIS AND GENERAL FEATURES

Class GRAPTOLITHINA Bronn, 1846

[Graptolithina BRONN, 1846, p. 149 (*nom. transl.* ELLES, 1922, p. 168)]

The Graptolithina are colonial, marine organisms which secreted a sclerotized exoskeleton with characteristic growth bands (fuselli) and growth lines. The thecae housing individual zooids are usually arranged in a single or double row along the branches (stipes) of the colony (rhabdosome), rarely in irregular aggregates. In most orders, the thecae are polymorphic and in three they are clearly related to an internal sclerotized stolon system. Rhabdosomes originate by a single bud from the initial zooid, housed in a conical sicula, producing simple, branched or rarely encrusting colonies. Sessile or pelagic. *M.Cam.-Carb.*

Six orders are now recognized: 1) Dendroidea NICHOLSON, 1872; 2) Tuboidea KOZŁOWSKI, 1938; 3) Camaroidea KOZŁOWSKI, 1938; 4) Crustoidea KOZŁOWSKI, 1962; 5) Stolonoidea KOZŁOWSKI, 1938; 6) Graptoloidea LAPWORTH, in HOPKINSON & LAPWORTH, 1875.

These are based principally on details of branch structure, which in turn reflects the nature and regularity of stolonal budding. An additional order, Dithecoidea, has been proposed by OBUT (1964), but since its branch structure has not been conclusively demonstrated, the genera concerned are here grouped together with others of uncertain taxonomic position under the general heading "Taxonomic Position Uncertain" (p. V54). Other groups of unknown affinity are the Graptovermida, Graptoblasti, and Acanthastida (p. V136-V139).

PIONEER WORK ON GRAPTOLITES

In the early days of paleontology, graptolites attracted comparatively little attention. Their remains were thought originally to be those of plants although LINNÉ believed them to be inorganic when bestowing the name *Graptolithus* upon *G. sagittarius* and *G. scalaris,* and it appears that WAHLENBERG (1821) was the first to recognize their animal nature. The generic name *Graptolithus* now has been suppressed by the International Commission on Zoological Nomenclature (Opinion 197), but it persists in the forms Graptolithina, Graptoloidea, and the anglicized version "graptolites."

The early phase of work by BRONN (1834), BECK (1839), and others, followed towards the middle of the century by publications of M'COY, BARRANDE, and SALTER, has given us a number of well-known generic names; but probably the first work of real insight and understanding is HALL's *Graptolites of the Quebec Group* (1865), where more than a dozen genera and over 50 species of graptolites (including dendroids) were described and beautifully figured.

Soon after this began the period of LAPWORTH's great contribution with a series of papers (extending mainly from 1870 to 1880) devoted not only to a more exact understanding of structure and morphology and a more precise determination of species, but above all to the demonstration of their stratigraphical value (see especially his *Geological Distribution of the Rhabdophora,* 1879-80). This phase of work on the group may be said to have culminated in the *Monograph of British Graptolites* (1901-18) where LAPWORTH was assisted by Miss ELLES and Miss WOOD to produce an exhaustive and superbly illustrated monograph which has been an indispensable aid to workers all over the world. Comparable work was being done in Sweden, at first by LINNARSSON (who published but little), later by TÖRNQUIST and HADDING, while RUEDEMANN's *Graptolites of New York* (1904-08) and his *Graptolites of North America* (1947) serve the same need for the North American continent.

Toward the end of last century a remarkable series of papers was published by HOLM (1890, 1895) and WIMAN (1895-1901), who may be said to have initiated the really detailed study of graptolite morphology, aided by novel techniques of solution and serial sectioning. After an interval of nearly 30 years, a revival of interest in the application of special techniques began with KRAFT's (1926) memoir on *Diplograptus* and *Monograptus,* and much of HOLM's work which was left unpublished at his death was completed in a series of papers by BULMAN (1932-36). The outstanding contribution of this character, however, was that of KOZŁOWSKI (1938, 1949), whose researches on the astonishingly well-preserved material from silicified nodules in Tremadocian rocks of Poland led to a new concept of the nature and affinities of the Graptolithina. Since then a steadily increasing emphasis (especially in Britain, Poland and Scandinavia) has been placed on detailed morphological description and analyses.

TECHNIQUES

PREPARATION OF SPECIMENS

Detailed information about structure and development of graptolites is obtained almost entirely from specimens which have been dissolved out of their matrix and rendered more or less transparent by the use of various oxidizing agents. The actual processes and reagents employed naturally depend upon the nature of the matrix and the degree of carbonization of the fossil.

Pure limestone matrix can be dissolved readily with hydrochloric or acetic acid, the latter being sometimes preferable with fragile material on account of its more gentle action. The concentration should be adjusted so that effervescence is not too brisk, and is maintained by repeated addition of drops of concentrated acid. The condition of preservation of the graptolite periderm is an important factor, and some limestone material otherwise suitable is rendered useless for treatment because the graptolite remains have been too highly carbonized and have become so brittle that they crumble to a powder when freed from matrix.

Impure limestone generally requires a double treatment, involving solution of the calcareous matter first and then (after washing out all trace of HCl) solution or disintegration of the arenaceous or argillaceous remainder with hydrofluoric acid. Repeated washing and decanting is necessary to remove all HF before the graptolite remains can be picked out with a pipette under low-power binocular. Much of the fine mud can be removed by elutriation, and some workers wash the whole through a series of sieves, although a greater risk of breakage is entailed in this process. Graptolites preserved in chert nodules of course can be dissolved out with HF without previous treatment.

Graptolites which have been dissolved out of calcareous rocks may contain bubbles of CO_2 which should be removed in a vacuum desiccator before further treatment. Clearing is most usually done in a watch-glass with potassium chlorate and concentrated nitric acid, but *eau de Javelle* and other bleaching reagents have been used. The period required varies with the thickness of periderm and the degree of carbonization, and can only be judged individually by constant observation through a low-power binocular; but the treatment cannot be prolonged, as a rule, beyond 20 minutes or half an hour without the specimens becoming too brittle to handle. Some workers prefer a much lower concentration over a correspondingly greater period of time. Quite a high proportion of material successfully dissolved from its matrix proves unsuitable for further treatment of this kind.

Specimens which cannot be cleared are best mounted dry if robust enough, because surface features are so much more easily seen than when mounted in a relatively high-refractive-index medium. They may be affixed in a cell between two glass slides with a minute drop of gum arabic. Transparencies may be mounted in Canada balsam or some proprietary mountant such as Euparal, which has the advantage of not requiring perfect desiccation in absolute alcohol and clearing in xylol, thus eliminating processes in which damage to the specimen may occur. Some workers prefer mounting in glycerine, which further eliminates the whole "alcohol series" and also enables the specimen to be rolled over (using a fine bristle) and viewed from different sides; but the technique of permanent mounting in glycerine presents many difficulties of its own. Storage of duplicate material, however, is always best in glycerine.

Some rhabdosomes which are too large (e.g., *Dictyonema*) or too delicate (e.g., *Rastrites*) to hold together on removal of the matrix may be cemented to a glass slide with Canada balsam or some proprietary cement after one side has been completely exposed, and when thus supported the rest of the matrix can generally be dissolved safely with HCl or HF. More recently, promising results have been obtained with blocks of polyester resins (e.g., Crystic 195 and Ceemar) in place of a glass backing (HUTT & RICKARDS, 1967). No transfer preparations can be cleared, however, as no mounting medium yet used has been found to withstand the effect of clearing reagents.

Shale material that is exceptionally well preserved (e.g., in relief in pyrite) may also be worth treating by one or other of the transfer methods described in the foregoing paragraph, and in some instances the graptolites may be sufficiently uncarbonized for complete isolation with HF and clearing, even though the stipe is completely flattened (SKOGLUND, 1961). In general, however, little can be done with specimens preserved in a shale or silt matrix beyond careful cleaning of the fossils with a fine needle under a medium-power binocular. It is sometimes an advantage, in order to gain greater contrast with the matrix, to varnish specimens after cleaning and for this purpose mastic varnish, Canada balsam, Euparal, or some similar substance may be used and can if necessary be removed later with xylol or alcohol. For subsequent examination and particularly for photography, it is desirable to cover the specimen with an ordinary microscope coverslip.

Dissolved graptolites can be embedded and sectioned with a microtome, and although HOLM satisfactorily used only paraffin wax embedding, better results can be obtained usually by double embedding in collodion and paraffin wax. Zoology technicians, much more accustomed to such procedure, usually are willing to undertake this part of the paleontologist's work.[1] Pyritized graptolites, and those preserved in limestone but too highly carbonized for any of the solution treatments, can be sectioned by serial grinding, and with a limestone matrix permanent transfers can be taken with collodion films. Restorations can be made from serial sections (microtome or grinding), either by a modification of the method of preparing block diagrams, or by drawing on glass plates, or better still as wax models, the thickness of the wax plates to be used being determined by the frequency of the sections and the magnification employed. Most generally useful, at least for proximal-end development, are reconstructions in the form of internal casts, made by cutting away a slightly exaggerated thickness of the internal and external walls, and assembling the resulting series of "thecal cavities." The result is something approaching a thecal diagram, as shown in Figures 15, 49, 62, and others.

ILLUSTRATION

The satisfactory illustration of graptolites has always presented a difficult problem. On account of their small size, enlarged figures are necessary to show details of structure and thecal form, but illustrations at natural size are so valuable as an aid to identification that the ideal is to have both.

Enlarged figures present no special difficulties, though it should be remarked that untouched photographs are rarely satisfactory; retouched protographs or wash drawings are far preferable (for example, see KRAFT, 1926, where both photographs of exceptionally high quality and wash drawings are reproduced; also excellent examples of retouched photographs in HOLM's plates, BULMAN, 1932-36, and of wash drawings in WIMAN, 1895-1901). At high magnifications, line drawings made with a camera lucida under the microscope, or with a Shadowmaster, often leave little to be desired and can be reproduced cheaply as text figures (for example, WALKER, 1953; URBANEK, 1966).

It is the natural-size figures which present the special problem, both on account of technical difficulties in reproduction and the high degree of artistic skill needed for the original drawings (for here again photographs are rarely satisfactory). The steel engravings accompanying HALL's *Graptolites of the Quebec Group*, and TÖRNQUIST's lithographic plates are alike admirable, but these methods of reproduction are now obsolete even if authors were able to emulate their drawings. LAPWORTH solved the problem *(Monograph of British Graptolites)* by photographic reduction of enlarged chalk and wash drawings, reproduced at natural size by a collotype process to which Messrs. BEMROSE devoted special care and attention. It is doubtful whether such plates could be produced today; but since modern zinc blocks cannot reproduce satisfactorily a line drawing of a graptolite rhabdosome at natural size, and the usual halftone screens are far too coarse, some form of collotype is essential for natural-size figures.

It cannot be overemphasized that there is no substitute for well-executed wash drawings or carefully retouched photographs. Innumerable examples of modern halftone (and even collotype plates) produced from unretouched photographs at natural size or even at small magnifications serve mainly to show the limitations of this quick and labor-saving method of illustration. Because of these technical difficulties, there is inevitably a tendency nowadays to discard natural-size figures. Line drawings, reproduced as text figures with magnifications of $\times 2$ to $\times 5$, probably will become the standard method of illustrating the general features of a graptolite rhabdosome in future; such figures can be drawn either with a camera lucida (at $\times 5$ to $\times 10$) or more readily drawn on an enlarged photographic print which is subsequently bleached and reduced for reproduction. The latter method is particularly useful with large or spreading rhabdosomes, and completely supersedes use of the Lapworth microscope (described in ELLES & WOOD, 1901-19).

If photography is used, better results commonly are obtained by immersing the specimen in alcohol or xylol, which reduces surface reflections and increases contrast between the graptolite and surrounding matrix.

---
[1] If any granules of pyrite are present in the rhabdosome, the sections are likely to be torn and there is danger of damaging the microtome knife.

STRUCTURE AND COMPOSITION OF PERIDERM

The periderm of the Dendroidea (KozŁOWSKI, 1938, 1949) consists of two layers, a main fusellar layer constructed of short transverse growth segments (fuselli) generally disposed with bilateral symmetry, and an outer, laminated, cortical layer (Fig. 5). A comparable structure has been demonstrated in other graptolithine orders. The fusellar layer corresponds closely in its appearance to the zooidal tubes of *Rhabdopleura* (Fig. 3) and *Cephalodiscus*. Cortical tissue is not present in *Rhabdopleura*, but is well developed in *Cephalodiscus*, and it varies greatly in amount in the Graptolithina; some dendroid and tuboid rhabdosomes exhibit so much "secondary thickening" that underlying structures are completely obscured, while even the growth lines of others remain clearly visible, and always more of it occurs at the base (proximal end) of a rhabdosome than at its distal extremities. To this extent it is a function of age.

In *Dictyonema*, the dissepiments are composed of cortical tissue. Also probably, but not certainly, the web and disc structures of some dichograptids and "float structures" of biserial graptolites are composed of cortical tissue. On the other hand, the microfusellar tissue associated with the apertures of some diplograptids and monograptids is believed to be a form of fusellar tissue. Thecal (apertural) spines are fusellar, as also is the virgella, but the exact nature of the lacinia and parts of the clathria remain obscure. Where periderm of the apertural region has been damaged or destroyed, it is replaced by tissue with normal fusellar structure, but if some area remote from an aperture is damaged, the regenerated tissue consists of a structureless membrane.

The chemical composition of the periderm of the Graptoloidea has been investigated by FLORKIN and his colleagues (1965). Using two species of *Pristiograptus* and one of *Climacograptus*, they have demonstrated its proteic nature and the complete absence of chitin. The presence of large amounts of serine (molecular fraction 10.6 to 22.8), alanine (6.3 to 9.5), glycine (20.1 to 23.4), aspartic acid (8.6 to 10.0) and glutamic acid (12.8 to 15.3)

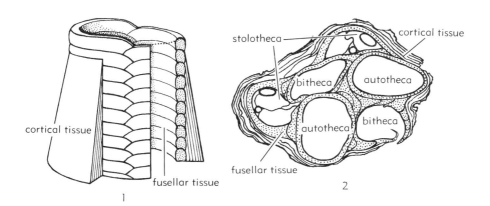

FIG. 5. Structure of graptolite periderm.

1. Diagram showing fusellar tissue laid down in alternating half rings, surrounded by laminated cortical tissue (114).
2. Transverse section of a *Koremagraptus* stipe (×130) showing cortical tissue surrounding fusellar tissue (stippled), which has the form of complete tubes for autothecae, and split tubes for bithecae and stolothecae; where growth bands are oblique (as in bitheca at lower right) many such bands are cut by the plane of the section (23).

suggests that these graptolite proteins are scleroproteic and such a composition is analogous to that of the cephalodiscoid coenoecium.

WETZEL (1958) and KRAATZ (1964, 1968) have carried out some preliminary examination of thin sections of graptolite periderm at high magnifications, using the electron microscope. WETZEL indicated certain differences between the periderm of the sicula of a *Diplograptus* species and that of a *Rhabdopleura* tube. and KRAATZ described several types of granular aggregates in the fusellar tissue of *Monograptus* and in the denser substance of the virgula and of a retiolitid meshwork. It is difficult at this stage to assess the significance of their findings.

GRAPTOLITE AFFINITIES

Because the graptolites are an extinct group of animals whose soft parts have left little or no trace upon the exoskeleton, their affinities always have been in dispute. Originally regarded as inorganic (LINNÉ, 1735) or of vegetable nature (VON BROMELL, 1727; BRONGNIART, 1828), at different times they have been assigned to the Cephalopoda (WALCH, 1771; WAHLENBERG, 1821; VON SCHLOTHEIM, 1822), Coelenterata (HALL, 1865; ALLMAN, 1872; NICHOLSON, 1872; LAPWORTH, 1873; BULMAN, 1932), Polyzoa (SALTER, 1866; ULRICH & RUEDEMANN, 1931), Pterobranchia (SCHEPOTIEFF, 1905), or considered to occupy an isolated position in the animal kingdom not clearly related to any living group of organisms (WIMAN, 1895; PERNER, 1895; RUEDEMANN, 1895; FRECH, 1897; ELLES, 1922).

However, reliable and detailed description of the histology of the periderm, of its chemical composition, and nature of the stolon system now point more decisively to pterobranch affinities. The existence of fusellar and cortical layers in the periderm and the arrangement of fuselli find close parallels in the skeletal tissues of the Pterobranchia, and no other living organism has such an organ as the pectocaulus (skeletal sheath of the stolon) to which the sclerotized stolon system of the graptolites is so closely comparable. In the order Crustoidea, it is even embedded in the lower wall of the stolotheca, as in *Rhabdopleura,* though in other graptolite orders it lies free in the stolothecal tube. The proteic nature of the periderm and the absence of chitin lends additional support to this view of their affinities.

The actual method of budding seems to have been somewhat different. In *Rhabdopleura* a permanent terminal bud *("blastozooide inachevé")* is present behind which successive individuals developed from the steadily lengthening stolon, which has not distally developed its sclerotized sheath (Fig. 3); whereas in the graptolites each stolotheca in turn seems to have represented the terminal bud of its branch (Fig. 6,*1,2*). In the Graptoloidea, evidence for the budding of successive thecal zooids from one another is even clearer; here the prothecal segment (Fig. 6,*3,4*) represents the stolotheca and the stolon, assumed by analogy to have existed, lacks any skeletal sheath. These differences from *Rhabdopleura* may not be very significant; the living *Cephalodiscus,* which is placed without question within the same class as *Rhabdopleura,* have no stolon system at all and is not a truly colonial organism; and the budding processes in *Rhabdopleura* on the one hand and graptolites on the other, are closely paralleled within a single order by the monopodial and sympodial budding of calyptoblastean hydroids. Certain other differences are mentioned in the section below.

For a detailed discussion of the question of graptolite affinities, reference may be made to KOZŁOWSKI, 1966.

NATURE OF GRAPTOLITE ZOOID

The nature of graptolite zooids is essentially conjectural, but by analogy with the pterobranchs it is now permissible to suggest a tentative restoration involving a bi-

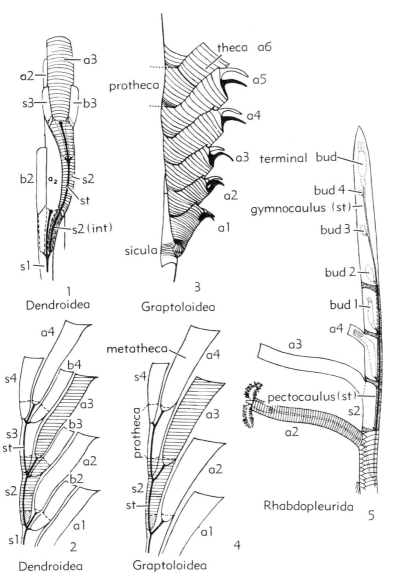

FIG. 6. Comparison of branch structure and mode of budding in Dendroidea (*1,2*), Graptoloidea (*3,4*), and Rhabdopleurida (*5*) (*29*).——*1*. Growing end of dendroid stipe (diagrammatic) with one complete autothecal unit shaded (internal portion of stolotheca stippled, external portion and daughter autotheca with growth lines).——*2-5*. Autothecal units shaded for comparison. [*a*, autotheca; *b*, bitheca; *(int)*, internal portion; *s*, stolotheca; *(st)*, stolon.]

lateral lophophore with two arms or groups of arms (Fig. 7). This in turn allows a convincing interpretation of some apertural modifications of the thecae, especially in the Graptoloidea, which may include spines, lobes, and even more or less tubular "hydrodynamic tunnels." Asymmetry in the apertural processes is extremely rare, but one group (cucullograptids) is characterized by the introduction of such asymmetry, and URBANEK (1966) has attributed to these highly-modified left-handed apertural structures of the extreme *Cucullograptus aversus rostratus* a hydrodynamic and supporting-protective role in relation to the hypertrophied left lobe of the lophophore (Fig. 7,5).

It is possible also to give some explanation, in terms of pterobranch affinities, of the polymorphism seen in many graptolite orders. Polymorphism in the coelenterate hydrosome is introduced by the presence of special reproductive or protective individuals in addition to the nourishing individuals. Likewise, the bithecae of Dendroidea at one time generally were regarded as housing such protective polyps. Dimorphism among higher organisms is prevailingly sexual and the occurrence of autothecae and bithecae on the stipes of many graptolites is now taken to indicate the presence of male and female zooids (KOZŁOWSKI, 1949, 1966b). In *Rhabdopleura* and *Cephalodiscus* males and females are usually indistinguishable externally, but certain species do show dimorphism. Thus, in *Cephalodiscus sibogae* the males are degenerate and bear an almost atrophied lophophore. The graptolite bithecae may be considered to represent such male zooids, the reduced state of their lophophore being reflected in the universal absence of apertural processes in the bithecal skeleton. The female zooids occupied the autothecae, which in several dendroid species have furnished traces of what are claimed to represent embryos (KOZŁOWSKI, 1949) and which probably possessed a well-developed lophophore as indicated by the varied apertural modifications commonly present, especially in the Graptoloidea. Disappearance of the bithecae in the Graptoloidea may imply a change to hermaphroditism, the autothecal females becoming hermaphrodite as the bithecal males were eliminated. The process may be actually visible in *Kiaerograptus*, a *Didymograptus*-like genus with bithecae regularly present distally but lacking in association with the first three of four autothecae of the rhabdosome (see Fig. 19,3). The existence of such forms provides a complete link between the dendroid Anisograptidae and graptoloid Dichograptidae.

In addition to these two thecal types, the Tuboidea may also exhibit microthecae, umbellate thecae, and conothecae; the first two appear to represent autothecae modified to a varying extent (and for an unknown purpose), but the conothecae differ more pronouncedly. Graptoblasts and cysts occur within the autothecae of the Crustoidea, and their interpretation also remains obscure.

It is improbable that the stolothecae contained separate zooids; indeed it is virtually

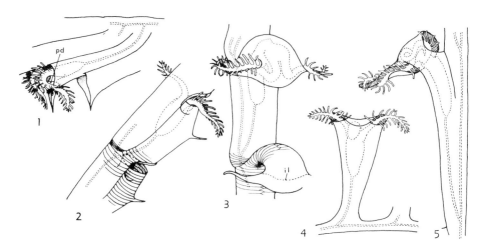

FIG. 7. Diagrammatic restorations of thecal zooids illustrating possible relations of lophophore to different types of apertural modification (Bulman, n).——*1.* Dichograptid.——*2.* Dicellograptid.——*3.* *Monograptus exiguus.*——*4.* Triangulate monograptid.——*5.* *Cucullograptus aversus rostratus.* [*il*, inner lip of aperture; *pd*, preoral disc.]

certain that in the Dendroidea each stolotheca was secreted by the same individual as the autotheca which succeeds it with continuity of periderm and without break or "unconformity" in the growth lines. Budding is thus essentially sympodial. Structural resemblances between stolothecae and bithecae (p. V28) are accountable on the assumption that the stolotheca was secreted by an immature autothecal zooid with the preoral lobe and lophophore still relatively undeveloped, while the bitheca was secreted by a "reduced" male in which these structures remained always undeveloped.

The laminated cortical tissue, to some extent at least invariably present among graptolites, is assumed to imply the existence of some extrathecal living tissue, possibly even enveloping the entire colony (KOZŁOWSKI, 1966b). A contrary view was expressed by BEKLEMISHEV (1951), who explained cortical tissue as the secretion of zooids which crept out of their thecal tubes just as do those of the living *Cephalodiscus,* despite the basal attachment of graptolite zooids to their stolon system. Although this is manifestly impossible in graptolites with strongly constricted thecal apertures, he contended that it could have been formed at an immature stage of thecal development while the apertural region (of the exoskeleton) was incomplete. This assumption overlooks the important fact that cortical tissue is not present distally, at the growing end of the stipe, but increases in amount proximally and can only have been deposited when the underlying thecae were mature.

The existence of this enveloping tissue constitutes a significant difference between graptolites and pterobranchs, and its relation to the body of the zooids is certainly difficult to visualize, bearing in mind that the fusellar layer of graptolites must surely have been secreted by some part of the preoral lobe. Something analogous is known to exist, however, in some Bryozoa,[1] producing an external thickening of the calcareous walls. It has been suggested (BULMAN, 1964) that this tissue may have played a role in the buoyancy of the Graptoloidea (p. V93), as well as in secretion and lengthening of the nema and development of "floats," webs and other extrathecal skeletal structures.

The nature of the sicular individual remains more obscure. The prosicula seems to represent the skeleton of a larva developed from a fertilized egg, originally exhibiting little trace of basal disc or nema, but possibly covered by some extrathecal membrane and either free-swimming or attached by a fleshy peduncle. This prosicular skeleton is so sharply distinct from that of the metasicula, however, that KOZŁOWSKI even believes them to have been secreted by different individuals. On this view, the prosicula corresponds to a fixed larval form, which on degeneration is replaced by a metasicular individual whose body occupied the entire cavity of the sicula to the apex of the prosicula; such a process finds some analogy in the embryonic stages of certain Polyzoa. The initial bud (or sicular stolotheca) likewise extends to the apex of the prosicula, perhaps originating as a bud from the peduncle of the metasicular zooid, and then growing up with the metasicular individual until it emerges generally through a foramen produced by resorption in the wall of the sicula in a manner comparable with the normal process of budding in *Rhabdopleura* (or less commonly through a notch as in the monograptids).

The possible role of siculozooid (the only sexually-produced individual in the colony) in controlling the pattern of rhabdosome development has been discussed by URBANEK (1960) and will be referred to more fully in the section on Graptoloidea.

DENDROIDEA

Order DENDROIDEA Nicholson, 1872

[*nom. transl.* RUEDEMANN, 1904, p. 578 (*ex* section Dendoidea NICHOLSON, 1872, p. 101)] [=suborder Cladophora HOPKINSON in HOPKINSON & LAPWORTH, 1875, p. 633]

Sessile Graptolithina, attached by apex of sicula, which is then generally more or less embedded in secondary cortical tissue forming a rootlike base, or more rarely attached by a nema; stipes composed of stolothecae, autothecae, and bithecae produced by regu-

---

[1] R. TAVENER-SMITH (1969) recently has inferred the existence of an external, colonial mebranous investment in fenestellids.

lar triad budding; rhabdosome typically erect, dendroid in habit of growth, developed by dichotomous or irregular branching, with anastomosis or dissepimental connection between adjacent stipes in many forms; pendent to horizontal (very rarely reclined) in forms with nema attachment. ?*M.Cam.*(Eng.-Nor.); *U.Cam.*(N.Am.-USSR)-*Carb.*(*Namur.*)(Eng.).

The Dendroidea are characterized essentially by their regular triad budding of the stolon on what has been termed the "Wiman rule," in which the autotheca constitutes the central individual at each division and the bithecae are produced alternately right- and left-handedly (Fig. 8). While this can be demonstrated conclusively in transparencies or in serial sections, it is more difficult to recognize in hand specimens, although often it can be inferred in material preserved in relief (e.g., pyritized specimens) where the triads can be detected owing to the dorsal position of the stolon system. In genera belonging to the Acanthograptidae, however, where the stipe is composed of elongate tubular individuals and the stolon system is commonly "internal," no outward indication of the budding mechanism may be seen. Exceptions to the regularly alternating triad budding are known, but are exceedingly rare.

The dendroid habit is not in itself diagnostic, since it is encountered also in the Tuboidea. Considerable uncertainty may arise therefore, both from lack of knowledge concerning fundamental generic characters and also from poor preservation of individual specimens.

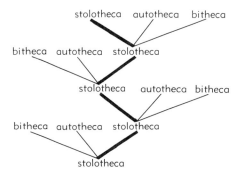

FIG. 8. Arrangement of dendroid thecae in alternating triads according to the "Wiman rule" (29).

MORPHOLOGY
THECAE
STOLOTHECAE

The stolothecae of Dendroidea, formerly called "budding individuals," form a continuous closed chain lying characteristically along the dorsal side of the branch, but, in the more complex acanthograptids and inocaulids, embedded in the stipe to a greater or lesser extent. Each stolotheca terminates distally against the base of the succeeding autotheca, liberating at the same level a bitheca on one side and another stolotheca on the other (Fig. 9,*2,3*). When only this much was known about branch structure, the term "budding individual" was not inappropriate. Internally, however, each stolotheca carries a section of the sclerotized stolon system, analogous to the pectocaulus of *Rhabdopleura*. Distally, each stolotheca encloses a thin-walled proximal extension of the daughter stolotheca and bitheca, together with a long stolon from the base of the autotheca lying centrally. Traced proximally, these unite near the mid-length of the stolotheca in a stolonal triad. Globular swellings (vesicular diaphragms) may occur at the points of origin of the three thecal stolons and where the thecal stolons join the bases of their respective thecae; such diaphragms also are seen in the Crustoidea (see Fig. 26) and in certain Tuboidea. Though usually well sclerotized, the stolon system appears to have been unsclerotized in the aberrant *Graptolodendrum*.

In a typical dendroid, the growth bands of the parent stolotheca pass uninterruptedly into the base of the daughter autotheca, though a marked "unconformity" delimits growth lines at the bases of free portions of the daughter stolotheca and bitheca (Fig. 9,*2,3*; Fig. 6,*1*), indicating that in effect each stolotheca is no more than the immature basal portion of the succeeding autotheca. Clearly, no reason exists for supposing the existence of a separate stolothecal zooid. In the Tuboidea, where budding is diad, the relations are not so simple and invariable. In the Graptoloidea no sclerotized stolon system is present, but the prothecal segment of the theca undoubtedly corresponds to the dendroid stolotheca (Fig. 6,*3,4*).

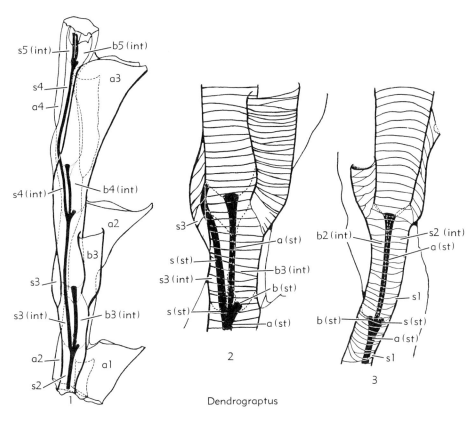

Fig. 9. Thecal constitution of a dendroid stipe (114).

1. *Dendrograptus regularis* Kozłowski (×40) viewed as transparency with growth lines omitted; stolon system in solid black, stolotheca and daughter thecae in heavy outline.
2. Portion of same with growth lines (×80).
3. *D. communis* Kozłowski (×80), distal end of branch showing immature stolotheca and bitheca. [*a*, autotheca; *b*, bitheca; *(int)*, internal portion; *s*, stolotheca; *(st)*, stolon.]

These relationships imply that, instead of the permanent terminal "leading bud" of *Rhabdopleura*, each autothecal zooid in turn has been the terminal zooid of its branch; the process is analogous to "sympodial budding" as compared with the "monopodial budding" of *Rhabdopleura*.

AUTOTHECAE

The autotheca, originally called "hydrotheca," is the largest and most conspicuous of the three types of dendroid thecae, and comprises a relatively long autothecal stolon and the theca proper. The stolon and the thin, rounded base of the theca are enclosed within the stolotheca of the preceding generation, but practically the whole of what is termed autotheca is external and its dorsal wall continues that of the stolotheca uninterruptedly. The midventral line is usually marked by the zigzag wedging out of growth bands laid down alternately to right and left, and where the whole distal region of the autotheca is isolated from the branch, a similar zigzag suture line is visible on the dorsal side as well. For the most part, the autothecae are practically straight, commonly provided with an apertural (ventral) process or spine, rarely with a dorsal spine, or both. In some forms (e.g., *Dendrograptus cofeatus*, Fig. 10,4), this ventral process is transversely enlarged and recurved over the aperture, but the dendroid autothecae rarely exhibit apertural

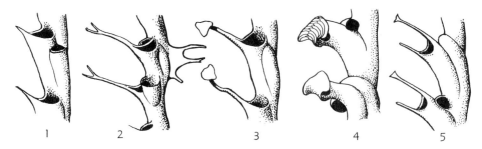

Fig. 10. Autothecal modifications in Dendroidea (29).

1. Apertural spine of denticle on hydrothecae of *Dictyonema flabelliforme* (EICHWALD).
2. Forked apertural spine of *D. cervicorne* HOLM.
3. Apertural spine with platelike termination in *D. peltatum* WIMAN.
4. Laterally expanded and reflexed ventral process (languette) shown by *Dendrograptus cofeatus* KOZŁOWSKI.
5. Ventral and dorsal spines in *Dictyonema rhinanthiforme* BULMAN.

modifications at all comparable with the more extreme types of elaboration shown by the Graptoloidea or Crustoidea. Some isolation of the distal end of the autotheca is by no means uncommon, accompanied by elongation of the theca reaching its extreme in the Acanthograptidae (see Fig. 21).

BITHECAE

The bithecae are shorter and as a rule narrower than the autothecae, and are commonly inconspicuous in external view, though in some species they form marked swellings along the branch, as in *Dictyonema cervicorne*, where they were for the first time recognized by HOLM in 1890. Their wall is incomplete along the side in contact with the branch and they are without apertural spines. They further resemble the stolothecae and differ from autothecae in possessing a very short stolon and in having a long, thin-walled proximal portion which is enclosed within the stolotheca of the preceding generation.

In its simplest form, the bithecal tube is nearly straight and, owing to its shortness in comparison with the autotheca, its aperture is normally situated beside that of the autotheca of the preceding generation. A common variant is for the bitheca to open into the cavity of the preceding autotheca, when it is practically invisible in external view; other modifications, usually involving some increase in length, are shown in Figure 11. In most species the behavior of the bithecae is constant for the species, but in a few it is variable and several different types occur together, in some (e.g., *Dictyonema falciferum*) characterizing a particular portion of the rhabdosome.

THECAL GROUPING

Regularity in triad budding is the distinctive feature of the Dendroidea, but instances of irregularity, though uncommon, are known. The Anisograptidae, essentially a transitional family linking the Dendroidea with the Graptoloidea, provide instances where the bithecae are in process of reduction or loss. Thus, bithecae are not developed in association with the proximal autothecae of the rhabdosome in *Kiaerograptus*, and in the aberrant genus *Graptolodendrum* (as in the tuboid? *Parvitubus*) the bithecae are mainly disposed along one or other side only of the branch.

The dendroid branch unit is the threefold association of autotheca, bitheca and stolotheca; and as the stolotheca does not open to the exterior (except at the growing tip of the branch) and was not inhabited by a distinct type of zooid, the effective unit is the autotheca combined with bitheca. Owing to their relative difference in length, this smaller unit is split, for the bithecal aperture is normally associated with the aperture of the autotheca of the preceding

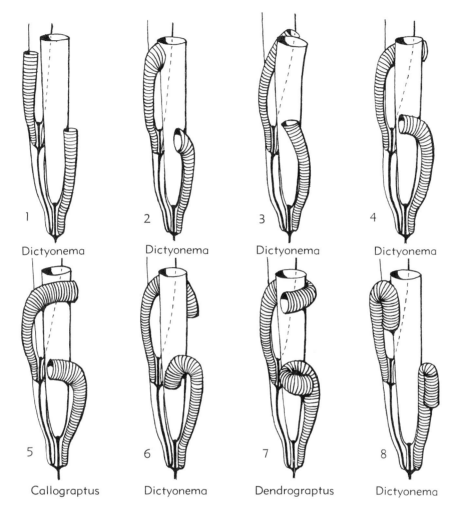

Fig. 11. Variations in form and relations of dendroid bithecae (shaded) (29).

1. *Dictyonema flabelliforme* (EICHWALD).
2. *D. peltatum* WIMAN.
3. *D. cotyledon* BULMAN.
4. *D. rarum* WIMAN.
5. *Callograptus infrabithecalis* KOZŁOWSKI.
6. *D. inconstans* BULMAN.
7. *Dendrograptus cofeatus* KOZŁOWSKI.
8. *Dictyonema wysoczkianum* KOZŁOWSKI.

generation (e.g., $bi2$ opens in association with $au1$). Exceptions to this generalization are connected with elongation of the thecal tubes.

In *Pseudocallograptus* (see Fig. 17,3) the adnate autothecae are elongated by some 50 percent as compared with normal *Callograptus,* but the bithecae are usually of normal length, so that $bi6$ opens adjacent to $au4$ instead of $au5$; but irregular variations in bithecal length also occur. It is probable that comparable increase in autothecal length, altering the normal apertural association, occurs in *Pseudodictyonema* (compared with *Dictyonema*), and possibly *Stelechocladia* (compared with *Dendrograptus*), and in some species of *Desmograptus* not yet generically separated.

A distinctive elongation of the thecae characterizes members of the Acanthograptidae and may be accompanied by surprisingly regular grouping of autothecae and bithecae. This is best exhibited in *Acanthograptus suecicus,* where four variously

associated thecae open together in groups or "twigs," each composed of two autothecae and two bithecae arranged as follows (Fig. 12,1):
twig 1: a2, a5, b4, b5,
twig 2: a4, a7, b6, b7,
twig 3: a6, a9, b8, b9,
and so on,
but the thecal composition of the twigs varies somewhat in different species. *A. czarnockii,* from the Tremadoc of Poland, and *A. divergens,* can be interpreted on a comparable basis though the thecal elongation is less pronounced. *A. musciformis* has more complex stipes, with more numerous thecae and more than one stolonal chain, and twigs may combine individuals derived from two or more lines of development. In *A. impar,* a cross section of the stipe may cut through as many as 30 to 35 individual tubes, of which five may be stolothecae, and though twigs are present, they tend to lose the regular fourfold grouping of simpler species. Finally, in

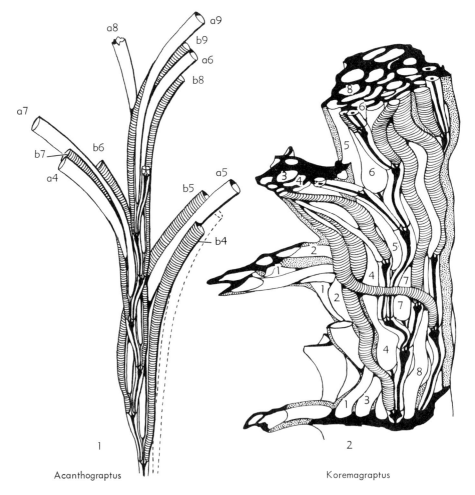

Fig. 12. Thecal grouping in the Acanthograptidae (29).

1. *Acanthograptus suecicus* (Wiman) showing regular association of 2 autothecae and 2 bithecae to form twigs.
2. Restoration from serial sections cut by Wiman of *Koremagraptus formosus* (Wiman), showing complex branch with numerous main stolons and larger, more irregular, twigs (bithecae shaded; numbering of some of the autothecae only to identify such thecae as can be traced throughout the series figured).

Koremagraptus, with its dominant but irregular anastomosis, branchlets and twigs show little if any regularity in their construction.

MODE OF BRANCHING

Preceding any bifurcation of a stipe, two main stolons must be produced and hence at a branching division or node two stolothecae must arise. Since the stolothecae are essentially the proximal portions of autothecae, it follows that the branching division entails production of two autothecae in place of autotheca and bitheca (i.e., the suppression of a bitheca). In spite of this, the external regularity in arrangement of the thecae along the stipe is not disturbed; slight adaptation in length of stolon and length of bitheca insures that the normal association of bithecal and autothecal apertures persists (Fig. 13,1), uninterruptedly.

The shape of a colony, especially a conical colony like that of *Dictyonema*, is largely dependent upon frequency of branching, but in most instances the branching divisions are uniformly scattered over the surface of the rhabdosome. However, in siculate species such as *Dictyonema flabelliforme* (Fig. 13,2) and in many anisograptids such as *Clonograptus* and *Anisograptus*, fairly regular "zones of branching" do appear, particularly proximally (Fig. 13,2), and relative lengths of various orders of branches may constitute one of the specific characters. Thus in *Anisograptus matanensis*, one of the three primary branches is about one quarter of the length of the other two, while in *A. richardsoni* all three are equal (though of different lengths from the corresponding branches of *A. flexuosus*). This suggests some degree of rhabdosomal control reminiscent of that so distinctive of the Graptoloidea. It is to be expected that branching occurs at the level of theca $n\pm1$ (as in cladia production of cyrtograptids) rather than that any particular order of branching has a precise absolute length.

Bifurcation of a complex branch containing several main stolons and elongate tubu-

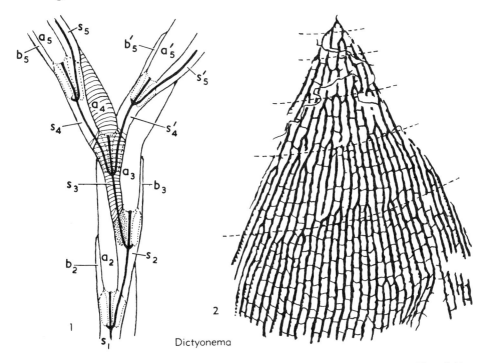

Fig. 13. Branching in *Dictyonema flabelliforme* (EICHWALD) (24).——*1*. Diagram of branching division with two stolothecae (*s4* and *s'4*) in place of stolotheca and bitheca; parent stolotheca *s3* and daughter autotheca *a4* shaded.——*2*. Approximate zones of branching in a rhabdosome of *D. flabelliforme*, ×1.

lar thecae (Fig. 12,2) may involve merely the separation of certain stolons, together with some associated autothecae and bithecae. Here, the technical "branching division" of the stolon involving the production of two stolothecae is not immediately related to the bifurcation of the stipe.

DISSEPIMENTS AND ANASTOMOSIS

In *Dictyonema*, the branches are united by transverse threads called dissepiments, which may be rather erratic in spacing and direction, but in certain species are extraordinarily regular. They also characterize *Ptiograptus* and a few may be developed in certain species of *Callograptus* and even *Dendrograptus*.

These dissepiments have been shown to be extrathecal in origin and composed of cortical tissue, secreted by the extrathecal living tissue responsible for secondary thickening in general. Some exhibit growth out from adjacent branches so as to meet and fuse in the center, and two closely adjacent dissepiments may be partially or completely united by a web of cortical tissue. Two different types of mesh, one coarse, with broad widely-spaced dissepiments and the other fine, with slender closely-spaced dissepiments, may occur in the same rhabdosome in *Dictyonema flabelliforme norvegicum*, but this is exceptional. The degree of variation in dissepimental structures and mesh characters suggests need for caution in their taxonomic use.

The complicated flange structure, which produces a honeycomb appearance on the dorsal (outer) side of the rhabdosome of *Dictyonema cotyledon*, appears to be related in some way to dissepimental structures. Though apparently composed of fusellar tissue (SKEVINGTON, 1963), its formation is difficult to explain and it may prove to be a variant of the pseudofusellar cortical tissue as developed in *D. wysoczkianum*. The terminal plates on the apertural spines of *D. peltatum* may fuse to a more or less continuous sheet, but it is extremely doubtful if stipes are ever connected by apertural spines; accordingly, RUEDEMANN's genus *Airograptus*, based on this assumption, is not here accepted. Bithecae may grow out sporadically along dissepiments, but again no evidence is offered that regular "pseudodissepiments" are formed by bithecae or autothecae. As in the Tuboidea, however, single thecae may connect adjacent branches obliquely in *Koremagraptus*, but this is a limiting case of anastomosis prevalent in this genus.

More or less regular anastomosis characterizes *Desmograptus* and several acanthograptid and inocaulid genera, and complicated but irregular transfer of thecae, singly or in groups, may occur particularly in forms with several stolonal chains present in their branches.

DEVELOPMENT[1]

The most detailed and complete account of dendroid development is that published by KOZŁOWSKI (1949) for *Dendrograptus communis*, which may serve as a type for the order (Fig. 14).

The prosicula is a thin-walled, almost cylindrical tube, closed and flattened at the base and usually developing a well-marked basal disc of attachment. Its walls are strengthened by a spiral thread *(Schraubenlinie, ligne helicoidale)* coiled indifferently in a clockwise or counterclockwise direction, but not exhibiting any longitudinal fibers comparable with those of the graptoloid prosicula (see Fig. 39).

From this initial prosicula, the metasicula is sharply differentiated by its closely set growth lines. These are not so strikingly regular as in the Graptoloidea, but a general zigzag line runs down the dorsal and ventral sides and the growth bands are for the most part alternating half rings, giving a definite bilateral symmetry to the metasicula. With continued growth, the metasicula of *Dendrograptus communis* begins to develop the dorsal and ventral apertural processes so conspicuous in the adult.

The initial bud or sicular stolotheca originates in a pore produced by resorption in the wall of the prosicula. After a short period of growth pressed against the wall of the prosicula and metasicula, it gives

---

[1] [The general term development has been retained, in preference to astogeny, chiefly because the expressions "dichograptid type of development," "diplograptid development," etc., are by now so well entrenched in the literature. Strictly speaking, moreover, the phase covered by the term includes the ontogeny of the sicula and the astogeny of the initial part of the colony, which begins with *th1'* and comprises a small but variable number of individual thecal ontogenies.]

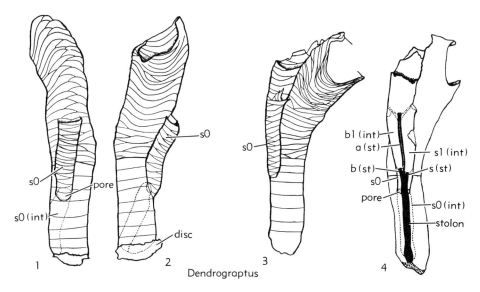

Fig. 14. Sicula and initial bud of *Dendrograptus communis* Kozłowski (114).——*1,2*. Prosicula with basal disc and spiral line, metasicula, and early stage of initial bud (stolotheca, *s0*) which extends internally *(s0int)* to base of prosicula.——*3,4*. Later stage of development showing the 3 descendants of stolotheca *0*. [All figures approx. ×45; letter symbols as in Fig. 9.]

rise distally to the first triad, consisting of autotheca, bitheca, and stolotheca; proximally it can be traced within the prosicula to the basal disc as a thin-walled tube containing the initial stolon. The first triad of thecae thus produced (Fig. 14,*4*) constitutes the initial part of the main stem of the colony, which in this species extends for four or five consecutive generations before beginning to branch.

Numerous other dendroid astogenies are known in varying degrees of detail, and most agree in all essentials with that just described. The list includes: *Acanthograptus suecicus* (STRACHAN, 1959), *Calyxdendrum graptoloides* (KOZŁOWSKI, 1960), *Dictyonema flabelliforme* (BULMAN, 1949) and various anisograptids (STUBBLEFIELD, 1929; BULMAN, 1950a, 1954; SPJELDNAES, 1963), *Dictyonema cavernosum* (BULMAN & RICKARDS, 1966), *Graptolodendrum mutabile* (KOZŁOWSKI, 1966) and *Rhipidodendrum samsonowiczi* (KOZŁOWSKI, 1949). In all of these, in contrast to the Graptoloidea, the initial bud appears to be produced from the prosicula, with the exception of *Graptolodendrum*, where it originates in the metasicula. Another difference from the Graptoloidea is the absence of longitudinal strengthening fibers in the prosicula; these may be related to the development of the nema and it is possible that they occur in *Dictyonema flabelliforme* and the anisograptids, but have not yet been detected since these prosiculae are not yet known in transparencies. STUBBLEFIELD (1929) noted their possible presence in *Adelograptus hunnebergensis*, but in the late, somewhat aberrant anisograptid *Calyxdendrum* they are absent. The relation of the initial bud and earliest-formed triad of thecae tends to be indifferently right- or left-handed.

Some irregularities and departures from the typical dendroid plan begin to appear in the anisograptids, where bithecal development may become irregular. To judge from *Kiaerograptus* (SPJELDNAES, 1963), the sicular bitheca from triad *1* persists after the loss of other proximal bithecae; *Adelograptus* (SPJELDNAES, 1963) can likewise be interpreted in terms of normal dendroid development but again only the sicular bitheca has been recorded. An extraordinary astogeny was described by LEGRAND (1963) for his genus *Choristograptus*, but it is impossible to interpret the proximal end from the photographic illustrations and the genus is provisionally included here as a synonym of *Adelograptus*.

Rhipidodendrum (Fig. 15,4), which here is provisionally retained in the family Dendrograptidae, differs from typical dendroids in that the first stolonal node is diad and produces only a bitheca and a stolotheca. This is usually succeeded by one normal triad and then a concentrated series of branching divisions; this genus shows much diversity, however, and in some colonies even the second node is a branching division.

No positive evidence can be cited of any significant (genetic) relationship between two or more rhabdosomes based on a common "root system," analogous to the synrhabdosomes of the Graptoloidea. The few examples figured (*Dictyonema cavernosum* in WIMAN, 1897a, and *Syrrhipidograptus* in POULSEN, 1924) are here ascribed to chance associations or a mistaken interpretation of root irregularities.

PALEOECOLOGY

The class Graptolithina includes four orders (Tuboidea, Camaroidea, Stolonoidea, and Crustoidea) which appear to have been sessile, but concerning the ecology of which practically nothing is known. The Dendroidea, however, are somewhat better understood.

That the Dendroidea, with their thickened "stems" and discoidal or ramifying basal organs, have the morphology of sessile organisms was long ago recognized by HALL (1865). Their remains occur with other shallow-water benthonic invertebrates, and their sporadic distribution is consistent with a sessile mode of life. Few instances of attachment to shells or pebbles have ever been recorded, but the restoration attempted by RUEDEMANN (1925) for the Gasport lens is probably quite typical of the ecology of the group. Here are the fossilized remains of a muddy channel in the

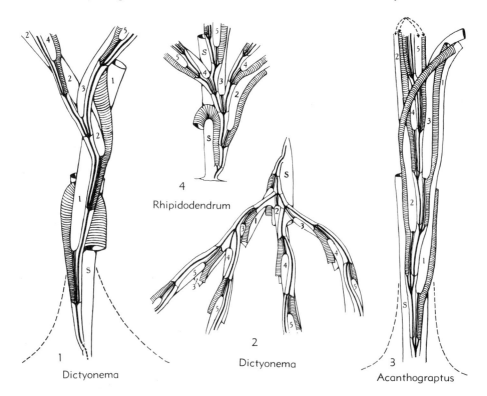

FIG. 15. Diagrams representing development of dendroid rhabdosomes (29).——*1. Dictyonema cavernosum* WIMAN.——*2. D. flabelliforme* (EICHWALD).——*3. Acanthograptus suecicus* (WIMAN).——*4. Rhipidodendrum samsonowiczi* KOZLOWSKI. [Autothecae numbered in order of appearance; bithecae shaded; stolon system in solid black; sicula, *S*.]

Lockport limestone sea, carpeted with a miniature forest of tough seaweed and bushy dendroid graptolites, while in the clearer water on either side up to the brink of the channel flourished a profusion of corals and crinoids, with associated brachiopods and mollusks.

This view of dendroid ecology recently has been questioned by BOUČEK (1957), who followed PRANTL in arguing that the Dendroidea were in fact epiplanktonic, living attached to large floating algae. For example, BOUČEK cited particularly the occurrence of very large rhabdosomes of *Dictyonema (Pseudodictyonema) graptolithorum* in ordinary (euxinic) graptolite shales associated with *Monograptus spiralis* and *Monoclimacis vomerina*. The rootlike structure at the proximal end of most dendroids clearly does not prove attachment to the sea bottom, and structurally such forms might equally well have been attached to large floating algae. But the occurrence of dendroids in true graptolite shales is somewhat unusual and an association with the remains of benthonic organisms is more normal. Of course benthonic dendroids might have been attached to seaweed as readily as to other objects on the bottom, and if such algae broke free and drifted away (as modern *Sargassum* does), this might account for examples suggestive of epiplanktonic association. BOUČEK attributed the more restricted distribution of dendroids (*vis-a-vis* Graptoloidea) to the comparably more restricted occurrence of particular species of large floating algae.

Whatever view may be taken of the mode of life of the Dendroidea as a whole, an epiplanktonic existence can be attributed convincingly to such Tremadocian dendroids as *Dictyonema flabelliforme* and the Anisograptidae. These colonies probably lived attached by their nemata to floating weed, "like a bell at the end of a rope" in the words of LAPWORTH (1897), and it has been claimed that in adopting an epiplanktonic mode of life, *D. flabelliforme* had taken the first step along the road leading to the Graptoloidea. Strong supporting evidence here lies as much in their geographical distribution as in morphology of the proximal end; their widespread occurrence, comparable with that of the Graptoloidea, is in marked contrast to that recorded for most dendroids. The *Staurograptus* rhabdosomes attached to the alga *Sphenophycus,* described by RUEDEMANN (1934), are suggestive, but the possibility of a drifted association cannot be overlooked. STØRMER (1933, 1935) has described and figured specimens of *Dictyonema flabelliforme* with a bladderlike structure at the proximal end; like the so-called "floats" of the Graptoloidea (p. *V*93), these structures are perhaps more likely to have supported vesicular tissue than to be themselves air bladders, but such forms would appear to have become truly planktonic.

GEOGRAPHIC DISTRIBUTION

Many dendroid genera (e.g., *Dictyonema, Dendrograptus, Desmograptus*) certainly have an extremely wide if not a world-wide distribution, but few species appear to have any notable geographic range. BOUČEK (1957) rightly pointed out that the dendroids are still very inadequately described and that few critical comparisons between materials from different areas are available, so that generalizations are dangerous. But of the rich dendroid fauna comprising some 50 species described by SPENCER (1884) and BASSLER (1909) from the Niagaran of Hamilton, Ontario, less than half have been reported from even nearby localities in the United States, and an extremely small proportion from other continents. BOUČEK (1957) monographed over 90 dendroid species from the Silurian of Bohemia, of which more than 80 are new (or can be referred to species already described by POČTA); only six of these are referred to Niagaran forms, and two to species from other European countries. In the present state of knowledge, it can be said surely that the geographical distribution of species of dendroid graptolites is restricted as compared with graptoloid species, and this would accord with a sessile, benthonic mode of life.

STRATIGRAPHIC DISTRIBUTION

A species attributed to *Dendrograptus* has been described by ÖPIK (1933) from the Middle Cambrian (*Paradoxides davidis*

Zone) of Norway, and an undescribed form is known from the same horizon in the Middle Cambrian of Shropshire, England. In addition to these, a number of Graptolithina (e.g., *Siberiodendrum, Dithecodendrum*) of uncertain taxonomic position (possibly representing a new order, but possibly dendroid or tuboid) have been described from the Middle Cambrian of Siberia, USSR, by OBUT (1964). Apart from these records, the earliest occurrence of the Dendroidea is in the Upper Cambrian; RUEDEMANN (1933) described a small fauna, including species of *Dendrograptus, Callograptus* and *Dictyonema*, from the Trempealeauan Stage of Wisconsin and a somewhat larger fauna (poorly illustrated) has been described from the Wilberns Formation of Texas by DECKER (1945). Other occurrences have been described from Quebec and from western Canada, and although it is possible that some of these records may be tuboid rather than dendroid, with little doubt dendroid graptolites were well established by Late Cambrian times. From then they persist with remarkably little conspicuous change to the Carboniferous, the highest dendroid being perhaps an undescribed species of *Dictyonema* from the Yoredale Series of Yorkshire, England. The time range of many individual genera is also extremely long; both *Dictyonema* and *Callograptus* extend from Upper Cambrian to Carboniferous, *Desmograptus* from Lower Ordovician (Arenig) to Carboniferous, and *Dendrograptus* from ?Middle or Upper Cambrian to Upper Silurian. Most genera range through the Ordovician and Silurian.

CLASSIFICATION

With the gradual accumulation of morphological detail, a satisfactory basis for classification of the Dendroidea may emerge, but at present, when the structure of certain species of a few genera is known in great detail while the majority are still "form genera," a conservative attitude has been adopted here in recognizing both families and genera. A large number of genera, including many newly described ones, have been relegated to *incertae sedis,* or, where it appears difficult to maintain their individuality on the basis of gross morphology, they have been classed as synonyms.

The distinction between the Acanthograptidae and Inocaulidae, accepted in the first edition of this *Treatise,* has now been abandoned. The structure of *Inocaulis* remains practically unknown (and even triad budding is as yet unproved), and from the point of view of gross morphology it appears to differ from *Acanthograptus* essentially in degree: the thecal tubes have become so slender as to be capillary. Again, the differences between *Palaeodictyota* and *Thallograptus* are scarcely such as to justify allocation to two separate families, and both can only be assigned familial and even ordinal position with reservation. To this extent, the broad classification adopted here agrees with that proposed by BOUČEK (1957), except that he erected a separate order for *Inocaulis* (to which he also assigned *Medusaegraptus* and *Palmatophycus*).

Indifferent preservation naturally creates uncertainty. A poorly preserved *Koremagraptus,* for example, could be indistinguishable from the tuboid *Reticulograptus;* the generic identification of particular species, like individual specimens, must remain provisional in perhaps the majority of instances.

The predominantly Tremadocian family Anisograptidae, which appeared originally to comprise a number of genera plausibly derived from perhaps a single species of siculate *Dictyonema,* has been extended to include the mid-Ordovician *Calyxdendrum,* which must have acquired its siculate character independently and at a much later date. Moreover, even the Tremadocian anisograptids now appear to be probably polyphyletic.

SYSTEMATIC DESCRIPTIONS

Family DENDROGRAPTIDAE Roemer in Frech, 1897

[Dendrograptidae ROEMER in FRECH, 1897, p. 568]

Rhabdosome conical, flabellate, or irregularly dendroid, usually with thecate or more or less thickened nonthecate stem terminating proximally in rootlike processes or disc of attachment, rarely attached by nema; branching generally dichotomous, stipes free or united by anastomosis or by

Dendroidea—Dendrograptidae

transverse dissepiments. Autothecae denticulate to tubular and isolate, commonly with unpaired apertural spine or process, inwardly facing in conical rhabdosomes; bithecae variable in form, usually inconspicuous externally; stolothecae situated on dorsal side of stipe. *?M.Cam., U.Cam.-Carb.*

The branches of dendrograptids are characteristically simple, with relatively short and denticulate autothecae, and even where these are more elongate and isolate,

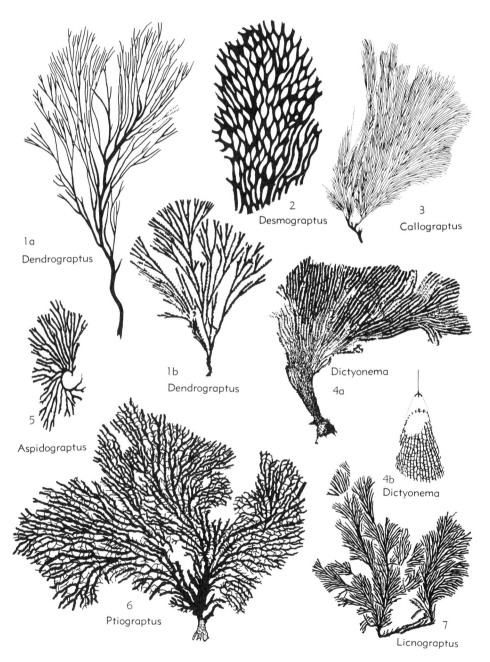

FIG. 16. Dendrograptidae (p. *V*38-*V*39).

their apertures are all directed ventrally in a single row, with the single stolonal chain on the dorsal side of the branch.

Dendrograptus HALL, 1858, p. 143 [*Graptolithus hallianus* PROUT, 1851, p. 189; SD HALL, 1862, p. 21] [=?*Ophiograptus* POULSEN, 1937, p. 24 (type, *O. inexpectans;* OD)]. Generally robust, shrublike in habit, branching irregularly, stipes usually divergent, unconnected, stem well developed, with basal attachment; autothecae denticulate, spined or with apertural processes. ?*M.Cam.,* *U.Cam.-Carb.,* almost worldwide.——FIG. 16,*1a*. *D. fruticosus* HALL, L.Ord.(Levis Sh.), Que.; ×1 (77).——FIG. 16,*1b*. *D. hallianus* (PROUT), U.Cam.(Trempeal.), Minn.; ×1 (209).

[It is probable that *Stelechocladia* POČTA, 1894, p. 206 (emend. BOUČEK, 1957, p. 35, type, *S. fruticosa* POČTA, 1894, p. 207; SD BOUČEK, 1957, p. 35 (=*Dendrograptus* (*Stelechocladia*) *suffruticosus* BOUČEK, 1957, nom. nov., p. 36); =?*Callodendrograptus* DECKER, 1945, p. 28, type, *C. sellardsi;* OD) is a subgenus bearing the same relation to *Dendrograptus* that *Pseudocallograptus* does to *Callograptus*.]

Aspidograptus BULMAN, 1934, p. 70 [*Clematograptus implicatus* HOPKINSON, 1875, p. 652; OD]. Similar to *Dendrograptus* but branching laterally from ?4 curved principal stipes; lateral branches close-set, irregularly produced, bifurcating repeatedly. *U. Cam.-Ord.,* Eu.-N.Am.-S.Am.-China.——FIG. 16,*5*. *A. implicatus* (HOPKINSON), L.Ord.(Arenig), Eng.; ×1 (18).

Callograptus HALL, 1865, p. 133 [*C. elegans;* SD MILLER, 1889, p. 175] [=*Odontocaulis* LAPWORTH, 1881, p. 175 (type, *O. keepingi;* OD); *Capillograptus* BOUČEK, 1957, p. 46 (type, *Callograptus dichotomus* POČTA, 1894, p. 182; M)]. Rhabdosome conical, flabellate or somewhat irregular, with the thecae stem (*Odontocaulis* condition) or more usually thickened nonthecate stem, branching dichotomously with some regularity, stipes subparallel to parallel, sporadically united by anastomosis or dissepiments. *U.Cam.-Carb.,* almost worldwide.

C. (Callograptus). Autothecae denticulate, rarely isolate distally, with normal length ratio; apertural processes in some developed into acute spines. *U.Cam.-Carb.,* almost worldwide.——FIG. 16,*3*. *C. elegans,* Levis Sh., Can.(Que.); ×0.75 (77).

C. (Alternograptus) BOUČEK, 1956, p. 131 [*C. (A.) holubi;* OD]. Proximal branching lateral, stipes alternating to left and right, distal branching normal, dichotomous. *L. Ord.(Arenig),* Eu.(Czech.).——FIG. 17,*1*. *C. (A.) holubi,* Klabava beds, Rokycany; proximal portion of rhabdosome, ×2 (11).

C. (Pseudocallograptus) SKEVINGTON, 1963, p. 19 [*Callograptus salteri* HALL, 1865, p. 135; OD]. Autothecae and bithecae elongate, producing "ropy" appearance of stipe and unusual association of thecal apertures; autothecae generally denticulate. *L.Ord.,* Eu.-N.Am.-S.Am.——FIG. 17,*3*. *C. (P.)* cf. *C. (P.) salteri* (HALL), Vaginatumkalk (Ontikan),Sweden(Öland); fragment of stipe showing thecal elongation, ×14 (214). [*a,* autotheca;; *bi,* bitheca; *st,* stolotheca].

Desmograptus HOPKINSON, in HOPKINSON & LAPWORTH, 1875, p. 668 [*Dictyograptus cancellatus* HOPKINSON, 1875, p. 668; M] [=*Rhizograptus* SPENCER, 1878, p. 460 (*pro Rhizograpsus* SPENCER, 1878, ICZN Opin. 650) (type, *R. bulbosus;* M); ?*Syrrhipidograptus* POULSEN, 1924, p. 1 (type, *S. nathorsti;* M)]. Rhabdosome conical, possibly flabellate rarely, stipes flexuous, united by regular anastomosis and rare dissepiments; autothecae denticulate to isolate. *L.Ord.(?Tremadoc)-Carb.,* Eu.-N.Am.——FIG. 16,*2*. *D. cancellatus* (HOPKINSON), Arenig Sh., S.Wales; ×1 (95).

[It is possible that species with strongly isolate thecae should be grouped together under the name *Syrrhipidograptus* POULSEN, 1924, bearing the same relation to *Desmograptus* that *Pseudocallograptus* does to *Callograptus*.]

Dictyonema HALL, 1851, p. 401 [*Gorgonia retiformis* HALL, 1843, p. 115; SD MILLER, 1889, p. 185] [=*Phyllograpta* ANGELIN, 1854, p. iv (type, *Gorgonia flabelliformis* EICHWALD, 1840, p. 207; M); *Rhabdinopora* EICHWALD, 1855, p. 453 (type, *Gorgonia flabelliformis* EICHWALD, 1840, p. 207; SD BULMAN, herein); *Graptopora* SALTER, 1858, p. 65 (type, *G. socialis;* M); *Dictyograptus* HOPKINSON, 1875, p. 667 (*pro Dictyonema* HALL); *Damesograptus* JAHN, 1892, p. 645 (type, *Dictyonema* sp. DAMES, 1873, p. 383; OD); *Dictyodendron* WESTERGÅRD, 1909, p. 62 (*pro Dictyonema ex D. flabelliforme*); *Dictyograptus* WESTERGÅRD, 1909, p. 63 (type, *Gorgonia flabelliformis* EICH-

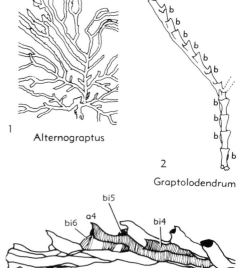

FIG. 17. Dendrograptidae [*a,* autotheca; *b, bi,* bitheca; *st,* stolotheca] (p. V38-V39).

WALD, 1840, p. 207; OD); *Airograptus* RUEDEMANN, 1916, p. 20 (type, *Dictyonema furciferum* RUEDEMANN, 1904, p. 606; OD); *Nephelograptus* RUEDEMANN, 1947, p. 196 (type, *N. rectibrachiatus;* OD)]. Rhabdosome conical, varying from almost cylindrical to almost discoidal, with thecate or nonthecate stem or rarely attached by nema; branching dichotomous, stipes straight, subparallel to parallel, united by transverse dissepiments, anastomosis rare; autothecae denticulate, commonly spined, rarely tubular and isolate; bithecae normally inconspicuous. *U.Cam.-Carb.*, almost worldwide.

D. (Dictyonema). Autothecae denticulate, with normal length ratio. *U.Cam.-Carb.*, almost worldwide.——FIG. 16,4a. *D. (D.) crassibasale* BASSLER, Sil.(Niag.), Hamilton, Ont.; ×1 (5).——FIG. 16,4b. *D. (D.) flabelliforme* (EICHWALD), L.Ord.(Tremadoc), Pedwardine, Eng.; ×1 (18).

D. (Pseudodictyonema) BOUČEK, 1957, p. 69 [*Dictyonema graptolithorum* POČTA, 1894, p. 196; OD]. Autothecae and bithecae elongate, producing "ropy" appearance of stipe; apertures denticulate or slightly isolate. *U.Sil.*, Eu. (Czech.).

Graptolodendrum KOZŁOWSKI, 1966, p. 4 [*G. mutabile;* OD]. Similar to *Dendrograptus* but with abnormal bithecae, mainly on one side of stipe only, and other abnormalities in detailed structure. *L.Ord.*, Eu.(Pol.).——FIG. 17,2. *G. mutabile,* glacial boulder (?*Glyptograptus teretiusculus* Zone); b indicates side (left or right) on which bithecae occur; ×5 (123).

Licnograptus RUEDEMANN, 1947, p. 196 [*L. elegans;* OD]. Several principal branches bearing fanlike groups of subparallel branches laterally and distally; thecal details unknown. *L.Ord.*, Can.(Que.-Newf.).——FIG. 16,7. *L. elegans,* Levis Sh., Que.; ×1 (209).

Ptiograptus RUEDEMANN, 1908, p. 175 [*P. percorrugatus;* OD]. Like *Dictyonema* but rhabdosome flabellate; dissepiments irregular, commonly angular; thecal details unknown. *Sil.-L.Carb.*, N.Am.-NW.Eu.——FIG. 16,6. *P. percorrugatus,* M.Dev., Ky.; ×1 (201).

Rhipidodendrum KOZŁOWSKI, 1949, p. 133 [*R. samsonowiczi;* OD]. Rhabdosome minute, fan-shaped, branching irregularly from 3 primary branches; autothecae and bithecae tubular, conspicuously curved. *L.Ord.-(Tremadoc)*, Eu.(Pol.).

Sagenograptus OBUT & SOBOLEVSKAYA, 1962, p. 74 [*S. gagarini;* M]. Somewhat similar to *Dictyonema*, but with coarse irregular meshwork. *L.Ord.*, USSR(Taimyr).

Family ANISOGRAPTIDAE Bulman, 1950

[Anisograptidae BULMAN, 1950, p. 79]

Rhabdosome siculate, pendent to horizontal or rarely reclined, quadriradiate, triradiate or bilateral; branching usually dichotomous, rarely lateral; stipes with typical dendroid structure, autothecae and bithecae characteristically simple, the latter reduced and partially absent in some, stolonal chains dorsal, superficial. *L.Ord. (Tremadoc), ?U.Ord.(Nemagraptus gracilis Zone).*

Anisograptus RUEDEMANN, 1937, p. 61 [*A. matanensis;* OD]. Rhabdosome triradiate, developed by dichotomous division to 6th order (usually 3rd or 4th) from 3 primary branches; typically horizontal but including declined and slightly reclined forms; autothecae denticulate, bithecae short, simple. *L.Ord.(Tremadoc)*, N.Am.-S.Am.-NW.Eu.-USSR-N.Afr.(Morocco).——FIG. 18,4. *A. matanensis,* Matane Sh., Que.; ×1 (25).

Adelograptus BULMAN, 1941, p. 114 [*Bryograptus? hunnebergensis* MOBERG, 1892, p. 92; OD] [=?*Choristograptus* LEGRAND, 1963, p. 52 (type, *C. louhai;* OD)]. Usually declined or almost horizontal, rarely pendent, commonly somewhat lax and flexuous, developed from 2 primary stipes by infrequent and irregular branching, apparently lateral rather than dichotomous; autothecae denticulate, bithecae and stolothecae in geologically early species. *L.Ord.(Tremadoc-Arenig)*, NW.Eu.-N.Am.-USSR-N.Afr.(Alg.)-N.Z.-Australia.——FIG. 18,3. *A. hunnebergensis* (MOBERG), Tremadoc, Eng.; ×1 (229).

Aletograptus OBUT & SOBOLEVSKAYA, 1962, p. 76 [*A. hyperboreus;* M]. Rhabdosome quadriradiate, comprising 4 undivided primary stipes; thecal structure unknown. *L.Ord.(Tremadoc)*, USSR (Taimyr).

Bryograptus LAPWORTH, 1880, p. 164 [*B. kjerulfi;* SD GURLEY, 1896, p. 64]. Rhabdosome pendent to declined, developed from 3 primary stipes by irregular and apparently lateral branching; stolothecae and bithecae present in geologically early species. *L.Ord.(Tremadoc-Arenig)*, NW.Eu.-N.Am.-N.Afr.-Australia-?China.——FIG. 18,2a. *B. kjerulfi,* Tremadoc, Sweden; ×1 (257).——FIG. 18,2b. *B. patens* MATTHEW, Tremadoc, Que.; ×1 (25).

Calyxdendrum KOZŁOWSKI, 1960, p. 109 [*C. graptoloides;* OD]. Rhabdosome minute, pendent, dendroid, branching at close intervals; autothecae conical, bithecae opening into autothecal cavities; prosicula lacking longitudinal fibers, but with relatively thick nema. *?U.Ord.(Nemagraptus gracilis Zone)*, glacial boulders, Eu.(Pol.).——FIG. 19,1. *C. graptoloides,* proximal end; ×35 (119). [a1, a2, etc., autothecae; b1, b2, etc., bithecae; n, nema; si, sicula; s0, s1, etc., stolothecae.]

Clonograptus NICHOLSON (ex HALL MS), 1873, p. 138 [pro *Clonograpsus* HALL & NICHOLSON, 1873, ICZN, Opin. 650] [*Graptolithus rigidus* HALL,

1858, p. 146; SD MILLER, 1889, p. 179] [=?*Herrmannograptus* MONSEN, 1937, p. 186 (type, *Graptolithus milesi* HALL, 1861, p. 372; OD)]. Rhabdosome bilateral, produced by dichotomous division, generally at steadily increasing intervals, to 8th or 9th order (usually 5th or 6th); branches diverging proximally, becoming subparallel, and in some species flexuous, distally; autothecae denticulate with moderate inclination, some species with low inclination and negligible overlap, some with exaggerated apertural spines; stolothecae and bithecae in geologically early species; central disc and web structures rare. [In view of the variability of *Clonograptus*, it seems scarcely feasible to maintain *Herrmannograptus* MONSEN, 1937, as a distinct genus.] L.Ord.(*Tremadoc-Llanvirn*), almost

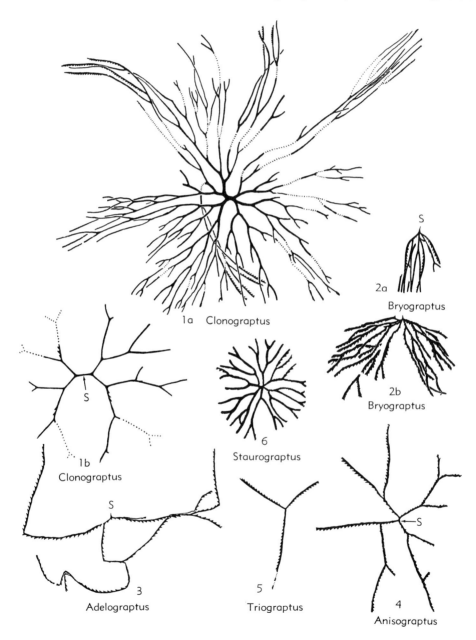

FIG. 18. Anisograptidae [*S*, sicula] (p. *V*39-*V*41).

worldwide.——FIG. 18,*1a*. *C. flexilis* (HALL), Levis Sh., Que.; ×0.5 (78).——FIG. 18,*1b*. *C. tenellus* LINNARSSON, Tremadoc, S.Sweden; ×1 (144).

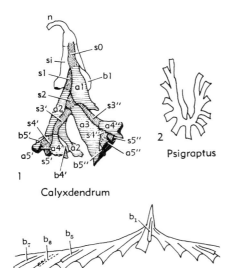

FIG. 19. Anisograptidae [*a*, autotheca; *b*, bitheca; *n*, nema; *s*, stolotheca; *si*, sicula] (p. *V*39-*V*41).

Kiaerograptus SPJELDNAES, 1963, p. 123 [**Didymograptus kiaeri* MONSEN, 1925, p. 172; OD]. Rhabdosome bilateral, composed of 2 undivided, horizontal stipes; autothecae of dichograptid type, bithecae present distally. L.Ord.(Tremadoc), Eu.(Norway)-?N.Am.——FIG. 19,*3*. **K. kiaeri* (MONSEN), Norway; [*b1*, etc., bithecae], ×5 (Bulman, n).

Psigraptus JACKSON, 1967, p. 314 [**P. arcticus*; OD]. Rhabdosome siculate, composed of 2 (or ?3) short, reclined stipes; autothecae distally isolate, stolothecae and bithecae believed present. L.Ord.(Tremadoc), N.Am.(Yukon).——FIG. 19,*2*. **P. arcticus*, Rock River; ×4 (105).

Radiograptus BULMAN (*ex* LAPWORTH MS), 1950, p. 89 [**R. rosieranus*; OD]. Rhabdosome triradiate, discoidal, composed of numerous branches dividing dichotomously, connected by sparsely developed dissepiments; thecal structure imperfectly known. L.Ord.(Tremadoc), N.Am.(Que.).

Staurograptus EMMONS, 1855, p. 108 [*pro Staurograpsus* EMMONS, 1885, ICZN, Opin. 650] [**S. dichotomus*; M]. Rhabdosome small, quadriradiate, developed by dichotomous division to about 4th order of 4 primary stipes, typically horizontal; bithecae imperfectly known. [Genus is commonly almost indistinguishable from discoidally preserved immature specimens of *Dictyonema* and it possibly has no validity.] L.Ord.(Tremadoc), N.Am.-Australia-?NW.Eu.——FIG. 18,*6*. **S. dichotomus*, Schaghticoke Sh., N.Y.; ×1 (201).

Triograptus MONSEN, 1925, p. 169 [**T. osloensis*; M]. Rhabdosome triradiate, composed of 3 horizontal, undivided stipes; stolothecae and bithecae present. L.Ord.(Tremadoc), NW.Eu.-N.Am.——FIG. 18,*5*. *T. canadensis* BULMAN, Matane Sh., Que.; ×1 (25).

Family PTILOGRAPTIDAE Hopkinson in Hopkinson & Lapworth, 1875

[Ptilograptidae HOPKINSON in HOPKINSON & LAPWORTH, 1875, p. 661]

Rhabdosome sessile, dendroid, with alternating pinnate arrangement of lateral branches. *L.Ord.-U.Sil.*

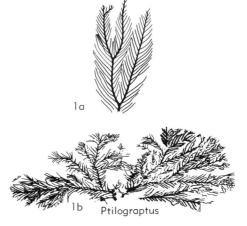

FIG. 20. Ptilograptidae (p. *V*41).

Ptilograptus HALL, 1865, p. 139 [**P. plumosus*; SD MILLER, 1889, p. 201] [=*Denticulograptus* SCHMIDT, 1939, p. 122 (type, *Ptilograptus acutus* HOPKINSON & LAPWORTH, 1875; OD)]. Rhabdosome with comparatively few main branches, bifurcating rarely and bearing closely set lateral branches arranged alternately on opposite sides; autothecae usually denticulate, but thecal details and constitution almost unknown. L.Ord.(Arenig)-U.Sil., Eu.-N.Am.-Australia. —— FIG. 20,*1a*. **P. plumosus*, L.Ord.(Levis Sh.), Que.; ×1 (77).——FIG. 20,*1b*. *P. delicatulus* RUEDEMANN, Ord.(Ottosee Sh.), Tenn., ×1 (209).

Family ACANTHOGRAPTIDAE Bulman, 1938

[Acanthograptidae BULMAN, 1938, p. 20]
[Incl. Inocaulidae RUEDEMANN, 1947, p. 230]

Rhabdosome sessile, conical to irregularly dendroid; stipes flexuous and anastomosing or rigid and irregularly branching, composed of elongate, tubular and in some forms almost capillary thecae, adnate proximally and isolate distally to varying extent, produced in normal dendroid triads but commonly showing distinctive grouping; stipes generally compound, with several stolonal chains enclosed within each branch. *?U.Cam., L.Ord.-M.Dev.*

Acanthograptus SPENCER, 1878, p. 461 [*pro Acanthograpsus* SPENCER, 1878, ICZN, Opin. 650] [**A. granti;* M] [=*?Boučekocaulis* OBUT, 1960, p. 148 (type, *Acanthograptus jubatus* OBUT, 1953; OD)]. Robust dendroid rhabdosome composed of rather stout branches bifurcating irregularly; very rarely anastomosing; thecae elongate, tubular, isolate distally to produce spinous appearance of branch; minor branches or "twigs" usually composed of 2 autothecae and 2 bithecae. *?U.Cam.,*

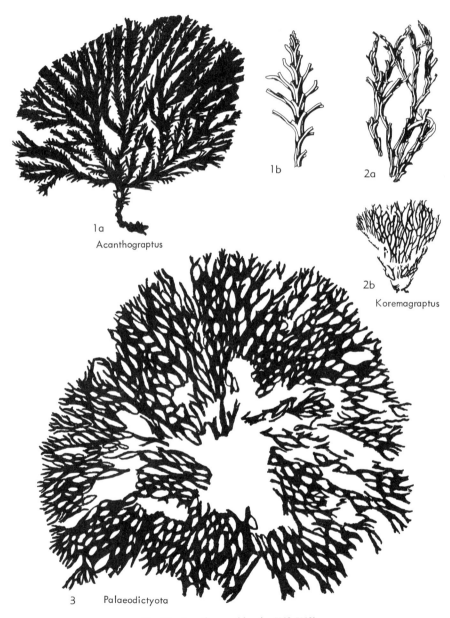

FIG. 21. Acanthograptidae (p. *V*42-*V*43).

Dendroidea—Acanthograptidae

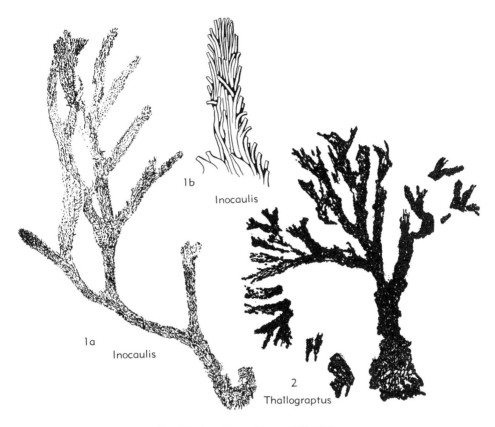

Fig. 22. Acanthograptidae (p. V42-V43).

L.Ord.-U.Sil., Eu.-N.Am.-Asia-Australia.——Fig. 21,1a. *A. granti, M.Sil.(Niag.), Ont.; ×1 (201). ——Fig. 21,1b. A. musciformis (Wiman), U. Ord., Baltic; distal part of branch, ×7 (267).

Inocaulis Hall, 1852, p. 176 [*I. plumulosa; M]. Very thick branches, bifurcating irregularly, composed of extremely fine, capillary thecae projecting distally as hairlike processes. Sil., Eu.-N.Am.——Fig. 22,1. *I. plumulosa, Niag., Ont.; 1a, ×1; 1b, distal fragment of stipe, ×5 (1a,5; 1b,201).

Koremagraptus Bulman, 1927, p. 345 [* K. onniensis; M] [=Coremagraptus Bulman, 1942, p. 285 (nom. van.); Trimerohydra Kozłowski, 1959, p. 217 (type, T. glabra; OD); Dyadograptus Obut, 1960, p. 147 (type, D. praecursor; OD); Archaeodictyota Obut & Sobolevskaya, 1967, p. 55 (type, A. dragunovi; OD)]. Rhabdosome conical or flabellate; branches complex, with several stolonal chains, particularly in species from higher stratigraphical levels; branches and "twigs" anastomosing irregularly; thecae very long, tubular, usually adnate for much of their length. ?U.Cam., L.Ord.-M.Dev., Eu.-USSR.——Fig. 21,2b. *K. onniensis, L.Sil.(U.Llandovery), Eng.; ×1 (17).——Fig. 21,2a. K. kozlowskii Bulman, U.Ord.(Balclatchie beds), S.Scot., showing tubular thecae and "twigs"; ×6 (23).

Palaeodictyota Whitfield, 1902, p. 399 [*Inocaulis ramulosus Whitfield (non Spencer, 1884), =I. anastomoticus Ringueberg, 1888, p. 131; SD Ruedemann, 1908, p. 200]. Resembling Koremagraptus, but without projecting thecae and twigs; branches probably composite but stolonal system unknown. Sil.-M.Dev., N.Am.-Eu.——Fig. 21,3. *P. anastomotica (Ringueberg), Sil. (Rochester Sh.), N.Y.; ×1 (201).

Thallograptus Ruedemann, 1925, p. 35 [non Thallograptus Öpik, 1928] [*Dendrograptus? succulentus Ruedemann, 1904, p. 581; OD] [=?Calyptograptus Spencer, 1878, p. 459 (pro Calyptograpsus Spencer, 1878, ICZN, Opin. 650) (type, C. cyathiformis; SD Miller, 1889, p. 175)]. Like Acanthograptus, but with more numerous and slender thecae, the isolate distal ends of which are rarely preserved; branch structure unknown. Ord.-Sil., N.Am.-Eu.-USSR.——Fig. 22,2. T. cervicornis (Spencer), M.Sil. (Rochester Sh.), N.Y.; ×1 (203).

TUBOIDEA

Order TUBOIDEA Kozłowski, 1938

[Tuboidea Kozłowski, 1938, p. 185] [Introduced by Kozłowski in 1938 without diagnosis but descriptive notes in text; defined by Bulman (21) in 1938, p. 92, but first adequately described by Kozłowski (114) in 1949, p. 140]

Sessile Graptolithina, with erect stipes and more or less dendroid rhabdosomes, or encrusting, with terminally erect thecae or sheaves of thecae arising from basal disc or thecorhiza; stolothecae less prominent than in Dendroidea, generally confined to basal disc in forms with thecorhiza; autothecae and bithecae present, budding commonly diad, with no regular succession and variably spaced nodes; specially modified autothecae (microthecae and umbellate thecae) and conothecae may occur. ?U.Cam., L. Ord.-Sil.

Like the Dendroidea, the Tuboidea are characterized by the presence of stolothecae, autothecae, and bithecae, but their association, arrangement and succession is far less regular. In the Dendroidea, with uniform triad budding, it is possible to regard the stolothecae as immature autothecae, but in the Tuboidea the relationship is not so simple; the terminal individual of a long chain of diad buds may be an autotheca, conotheca, or bitheca.

The two families recognized by Kozłowski in 1949 are possibly not so sharply delimited as they originally appeared to be, and *Dendrotubus? erraticus* Kozłowski and *Parvitubus* Skevington are to some degree intermediate. It is probable that several families are involved here, for the range of forms assigned to the Tuboidea is increasing and many now classed as dendroids (or *incertae sedis*) may prove to be tuboid. However, structural details are as yet insufficiently known to provide a reliable classification and the two original families are provisionally retained here; distinction between them rests mainly upon the dominantly dendroid habit of the Tubidendridae and the discoidal encrusting nature of the Idiotubidae.

MORPHOLOGY

The general form of the entire rhabdosome varies from an essentially dendroid, possibly flabellate, form to an encrusting assemblage of thecae (thecorhiza) from which tubular autothecae arise singly or in groups or sheaves.

In dendroid forms adjacent stipes may be connected by anastomosis or by transfer of single thecae (autothecae or bithecae) simulating dissepiments. Autothecae commonly open on one ("inner") side. Bithecae are more abundant than autothecae and the spacing and positioning of their apertures is less regular. Conothecae, when present, are quite irregular and development of the tubidendrid rhabdosome is as yet unknown.

In encrusting forms, stolothecae and bithecae are typically confined to the thecorhiza, but in some forms provisionally included in the family, sheaves of thecae (branches) may include stolothecae and bithecae; such sheaves divide infrequently, however, and are not of dendroid habit.

THECAE

STOLOTHECAE

In *Tubidendrum,* the stolon system is well developed, the stolons being provided with thick, well-sclerotized and strongly pigmented walls; but in *Reticulograptus,* sclerotized stolons have not been observed within the stolothecae, even though autothecal and bithecal stolons can be recognized. Instead of occupying the external position characteristic of typical Dendroidea, the stolothecae are commonly embedded in the stipe and several stolonal chains are usually present in a single stipe. Stolonal budding is diad (Fig. 23) in a manner associating any pair of thecae except two autothecae; but all divisions producing a conotheca appear to mark the end of a particular stolonal chain, the other individual being a bitheca. Stolothecae are of variable length and no regular budding rhythm is observed. Their distribution in thecorhizate forms appears to be quite irregular.

AUTOTHECAE

The autothecae are elongate and tubular, varying considerably in length and produced from autothecal stolons which also vary greatly in length. Despite this, the

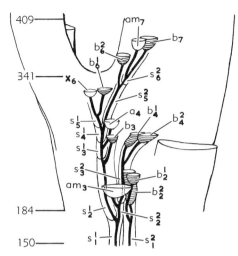

Fig. 23. Stolon system of *Tubidendrum* (solid black, with bases only of daughter thecae) reconstructed from and lettered as in serial sections figured by Kozłowski (114, pl. 21). [*a*, autotheca; *am*, microtheca; *b*, bitheca; *s*, stolotheca; sections 150-409.]

autothecal openings on branches or sheaves may exhibit surprising regularity in spacing. Ventral and, less commonly, dorsal apertural spines or processes may be present and the whole apertural region may be distinctly isolate.

In *Tubidendrum*, the autothecae are remarkable for the coiling of their central portions (Fig. 24) into a helical spiral with some seven or more turns. This coiling,

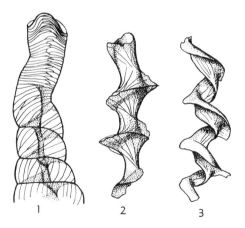

Fig. 24. Structural features of *Tubidendrum* (114).——*1*. Coiled and distal portions of an autotheca, ×40.——*2,3*. Examples of columella from the helicoidal parts of 2 autothecae, ×90.

either right- or left-handed, is so tight as to produce a columella analogous to that of a turreted gastropod shell. In one measured example, the autothecal stolon was 175μ long, the proximal portion of the autotheca $1,315\mu$, the coiled portion (measured along the axis) 620μ, and the straight distal portion 250μ. These coiled thecae, nevertheless, are enveloped in the stipe. Autothecal dimorphism occurs in *Tubidendrum*, one type (called **microthecae**) having a narrow terminal portion about one-third the diameter of a normal autotheca, with an oblique aperture which faces the opposite side of the stipe from that of the normal autotheca; in other respects, form and dimensions are comparable.

In the encrusting forms, each autotheca characteristically comprises two portions, a proximal adnate part incorporated in the thecorhiza and an erect distal part, free or associated in a sheaf with other thecae. Probably merely because both organisms are encrusting, the general characters are not unlike those of *Rhabdopleura*. The basal portion incorporated in the thecorhiza consists of regular growth bands of fusellar tissue on its upper surface, but the lower surface is a structureless membrane. The erect portion, however, is composed of regular growth bands disposed right and left, forming two (dorsal and ventral) zigzag sutures. Near the base of this free portion, the autotheca may show some helical coiling (see Fig. 28,*1*) comparable on a smaller scale with that of *Tubidendrum*. Autothecae produced from stolothecae in the branches *(Dendrotubus? erraticus)* are of a generalized cylindrical form. In *Galeograptus*, autothecae are also dimorphic, those on the proximal portions of the branching sheaves possessing elaborate apertural modifications; these **umbellate** thecae (Fig. 25,*1*) develop an umbrella-shaped structure shielding the aperture of the preceding theca, and the shields fill the cavity formed by the ring of stipes in the proximal region with a vesicular mass of skeletal tissue.

BITHECAE

It is a distinctive feature of the branching tuboids that the bithecae are about twice as numerous as the autothecae and are irregularly positioned. The majority

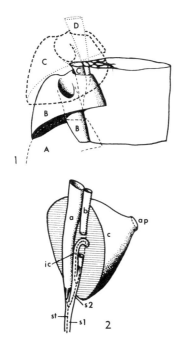

Fig. 25,1,2.——1. Diagrammatic figure of portion of stipe of *Galeograptus* showing umbellate thecae A, B, C, each with apertural process shielding the preceding thecal aperture; D is the proximal portion of the 4th theca; approx. ×25 (37).——2. Diagrammatic figure of conotheca of *Reticulograptus*, approx. ×30 (37). [*a*, autotheca; *ap*, aperture; *b*, bitheca; *c*, conotheca; *ic*, internal portion of conotheca; *s1*, *s2*, stolothecae; *st*, main stolon.]

have small relatively inconspicuous apertures almost flush with the surface of the stipe, but slight isolation may occur and some few may constitute "pseudodissepiments," transferring to and opening on an adjacent branch. Bithecal stolons vary in length, but are generally considerably longer than in the Dendroidea. In encrusting rhabdosomes, bithecae in general are limited to the thecorhiza and their numerical relation to autothecae is unknown.

CONOTHECAE

Conspicuous conelike bodies with a small aperture at the apex of the cone occur in several tuboid genera (Fig. 25,2). Their detailed morphology is imperfectly known, but they arise from a thin-walled cylindrical proximal portion, enclosed within the cone, and no sclerotized stolon has been detected. In branched rhabdosomes (e.g., *Reticulograptus*) they occur at irregular and often widely-spaced intervals, and the stolonal division producing a conotheca seem to terminate the stolon chain, since the other theca appears invariably to be a bitheca. Conothecae also have been detected on the thecorhiza of *Discograptus* and *Idiotubus*.

FORM OF RHABDOSOME AND THECAL GROUPING

A graded morphological series leads from *Idiotubus*, with autothecae arising singly from the upper surface of the thecorhiza, through *Dendrotubus* and *Discograptus*, with autothecae concentrated into groups, to *Galeograptus* and *Cyclograptus*, with a peripheral concentration of large sheaves of autothecae and bithecae. The same series illustrates also a progressive increase in regularity with which the thecae are distributed. In *Idiotubus*, the erect portions of the autothecae appear to have been distributed quite haphazardly over the surface of the thecorhiza; in *Dendrotubus* the arrangement is generally irregular but with a tendency towards greater regularity at the distal ends of the thecal bundles. In *Discograptus*, the thecae are arranged more precisely along several radii, steadily increasing in height peripherally, and the outer circle of sheaves comprise a series of autothecae regularly increasing in length and with regularly spaced apertures. In *Galeograptus* and *Cyclograptus*, the stipes are confined to the periphery and are composed of very numerous thecae with bithecal apertures concentrated around their bases or occurring on the branches themselves. Differences in the stolonal systems underlying these varied groupings are at present unknown. *Dendrotubus? erraticus* shows a tendency toward an irregularly branched tubidendrid rhabdosome rather than the unbranched radial groupings of *Galeograptus*.

DEVELOPMENT[1]

Development of a tuboid rhabdosome recently has been described (KOZŁOWSKI, 1963) in a form provisionally referred to *Dendrotubus* (Fig. 26). A slender, erect metasicula arises from a prosicula shaped apically like a conical flask. A pore, formed

---
[1] See footnoe on page V32.

in the prosicula after formation of the metasicula is complete, transmits the stolon system, and the astogeny is spiral, either right- or left-handed. Where the initial "prostolon" reaches the prosicular wall two almost spherical vesicular diaphragms originate the first stolonal and thecal stolons, and this first theca is apparently a bitheca. Subsequent diad divisions, always with vesicular diaphragms, are shown diagrammatically in Figure 26. No other tuboid astogeny is known, but the erect cylindrical metasicula has been identified in a central position in *Discograptus*, where some indication of a comparable spiral development also occurs.

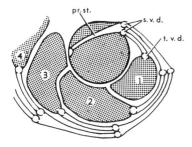

Fig. 26. Diagrammatic illustration of astogeny of *Dendrotubus* (122). [*pr.st.*, prostolon in circular basal portion of sicula; *s.v.d.*, vesicular diaphragms at base of stolons; *t.v.d.*, vesicular diaphragms at base of thecae; *1,2,3,4*, adnate basal portions of successive thecae.].

SYSTEMATIC DESCRIPTIONS

Family TUBIDENDRIDAE Kozłowski, 1949

[Tubidendridae Kozłowski, 1949, p. 160]

Rhabdosome erect, conical or ?flabellate; stipes dividing irregularly and anastomosing or united by single thecae, comprising at any given level numerous thecae of several generations; stolothecae more or less embedded in the stipe; autothecae tubular or spirally coiled in middle portion, dimorphism in one genus, one type (microthecae) with contracted apertural region; conothecae present in one genus; bithecae tubular with stolons of variable length; stolon system well developed and sometimes highly sclerotized. *L.Ord.(Tremadoc)-Sil.*

Tubidendrum Kozłowski, 1949, p. 160 [**T. bulmani;* OD]. Rhabdosome an irregular network, branches connected by tubular thecae, especially bithecae; autothecae helically coiled, dimorphic (autothecae and microthecae). *L.Ord.(Tremadoc),* Eu.(Pol.).——Fig. 24,*1-3*. **T. bulmani; 1,* autotheca, ×40; *2,3*, columellae, ×90 (114).

Reticulograptus Wiman, 1901, p. 189 [**Dictyonema tuberosum* Wiman, 1895; M] [=*Multitubus* Skevington, 1963, p. 51 (type, *M. spinosus;* OD); =?*Marsipograptus* Ruedemann, 1936, p. 385 (type, *M. bullatus;* OD)]. Branches anastomosing or connected by tubular thecae; autothecae with regularly spaced apertures; conothecae commonly present. *L.Ord.-Sil.(Wenlock),* Baltic-N.Am.——Fig. 27,*1*. **R. tuberosus* (Wiman), U.Ord. boulder, Gotl.; thecal composition diagram of portion of stipe; ×20 (37).——Fig. 27,*2,3*. *R. thorsteinssoni* Bulman & Rickards, Sil., Canad.Arctic; *2*, portion of stipe with numerous bithecae; *3*, portion of stipe with conothecae; ×20 (37).

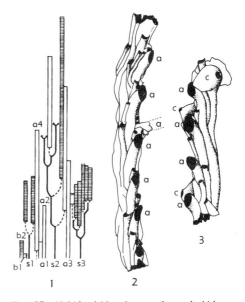

Fig. 27. Tubidendridae [*a,* autotheca; *b,* bitheca; *c,* conotheca; *s,* stolotheca] (p. V47).

Family IDIOTUBIDAE Kozłowski, 1949

[Idiotubidae Kozłowski, 1949, p. 144]

Rhabdosome an encrusting, more or less discoidal assemblage of thecae (thecorhiza) from which tubular thecae arise singly or in groups or sheaves; stolothecae mainly confined to thecorhiza; proximal portion of at least initial autothecae incorporated in

Graptolithina

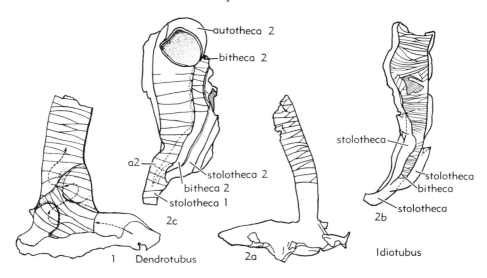

FIG. 28. Idiotubidae (p. V48).

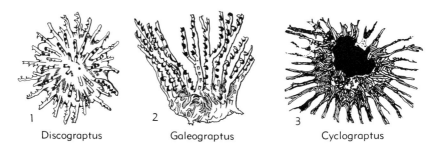

FIG. 29. Idiotubidae (p. V48-V49).

thecorhiza, distal portion tubular, erect; umbellate thecae present in one genus, conothecae in two genera; bithecae limited to thecorhiza or extending into thecal sheaves, rarely originating on branches. ?*U.Cam.*(USSR), *U.Ord.-Sil.*

Idiotubus KOZŁOWSKI, 1949, p. 144 [*I. typicalis*; OD]. Irregularly distributed erect portions of autothecae arising singly from surface of thecorhiza. *L.Ord.(Tremadoc)*, Eu.(Pol.).——FIG. 28,2. *I.* sp.; *2a*, autotheca and fragment of thecorhiza, ×25; *2b*, bitheca with stolothecae, ×25; *2c*, autotheca with associated bitheca and stolothecae, ×50 (114).

Calycotubus KOZŁOWSKI, 1949, p. 156 [*C. infundibulatus*; OD]. Autothecae fused by their lateral walls into irregular groups. *L.Ord.(Tremadoc)*, Eu.(Pol.).

Conitubus KOZŁOWSKI, 1949, p. 159 [*C. siculoides*; OD]. Known only by conical autothecae. *L.Ord.(Tremadoc)*, Eu.(Pol.).

Cyclograptus SPENCER, 1883, p. 365 [*C. rotadentatus* SPENCER, 1884, p. 42; M] [=*Rhodonograptus* POČTA, 1894, p. 205 (type, *R. astericus*, =*Sphaerococcites scharyanus* GOEPPERT, 1860, p. 454; M)]. Rhabdosome discoidal, erect portions of autothecae grouped into 20 to 30 peripheral sheaves bifurcating at their mid-length. *M.Sil. (Niag.-Wenlock)*, N.Am.-Eu.(Czech.).——FIG. 29,3. *C. rotadentatus*, Hamilton, Ont.; ×2 (26).

Dendrotubus KOZŁOWSKI, 1949, p. 153 [*D. wimani*; OD]. Erect portions of autothecae forming irregularly distributed groups, central portions commonly coiled into helical spiral. *L.Ord. (Tremadoc)*, Eu.(Pol.).——FIG. 28,1. *D. wimani*; basal part of autotheca showing spiral coiling; ×65 (114).

Discograptus WIMAN, 1901, p. 191 [*D. schmidti*; M]. Rhabdosome discoidal, erect portions of autothecae in more or less radially arranged groups on upper surface of thecorhiza; bithecae and conothecae confined to thecorhiza. *U.Ord.*, Baltic. ——FIG. 29,1. *D. schmidti*, silicified boulders, Gotland; ×3 (267).

Epigraptus EISENACK, 1941, p. 24 [*E. bidens*;

M]. Similar to *Idiotubus*. Ord.(*Wesenberg F.*), Eu.(Estonia).

?**Fasciculitubus** OBUT & SOBOLEVSKAYA, 1967a, p. 56 [*F. tubularis;* OD]. Robust thecae arising in irregular groups from the thecorhiza. *U.Cam.,* USSR (Sib.).

Galeograptus WIMAN, 1901, p. 189 [*G. wennersteni;* M]. Rhabdosome discoidal, erect portions of autothecae associated in comparatively few (8 to 10) peripheral branches bifurcating usually once near their mid-length; proximal autothecae with umbellate apertural processes; bithecae extending along the branches. *U.Ord.-M.Sil.(Wenlock),* Baltic-Eng.——FIG. 29,2. *G. wennersteni,* lateral view of rhabdosome; silicified boulder, Sweden (Gotland); ×3 (267).

?**Parvitubus** SKEVINGTON, 1963, p. 47 [*Azygograptus? oelandicus* BULMAN, 1936, p. 46; OD]. Erect, undivided branches comprising stolothecae, autothecae and bithecae, possibly grouped on a thecorhizal base; bithecae restricted to one side of stipe, opening into autothecae. *L.Ord.(Vaginatumkalk, Ontikan,* Sweden(Öland).——FIG. 30, 1-2. *P. oelandicus* (BULMAN); *1,* basal portion of stipe; *2,* distal portion of stipe [*a,* autothecae; *bi,* bithecae], ×7.5 (214).

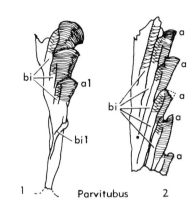

FIG. 30. Idiotubidae [*a,* autotheca; *bi,* bitheca] (p. V49).

CAMAROIDEA

Order CAMAROIDEA Kozłowski, 1938

[Introduced by KOZŁOWSKI in 1938 (p. 185) without diagnosis but descriptive notes in text; defined by BULMAN (21) in 1938, p. 92, but first adequately described by KOZŁOWSKI (114) in 1949, p. 170]

Encrusting Graptolithina comprising autothecae and indistinct stolothecae, bithecae present in some; autothecae strongly differentiated into two parts, an inflated basal vesicle (camara) and a free tubular distal portion (collum); bithecae tubular; stolothecae forming bifurcating network above camarae or represented by extracamaral tissue surrounding stolons. *Ord.*

MORPHOLOGY

The shape of the complete rhabdosome is unknown, as is its proximal end and mode of development. As in the Graptoloidea, the dominant element is the autotheca, but sclerotized stolons are present invariably, some enclosed in stolothecae, and bithecae characterize one genus.

AUTOTHECAE

The autothecae are very sharply differentiated into distinct proximal and distal portions. The **camara** (proximal portion) is a more or less inflated vesicle or cell, whose upper wall exhibits characteristic fusellar structure in contrast to the lower surface which is structureless; the camarae, commonly embedded in extra-camaral tissue, form a sort of encrusting mosaic. At the proximal extremity of each is attached the autothecal stolon, separated from the cavity of the camara by one or more transverse partitions or septa. From one extremity also, although not necessarily the distal one, arises the slender tubular **collum,** which is morphologically equivalent to the free portion of the autothecae in the Tuboidea. Its wall shows a somewhat irregular fusellar structure and terminates in an apertural process (corresponding to the ventral process of Tuboidea and other graptolites) or more typically thins out distally to terminate in a sharp jagged edge; it has been suggested that this latter feature indicates a gradual transition in life from the sclerotized thecal wall into the soft skin of the zooid. Rarely (as in *Cysticamara accollis*). the collum is absent altogether and the thecal aperture is situated on the upper surface of the camara. Occlusion of the autothecae by a sclerotized diaphragm, generally near the base of the collum, is of very common occurrence.

BITHECAE

Where present *(Bithecocamara)*, the bithecae are relatively inconspicuous tubular thecae (as in Tuboidea) which are distributed irregularly and unevenly on or near the surface of the colony.

STOLON SYSTEM

Stolothecae are present in some genera *(Bithecocamara)* as a bifurcating network of tubes near the upper surface of the colony, and even here they are never very clearly differentiated; in other genera, a stolon system occurs more or less embedded in extracamaral tissue which forms a sort of sheath, perhaps representing modified vestiges of original stolothecae. Apart from bifurcations of the stolons (and a corresponding bifurcation of the stolothecal tubes where present) no details of budding are known; it was probably irregular, and the autothecae at least possessed long autothecal stolons.

SYSTEMATIC DESCRIPTIONS

Family BITHECOCAMARIDAE Bulman, 1955

[Bithecocamaridae BULMAN, 1955, p. 42]

With all three types of thecae; autothecae with well-developed collum. *L.Ord.*

Bithecocamara KozŁowski, 1949, p. 176 [*B. gladiator;* OD]. Only genus. *L.Ord.(Tremadoc),* Eu.(Pol.).——FIG. 31,*1. B.* sp.; reconstr., approx. ×80 (Bulman, n).

Family CYSTICAMARIDAE Bulman, 1955

[Cysticamaridae BULMAN, 1955, p. 42]

Bithecae absent, stolothecae obscure or absent, stolons usually embedded in extracamaral tissue. *L.Ord.*

Cysticamara KozŁowski, 1949, p. 183 [*C. accollis;* OD] [=*Syringataenia* OBUT, 1953, p. 54 (type, *S. bystrowi;* M]. Stolons embedded in extracamaral tissue with fusellar structure. *L.Ord. (Tremadoc,* Eu.(Pol.); *Ontikan(Orthoceras* Ls.*),* Öland-NW.USSR).

Flexicollicamara KozŁowski, 1949, p. 182 [*F. bryozoaeformis;* OD]. Collum strongly bent back ventrally and fused to upper wall of camara. *L. Ord.(Tremadoc),* Eu.(Pol.).

Graptocamara KozŁowski, 1949, p. 187 [*G. hyperlinguata;* OD]. Collum provided with conspicuous apertural process. *L.Ord.(Tremadoc),* Eu.(Pol.); *Ontikan(Vaginatumkalk),* Öland.

Tubicamara KozŁowski, 1949, p. 188 [*T. coriacea;* OD]. Funnel-shaped collum with ventral apertural process; abundant cortical tissue. *L.Ord. (Tremadoc),* Eu.(Pol.).

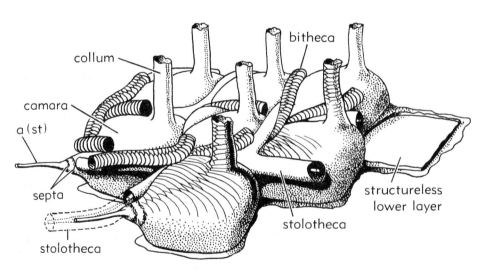

FIG. 31. Diagrammatic restoration of a camaroid, based on *Bithecocamara* KozŁowski, approx. ×80 (29). [*a(st),* autothecal stolon.]

CRUSTOIDEA

Order CRUSTOIDEA Kozłowski, 1962

[Order Crustoidea Kozłowski, 1962, p. 31]

Encrusting Graptolithina with colonies comprising autothecae, stolothecae and bithecae produced in triads; autothecae with erect (isolate) distal neck showing distinctive apertural modifications; bithecae cylindrical, adnate throughout their length; stolothecae tubular, with conspicuous well-sclerotized stolons; lower (adherent) wall of all thecae membranous, structureless; upper wall fusellar. *L.Ord.-U.Ord.*

MORPHOLOGY

The development and form of the complete colony of Crustoidea is unknown, for remains are fragmentary, but they are known to be encrusting, probably irregular and irregularly spreading; where the thecae are not in lateral contact, a thin structureless interthecal membrane is usually adherent to the substrate and continuous with the lower membrane of the thecae.

Some resemblance between the Crustoidea and Camaroidea is observed, but the former are less compact, have a more regular stolon system, and exhibit more highly elaborated apertural processes of the autothecae. *Hormograptus,* here provisionally included as an aberrant crustoid, may prove to be synonymous with *Chaunograptus* and provide a link with a number of obscure adherent forms, but the encrusting habit is of course represented in many unrelated phyla. Of all the Graptolithina, the Crustoidea constitutes the order most closely resembling morphologically the Rhabdopleurida.

AUTOTHECAE

The autothecae comprise a proximal portion in contact with the substrate, some in contact with one another laterally, with an isolate, erect, distal neck. The proximal portion is generally more or less inflated but may be almost cylindrical. It passes imperceptibly into the neck, from the distal end of which the apertural lobe is developed. This latter may be considerably enlarged and elaborated to produce a median fold and two auricular lateral folds (Fig. 32).

BITHECAE

The bithecae are slender and tubular, but vary much in length; in consequence, they may open beside, behind, or considerably in front of the autothecal aperture of the same generation. A tendency toward a right- and left-hand alternation in successive generations is seen, but this is not so regular as in the Dendroidea. Bithecal apertures are devoid of any special modification.

STOLOTHECAE

The stolothecae are slender, commonly sinuous, and distally each passes imperceptibly into the base of daughter autotheca. Externally they are distinguishable from the bithecae by their much more regular fusellar structure, with conspicuous zigzag suture. Stolons are well sclerotized, 20 to 35 microns in diameter and marked with fine transverse annulations. They exhibit well-developed nodes at the points of origin of the stolonal triads, as in the Dendroidea, but a distinctive feature is their attachment to the structureless basal wall of the stolothecae. The autothecal stolon is appreciably shorter than the bithecal stolon, and it does not terminate at the base of the theca, but penetrates to its interior. Indications of supernumerary and secondary stolons have been observed in some fragments, but their role is unknown.

GRAPTOBLASTS AND CYSTS

Vesicles or cysts may occur inside autothecae, varying greatly in size and shape, rarely filling the entire cavity of the autotheca. They may or may not be connected with the stolon, and possess blackish, structureless walls. It is possible that they represent the envelopes of degenerate zooids.

Graptoblasts (see also p. *V*136) may also occur within the autothecae of Crustoidea, completely filling the autothecal cavity, and such autothecae are devoid of normal apertural modifications. Their relationship is at present unresolved.

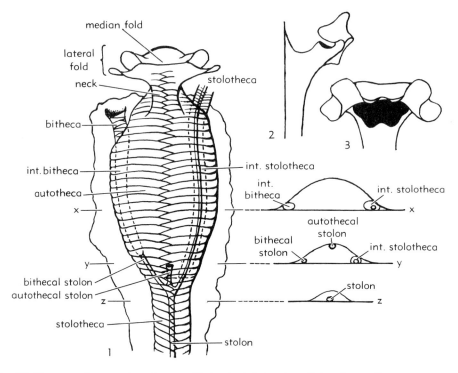

Fig. 32. Diagrammatic restoration of a crustoid thecal triad, based on *Bulmanicrusta* Kozłowski, approx. ×60.——*1*. Dorsal view of autotheca (*xx, yy, zz*, transverse sections).——*2,3*. Lateral and apertural views of autothecal apertural modification in *B. latialata* (Bulman, n).

SYSTEMATIC DESCRIPTIONS

Family WIMANICRUSTIDAE Bulman, n. fam.

Encrusting rhabdosomes comprising autothecae, bithecae, and stolothecae developed by regular triad budding; autothecae with erect distal neck and conspicuous apertural modifications; bithecae and stolothecae tubular, adnate, the latter with heavily sclerotized stolons attached to the adherent basal wall. *L.Ord.-U.Ord.*

Wimanicrusta Kozłowski, 1962, p. 42 [*W. urbaneki;* OD]. Apertural neck short or absent, apertural lobe linguiform. *L.Ord.* (glacial boulders, *?Glyptograptus teretiusculus* Zone), Eu. (Pol.).

Bulmanicrusta Kozłowski, 1962, p. 31 [*B. latialata;* OD]. Autothecae with or without well-developed neck, very large apertural lobe with median and auriculate lateral folds. *L.Ord.* (glacial boulders, *?Glyptograptus teretiusculus* Zone), Eu.(Pol.).——Fig. 32,*1-3*. *B. latialata; 1-3*, reconstr., approx. ×60 (Bulman, n).

Ellesicrusta Kozłowski, 1962, p. 38 [*E. longicollis;* M]. Autothecae with elongate neck, apertural lobe with slight lateral folds. *L.Ord.* (glacial boulders, *?Glyptograptus teretiusculus* Zone), Eu. (Pol.).

Holmicrusta Kozłowski, 1962, p. 41 [*H. sombrero;* M]. Autothecae with long neck and large flattened apertural lobe. *L.Ord.* (glacial boulders, *?Glyptograptus teretiusculus* Zone), Eu. (Pol.).

Lapworthicrusta Kozłowski, 1962, p. 44 [*L. aenigmatica;* M]. Slender autotheca without interthecal membrane and with only slight apertural modifications. *L.Ord.* (glacial boulder, *?Llanvirn*), Eu.(Pol.).

Ruedemannicrusta Kozłowski, 1962, p. 39 [*R. geniculata;* M]. Autothecae with long curved neck bearing strong internal ridges. *L.Ord.* (glacial boulders, *?Glyptograptus teretiusculus* Zone), Eu.(Pol.).

Family HORMOGRAPTIDAE Bulman, n. fam.

?Aberrant Crustoidea. Rhabdosome encrusting, irregularly branching; stolon system well developed, with triad budding commonly related to two stolothecae and an autotheca; autothecae adherent proximally, distally unknown; bithecae possibly absent or irregularly developed. *U.Ord.*

Hormograptus ÖPIK, 1930, p. 8 [*pro Thallograptus* ÖPIK, 1928, p. 35, *non Thallograptus* RUEDEMANN, 1925, p. 35] [**Thallograptus sphaericola* ÖPIK, 1928, p. 39; OD]. Only genus. *U.Ord. (Kukruse, Nemagraptus gracilis Zone),* Eu.(Est.).

STOLONOIDEA

Order STOLONOIDEA Kozłowski, 1938

[Introduced by KOZŁOWSKI in 1938, p. 185, without diagnosis but descriptive notes in text; defined by BULMAN (21) in 1938, p. 92, but first adequately described by KOZŁOWSKI (114) in 1949, p. 191]

Sessile or encrusting Graptolithina composed essentially of stolothecae and ?autothecae; stolothecae containing an exaggerated development of stolons dividing at irregular intervals and quite irregular in form. *Ord.*

MORPHOLOGY

This order is represented by extremely fragmentary remains and only an imperfect account of morphology can be given. It is, however, clearly distinguished from all other orders of Graptolithina by extraordinary development of the stolons themselves. These divide quite irregularly, in some forming whole interlacing groups (Fig. 33,*1*) and in others giving off single branches. Their course is erratic, vermiform, and they vary greatly in diameter from 50 to 350 microns. They have thick walls and the central lumen usually is filled with secondary deposit. The stolons are included in stolothecal tubes, either singly or in sheaves or groups, but these appear to have been extremely fragile and are preserved only rarely. The stolothecal tubes possess a normal though somewhat irregular fusellar structure, except on the lower surface of the encrusting forms where the thecal wall is structureless (as in Camaroidea).

The stolons appear to leave the stolothecae by pores produced by resorption, and on leaving the parent stolotheca give rise to new stolothecae or to what appear to be autothecae. Unlike most Graptolithina, where a sudden passage occurs from stolon to base of theca, the stolon here steadily increases in diameter until at a certain point the structureless substance of the stolon gives place to the fusellar structure of the theca proper. Autothecae are no more commonly preserved than stolothecae but were evidently tubular, straight or more commonly curved, with fusellar walls, opening on the surface of the colony with apertures devoid of any apertural processes.

SYSTEMATIC DESCRIPTIONS

Family STOLONODENDRIDAE Bulman, 1955

[Stolonodendridae BULMAN, 1955, p. 43]

Characters of the order. *Ord.*

Stolonodendrum KOZŁOWSKI, 1949, p. 194 [**S. uniramosum;* OD]. Large, irregularly vermiform stolons enclosed in thin-walled stolothecal tubes. *L.Ord.(Tremadoc),* Eu.(Pol.); (?*Ontikan*), Öland.——FIG. 33, *1,2. S.* sp., Tremadoc, Pol.; fragments of stolons, ×20 (114).

?**Melanostrophus** ÖPIK, 1930, p. 10 [**M. jokini;* OD]. Long, irregularly bent and coiled tubes in confused association; presence of growth lines demonstrated by EISENACK. *L.Ord.-U.Ord.,* Eu. (Baltic).

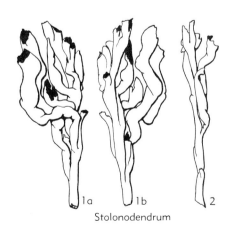

FIG. 33. Stolonodendridae (p. *V*53).

DENDROIDEA, TUBOIDEA, CAMAROIDEA, CRUSTOIDEA, STOLONOIDEA

TAXONOMIC POSITION UNCERTAIN

A considerable number of graptolite genera are not known in sufficient detail to be placed in the foregoing taxonomic sections. Many (e.g., *Chaunograptus*) have been described as dendroid; some (e.g., *Archaeocryptolaria*) as coelenterate (hydroid); and for others (e.g., *Dithecodendrum*) a distinct order has been proposed.

Definite pterobranch and hydroid remains are known from the Lower Ordovician and recently KOZŁOWSKI (1967) has described plausible representatives of phoronideans, Pogonophora, and what appear to be thecae of scyphozoan scyphistomae from glacial boulders of Ordovician rocks. The gross morphology of all such organisms, when poorly preserved, is little guide to their true affinity; and in the absence of detail concerning such features as the stolon system and the presence of fusellar tissue, it seems preferable to accept a large group of unclassified genera, some members of which may ultimately prove not to be Graptolithina at all.

The genera listed below vary greatly in character and in size. The majority are more or less dendroid in habit, but several are encrusting. The Dithecoidea of OBUT (1964) appear to comprise autothecae only, but no stolon system has been demonstrated and they compare superficially with such genera as *Archaeocryptolaria* and others which until now have been provisionally classed as hydroids. But among these

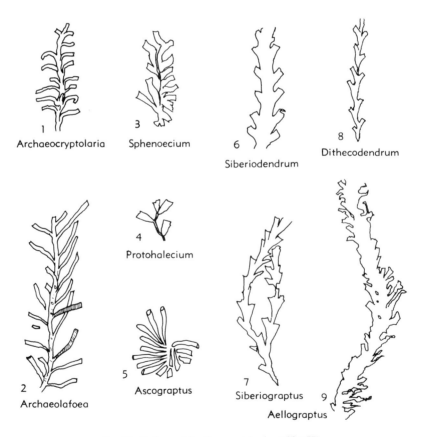

FIG. 34. Order and family uncertain (p. *V55-V57*).

Taxonomic Position Uncertain

latter, *Archaeolafoea* and *Sphenoecium* exhibit inconclusive traces of what appears to be fusellar structure, which would suggest a generally "graptolitic" affinity. Others, like *Leveillites* or *Diplospirograptus*, present superficial resemblance to algal forms, though it is perhaps doubtful whether such delicate filamentous algae could survive under the given conditions of preservation of these fossils.

Aellograptus OBUT, 1964, p. 306 [*A. savitskyi;* OD]. Elongate, unbranched or sparsely divided rhabdosomes with numerous short, projecting, cylindrical thecae. U.Cam., USSR(Sib.).——FIG. 34,9. *A. savitskyi;* ×3 (172).

Archaeocryptolaria CHAPMAN, 1919, p. 392 [*A. skeatsi* CHAPMAN; SD BULMAN, herein]. Rhabdosome slender, branching infrequently; thecae tubular, widely-spaced, adnate proximally with conspicuous isolate distal portions. M.Cam., Australia(Victoria).——FIG. 34,1. *A. skeatsi,* Lancefield, Victoria; ×2.5 (42).

Archaeolafoea CHAPMAN, 1919, p. 390 [*A. longicornis;* M]. Rhabdosome slender, flexuous, branching infrequently; thecae elongate, isolate, narrowing towards their bases, with indistinct traces of fusellar structure. M.Cam., Australia (Victoria).——FIG. 34,2. *A. longicornis,* Lancefield, Victoria; ×2.5 (42).

Ascograptus RUEDEMANN, 1925, p. 18 [*A. similis;* OD]. Relatively large, conical thecae arranged spirally along an unbranched axis. Sil., N.Am. FIG. 34,5. *A. similis,* Lockport Ls., N.Y.; ×7 (209).

Cactograptus RUEDEMANN, 1908, p. 196 [*C. crassus;* OD]. Elongate branches apparently dendroid, with prominent thecae projecting on both sides. ?Cam., Australia, Sil., N.Am.——FIG. 35,3. *C. crassus,* Sil.(Clinton Sh.), N.Y.; ×1 (201).

Ceramograptus HUDSON, 1915, p. 129 [*C. ruedemanni;* OD]. Stipes apparently multiserial. U.Ord., N.Am.(Can.).

Chaunograptus HALL, 1883, p. 58 [*Dendrograptus (Chaunograptus) novellus;* M]. Minute dendroid rhabdosome, usually encrusting, with short, conical thecae. ?Cam., Ord.-Dev., N.Am.-Eu. (Czech.).——FIG. 35,2. *C. contortus* RUEDEMANN, U.Ord. (Richmond beds), Ind.; ×3.5 (209).

Coelograptus RUEDEMANN, 1947, p. 266 [*Inocaulis problematica* SPENCER, 1878; OD]. Similar to *Chaunograptus,* but coarser, with no recognizable thecae. U.Sil.(Niag.), N.Am.(Can.).

Crinocaulis OBUT, 1960, p. 148 [*C. flosculus;* OD]. Resembling *Palmatophycus,* but with differently arranged terminal filaments. L.Sil., Eu. (Est.).

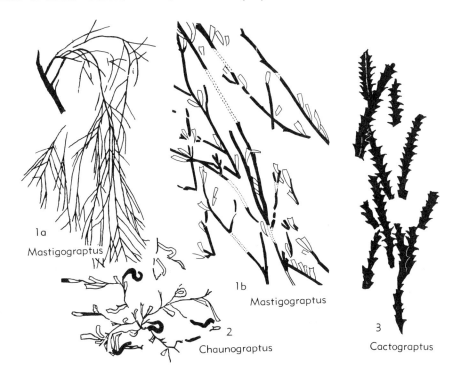

FIG. 35. Order and family uncertain (p. V55-V56).

Diplospirograptus RUEDEMANN, 1925, p. 34 [*D. goldringae;* OD]. Rhabdosome bifurcating near base to produce spirally coiled branches bearing distally a close-set brush of filamentous thecae. *Sil.,* N.Am.

Dithecodendrum OBUT, 1964, p. 306 [*D. sibericum;* OD]. Slender, elongate rhabdosome; autothecae cylindrical, isolate distally, arranged biserially. *M.Cam.,* USSR(Sib.).——FIG. 34,8. *D. tenuiramosum* OBUT; ×3 (172).

Estoniocaulis OBUT & RYTSK, 1958, p. 137 [*Inocaulis järvensis* ROSENSTEIN MS, OBUT & RYTSK, 1958; OD]. Resembling *Diplospirograptus,* but with smaller terminal tufts and branched stem not spirally coiled. *L.Sil.,* Eu.(Est.).

Haplograptus RUEDEMANN, 1933, p. 323 [*H. wisconsinensis;* OD]. Elongate conical or vermiform thecae associated to form an irregularly dendroid rhabdosome. *Cam.-Ord.,* N.Am.

Leveillites FOERSTE, 1923, p. 62 [*L. hartnageli;* M]. Lateral branches set with numerous small tufts of filaments. *L.Sil.,* N.Am.(Ont.).

Mastigograptus RUEDEMANN, 1908, p. 210 [*Dendrograptus tenuiramosus* WALCOTT, 1883; OD]. Much-branched, dendroid rhabdosome with slender, dense-walled stipe (?stolothecae) and thin-walled, conical thecae apparently including autothecae and bithecae; budding in triads but arrangement uncertain and little difference between

FIG. 37. Order and family uncertain (p. V56).

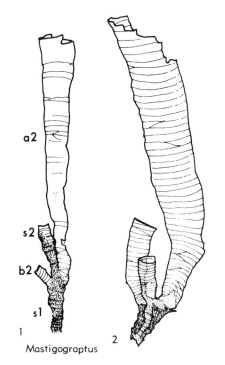

FIG. 36. Order and family uncertain [*a*, autotheca; *b*, bitheca; *s*, stolotheca] (p. V56).

auto- and bithecae; fusellar structure with complete rings and oblique sutures. *Cam.-Sil.,* N.Am.-Eu.-?Australia.——FIG. 35,1. **M. tenuiramosus* (WALCOTT), Ord.(Utica Sh.), N.Y. *(1a);* (Eden Sh.), Ky.*(1b); 1a,* fairly complete rhabdosome, ×1; *1b,* fragment showing thecae, ×5 (209).——FIG. 36,1,2. *M.* sp., Ord.(glacial boulder), N.Ger.; *1,2,* two thecal groups showing fusellar structure and triad budding; approx. ×170, ×130 [*a,* autothecae; *b,* bithecae] (Andres, 1961).

Medusaegraptus RUEDEMANN, 1925, p. 29 [**M. mirabilis;* OD]. Thick main branch ending in a blunt point at the base and terminating distally in a dense mass of long, filamentous thecae. *Sil.,* N.Am.-Eu.(Czech.).——FIG. 37,1. **M. mirabilis,* Lockport Dol., N.Y.; ×1 (203).

Palmatophycus BOUČEK, 1941 (revised, BOUČEK, 1957, p. 148) [**P. kettneri;* OD]. Stipe terminating distally in a crown of lateral branches, each furnished with a dense mass of filamentous thecae. *M.Sil.,* Eu.(Czech.).

Polygonograptus BOUČEK, 1957, p. 151 [**Palaeodictyota sokolowi* OBUT, 1953; OD]. An irregular network of mainly pentagonal or hexagonal meshes; thecal structure completely unknown. *U. Ord.-Sil.*, Eu.(USSR-Czech.).

Protohalecium CHAPMAN & THOMAS, 1936, p. 203 [**P. hallianum;* M]. Branching rhabdosome provided terminally with conical thecae. *M.Cam.*, Australia(Victoria).——FIG. 34,*4*. **P. hallianum*, Heathcote, Victoria; ×2 (43).

Rhadinograptus OBUT, 1960, p. 151 [**R. jurgensonae;* OD]. Similar to *Mastiograptus*, but with bundle-like accumulations of slender, conical thecae. *L.Sil.*, Eu.(Est.).

Ruedemannograptus H. & G. TERMIER, 1948, p. 174 [*pro Streptograptus* RUEDEMANN, 1947, *non* YIN, 1937] [**Streptograptus tenuis* RUEDEMANN, 1947; SD BULMAN, herein]. Irregularly branched dendroid rhabdosome with projecting, ?tubular thecae. *Ord.*, N.Am.(Tenn.)-N.Afr.(Morocco).

Siberiodendrum OBUT, 1964, p. 306 [**S. robustum;* OD]. Rhabdosome robust, with short, wide, cylindrical thecae isolate distally. *M.Cam.-U. Cam.*, USSR(Sib.).——FIG. 34,*6*. **S. robustum;* ×1.5 (172).

Siberiograptus OBUT, 1964, p. 306 [**S. kotujensis;* OD]. Rhabdosome slender, branching, with large, distally isolate thecae arranged biserially. *U.Cam.*, USSR(Sib.).——FIG. 34.7. **S. kotujensis;* ×1.5 (172).

Sphenoecium CHAPMAN & THOMAS, 1936, p. 205 [*pro Sphenothallus* CHAPMAN, 1917, *non* HALL, 1848] [**S. filicoides;* SD BULMAN, herein]. Robust, almost cylindrical thecae arranged radially (like *Fasciculitubus* OBUT & SOBOLEVSKAYA) or serially, with faint traces of fusellar strucure. *M. Cam.* Australia(Victoria).——FIG. 34,*3*. **S. filicoides*, Heathcote, Victoria; ×2 (43).

GRAPTOLOIDEA

Order GRAPTOLOIDEA Lapworth, 1875

[*nom. transl.* RUEDEMANN, 1904, p. 573 (*ex* section Graptoloidea LAPWORTH, in HOPKINSON & LAPWORTH, 1875, p. 633)] [=suborder Rhabdophora ALLMAN 1872, p. 380]

Planktonic or epiplanktonic Graptolithina; rhabdosome generally of few stipes, always comprising only one type of theca (autotheca) without sclerotized stolons; sicula pendent in relation to apical nema, stipes pendent to scandent, uniserial or biserial, very rarely triserial or quadriserial. *L. Ord. (?Tremadoc, Arenig) - L. Dev. (Siegen., ?Emsian)*.

MORPHOLOGY

GENERAL FEATURES

The Graptoloidea may be regarded best as simplified Dendroidea, to which order they are closely related through the family Anisograptidae. Such simplification, accompanying change to a pelagic mode of life, involves loss of bithecae and loss of a sclerotized stolon system; accordingly no sign is found in a graptoloid branch of the triad budding on the "Wiman rule" which distinguishes the Dendroidea. By analogy with other orders, the presence of a stolon may be inferred, but in the Graptoloidea this is not enclosed in a skeletal sheath and thus presumably was comparable with the **gymnocaulus** of *Rhabdopleura*. The proximal portion of the autotheca is morphologically equivalent to the dendroid stolotheca. The transition from one order to the other was probably gradual and may have occurred in several independent lines, and the placing of the Anisograptidae is arbitrary. Grouping this family (which includes many genera known to possess typical dendroid branch structure) with the Dendroidea leaves the typical graptoloid branch to be defined as composed of autothecae only. Following the loss of bithecae, the stipe shows a very general tendency to increase in breadth to a distal maximum.

The **sicula** gives rise laterally to a single initial bud from which ultimately the entire rhabdosome develops, and the apex of the sicula is prolonged as a slender thread known as the **nema** (or **virgula** in scandent forms) by the distal end of which the rhabdosome in general was attached. The relation between direction of growth of the branches of the rhabdosome and the nema has afforded a basis of subdivision among the Graptoloidea, and varies (Fig. 38) from pendent, through horizontal and reclined, to scandent, a general tendency in history of the group being toward attainment of a scandent direction of growth. There is likewise a general tendency toward reduction in the number of branches, the earliest genera being for the most part multiramous forms, whereas the youngest genera are composed of but one or two scandent stipes (Diplograptidae, Monograptidae). In these scandent forms it was not at

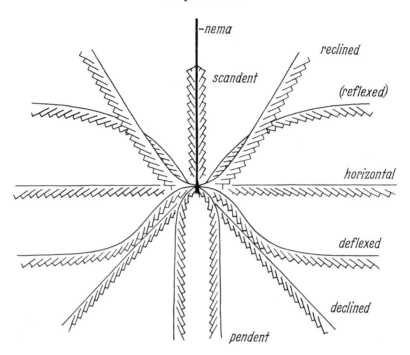

Fig. 38. Diagram illustrating relations of graptolite stipes to nema and terminology applied (21).

first recognized that the virgula is in fact identical with the nema of pendent forms. FRECH's suggested orders, Axonolipa and Axonophora, based essentially on absence or presence of an "axis," really reflect the degree of scandency attained by the rhabdosome, following the reduction in stipe number to two or one. No hard and fast line can be drawn between them, however, and the dicellograptids and dicranograptids were placed by FRECH in his Axonophora while RUEDEMANN, adopting the orders with slight modification, included them in the Axonolipa.

Biserial (axonophorous) rhabdosomes show two basically different structures. More commonly (in the Diplograptina and the Dicranograptidae) the two component stipes are oriented "back-to-back," with their dorsal sides in contact, and such rhabdosomes are termed **dipleural**. In the Glossograptina the two stipes lie "side-by-side," with one lateral wall in contact, and this structure is termed **monopleural** (JAANUSSON, 1960).

In uniserial scandent rhabdosomes (Monograptina), the stipe is commonly curved and spiral in some (plane or helical); in certain forms second- and higher-order rhabdosomes called **cladia** are developed from modified thecae or from the aperture of the sicula itself.

Loss of bithecae and particularly reduction in number of stipes results in a striking reduction in the total number of thecae in a rhabdosome. From as many as 20,000 to 30,000 in a large *Dictyonema flabelliforme,* the total falls to 2,000 or 3,000 in some *Clonograptus* species, and perhaps in a large multiramous dichograptid; averages of 100 to 200 occur in most leptograptids, dicellograptids, and diplograptids; 50 to 100 in a typical *Monograptus,* but in some no more than 10 to 20. It is also accompanied by a pronounced tendency toward thecal (apertural) elaboration. Recent morphological studies of the Graptoloidea have led to a greater appreciation of the range of complex thecal modifications, and considerable attention is now being paid to patterns of rhabdosome development. This order is outstanding for regularity in astogeny, exemplified in branching patterns, in cladia production, and in precisely controlled thecal sequences, which are discussed further in the sections on "Thecae" (p. V66) and on "Development of the Rhabdosome" (p. V71).

MUSCLE SCARS. The occurrence of muscle scars in specimens of *Climacograptus typicalis* and *Orthograptus quadrimucronatus* has been claimed by ULRICH & RUEDEMANN (1931), who used this as an argument in favor of polyzoan affinities, and later by HABERFELNER in *Monograptus* (1933). The published figures are not convincing, and we agree with KOZŁOWSKI in the view that the presence of muscle impressions in graptolites is by no means established.

GRAPTOGONOPHORES. Extrathecal structures, variously termed "ovarian vesicles," "reproductive sacs," and "graptogonophores," have been figured and described, first by HALL (1859, 1865) in *"Graptolithus whitfieldi"* (referred by RUEDEMANN to *Hallograptus bimucronatus*) and later by HOPKINSON (1871) and NICHOLSON (1866, 1872) in *Diplograptus* spp. These early figures are schematic; some represent the scopulae of lasiograptids, a peculiar spinous or fibrous development from the edges of the median septum comparable with the lacinia, whereas others probably were vesicles, having a firm outline, and seem to have been attached to the apertural region of otherwise normal thecae. Siculae and early growth stages may be abundantly associated with such specimens, and in some they appear to be attached to the vesicles by their nemata, though it is impossible to assert that they are not drifted into juxtaposition. Such capsules were claimed to represent ovicells, but their rarity and restriction to biserial graptolites is curious, and no convincing interpretation has yet been given.

SICULA

The sicula of the Graptoloidea is in general more conspicuous than that of the typical dendroids. Its significance was first recognized (LAPWORTH in HOPKINSON & LAPWORTH, 1875) and its morphology worked out (WIMAN, 1893; KRAFT, 1926; and others) in the Graptoloidea, and its role in the Dendroidea was only more recently appreciated.

The walls of the **prosicula** are strengthened not only by the spiral thread (*Schraubenlinie* of KRAFT) but also by longitudinal rods or fibers *(Längsverstärkungsleisten)* (Fig. 39,1). Three or four of these merge into the base of the nema, where it grows out as a hollow tube from the apex of the prosicula, and clearly serve as an anchorage for this; the remainder seem to grow from the apertural margin of the prosicula, thinning and disappearing when traced toward the apex.

The **metasicula** is sharply differentiated from the prosicula by its close-set regular growth lines, meeting in a zigzag suture down the two opposite sides (dorsal and ventral); on the virgellar or ventral side is embedded the **virgella** (Fig. 39,2), usually a conspicuous spine involving deposition of secondary skeletal tissue; while on the dorsal side, symmetrically placed apertural spines (or rarely, one spine) may arise when growth is complete (see Fig. 48,6), as also may broad lateral lappets.

In size, the sicula varies between wide limits. Usually 1.5 to 2.5 mm. in length; it attains 5 or 6 mm. in some monograptids *(Monograptus gregarius, M. acinaces)* and it may be a centimeter or more in *Cystograptus* and *Corynoides* (Fig. 39,6). In shape, however, the sicula varies little, apart from slight differences in length-breadth ratio, or a gentle curvature; apertural spines are not uncommon and more elaborate processes occur in a few species (Fig. 39,5). In *Linograptus,* an umbrella-shaped structure called the **virgellarium** occurs at the tip of the virgella and may be connected with buoyancy (Fig. 39,7); in *Climacograptus baragwanathi,* the virgella breaks up distally into a mass of reticulating fibers.

More or less regularly spaced internal rings (annuli) are present in the siculae (Fig. 39,4) and more rarely the proximal thecae of various species of *Monograptus,* and in *Linograptus* the first ring may even be situated on the prosicula. Except for the ring separating the prosicula from the metasicula, and that related to the initial bud, they are not apertural and do not correspond to periods of arrested growth; in fact, the annuli are commonly oblique to the fuselli. Their significance is not yet known; ringed and ringless forms occur in the same species and the number of rings, when present, is variable.

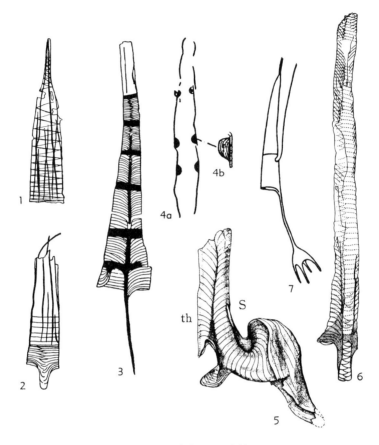

Fig. 39. Sicula in Graptoloidea.

1. Prosicula *(Diplograptus* sp. *cf. D. maxwelli)* with spiral and longitudinal lines, ×55 (255).
2. Early stage in metasicula of same, showing origin of virgella, ×55 (255).
3. Nearly adult metasicula of *Monograptus* with annuli, ×35 (256).
4. Longitudinal section *(4a)* of same showing internal position and structure *(4b)* of annuli, ×80 (256).
5. Apertural modifications of sicula of *Corynites wyszogrodensis*, ×35 (117). [*S*, sicula; *th*, theca.]
6. Elongate metasicula of *Corynoides curtus*, ×30 (23).
7. Sicula of *Linograptus posthumus* with virgellarium, ×20 (252).

In the Archiretiolitinae, the sicula is usually fully sclerotized although the rhabdosomal walls are reduced to a reticulum (and a comparable relation occurs in the dichograptid *Dinemagraptus*); but in the true retiolitids the sicula is either not sclerotized or is only represented by the delicate prosicular portion.

Further details are given below under the heading "Development."

THECAE

GENERAL RELATIONS

The form of the thecae, particularly in the apertural region, varies greatly and in fact constitutes one of the most valuable bases for the recognition of species and even genera. Outwardly, the simplest type is a straight, almost cylindrical tube partly overlapped by that which precedes it and partly overlapping that which follows it. In the past, much confusion has existed regarding the precise use of such terms as "thecae," "common canal," "interthecal septum," and others, but largely through the work of WALKER (1953), URBANEK (1953, 1958) and others, it now seems possible to define the parts with some precision. So far as concerns the periderm, each

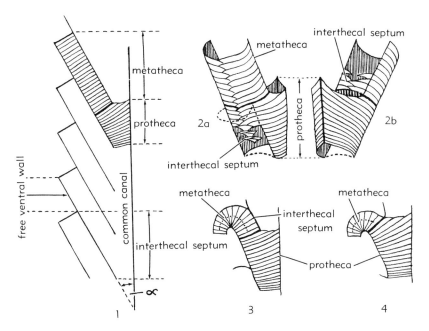

Fig. 40. Diagrams illustrating terminology of a graptolite stipe and thecae (α = angle of inclination) (29).

theca develops from the preceding theca and usually a slight but definite break or "unconformity" occurs between growth-lines of one theca and those of the next. Each theca can thus be divided into two parts, a **protheca** corresponding to the stolotheca of the dendroids and a **metatheca** corresponding to the autotheca of the dendroids[1] (Fig. 40). In the description of some highly elaborated thecae, it may be convenient to distinguish between a subapertural and an apertural portion of the metatheca (URBANEK, 1966), and in diverse graptolites, the initial portion of each protheca constitutes a distinct node which has been called the **prothecal fold**. The sum of the prothecae of a stipe, whether uniserial or biserial, constitutes the **common canal** of earlier authors (Fig. 40,1) and corresponds to the chain of stolothecae along the dorsal wall of a dendroid stipe. Although the common canal has thus no real entity as a structural unit, it is sometimes a useful descriptive term. A peculiar structure called the **appendix** occurs in *Gothograptus* (see Fig. 95,10b) and some other members of the Plectograptinae, where what appears to be a continuation of the common canal extends as an open tube beyond the thecate portion of the rhabdosome.

Normally, each thecal zooid buds off a single descendant zooid, but all bilateral rhabdosomes include at the proximal end at least one **dicalycal theca** (JAANUSSON, 1960) which produces two descendants, and in branched graptoloid rhabdosomes every branching division must relate to a dicalycal theca. This is further discussed in the sections headed "Development" (p. V71) and "Branching" (p. V82). The occurrence of dicalycal thecae at other points than the proximal end of a rhabdosome has not yet been investigated owing to lack of suitable material. By its nature, the dicalycal theca implies the presence of two prothecal segments, or what may be termed a prothecal and a mesothecal segment.

With minor exceptions, such as the apertural processes of certain cucullograptids, a theca remains bilaterally symmetrical in the plane of the stipe, whatever modifications in form it may undergo.

Each theca is composed of alternating right and left half-rings or fuselli, which basically meet in a zigzag suture (as in the

---

[1] The term "theca" thus used corresponds to the "thecal segment" of TÖRNQUIST, 1899; the protheca corresponds to the "semitubus" and "metatubus" of MÜNCH, 1938, and metatheca to his "thecatubus."

sicula) along the dorsal and ventral sides; but irregularities are not uncommon and in particular, complications tend to be introduced where the thecae overlap—that is to say, on the interthecal septa and the median septum, where a single peridermal wall separates two distinct zooids.

INTERTHECAL SEPTUM

An interthecal septum comprises the dorsal wall of one theca, n, and a portion of the ventral wall of the succeeding (overlapping) theca, $n+1$. In general, the fuselli of the proximal portion are continuous with the fuselli of theca n, while those of the distal portion are continuous with the fuselli of theca $n+1$ (Fig. 40,2). Commonly intercalary fusellar segments occur, the origin of which remains uncertain; WALKER (1953) considers them a secretion of theca $n+1$, whereas URBANEK (1958) associates them with theca n. In some examples, ranging from didymograptid at one extreme to monograptid at the other, the growth of theca n is so far in advance of theca $n+1$ that the entire septum must clearly be secreted by theca n; but THORSTEINSSON (1955) has contended that in some cyrtograptids, the septum is wholly secreted by theca $n+1$. Finally, SKEVINGTON (1965) has claimed that in *Glyptograptus dentatus* the interthecal septa, although marked externally by grooves separating the fuselli of adjacent thecae, are absent internally, and he has attributed this to the very slight angle of inclination and close proximity of adjacent zooids; but the interthecal septa are clearly complete in known climacograptids with an even lower angle of inclination.

MEDIAN SEPTUM

The dorsal wall, where it is prothecal, is usually quite regular, but may show intercalary fuselli; it may be nonexistent where the theca (or some portion of it) is adherent to a previously secreted part of the rhabdosome such as the sicula. In dipleural biserial rhabdosomes, the dorsal wall of each stipe is represented by the median septum. Information regarding the structure of this remains scanty, but although it marks the apposition of two distinct stipes, no positive evidence indicates that it comprises more than one layer. URBANEK (1959) has related this to the fact that the terminal theca of one or other series grows slightly ahead and produces the median septum as its dorsal wall. The fuselli are close-set, inclined distally and inward toward the central nema and the structure is somewhat irregular, with numerous intercalary fuselli (see Fig. 46,2).

Where the median septum is complete, it arises between $th2^2$ and $th3^1$ (the first four thecae being alternating in origin), but in some biserial graptolites the septum tends to be delayed as more and more of the proximal thecae alternate in origin. The process culminates in a wholly aseptate rhabdosome which in one sense is the only truly biserial condition. Progressive delay in the origin of the septum has been shown by WAERN (1948) to have stratigraphical value in a series of forms related to *Climacograptus scalaris*. Little general stratigraphical significance is seen in the tendency, however, since many late (Silurian) forms have a complete septum, whereas aseptate forms or those with a notably delayed (incomplete) septum may occur well down in the Ordovician.

The term **cryptoseptate** has been suggested by URBANEK (1959) for the condition in *Gymnograptus linnarssoni*, for example, where the nema is attached to the rhabdosome walls by peridermal rods but a peridermal septal membrane appears to be lacking.

In other instances again, the septum may be present at first on one side of the rhabdosome only (apparently usually the obverse), thecae on the opposite side being to all appearances alternating in origin; such a condition is referred to as a **partial septum**. A stage has been reached in *Cephalograptus cometa* where the partial septum is reduced to a mere ridge on the periderm of the obverse side.

No details are yet available concerning the structure of the median septum in monopleural rhabdosomes (Glossograptina) (see Fig. 90).

PRINCIPAL TYPES

The basic types of graptolite thecae may be defined as dichograptid, leptograptid, dicellograptid, climacograptid, triangulate (and isolate), hooked, lobate and auriculate; but the extent of variation now being revealed within almost any of these types makes precise definition difficult. In the majority of these, and commonly as a result of differential growth, the main axis of the theca becomes curved and the more extreme forms involve modification of the entire apertural region. An introverted theca is one with the aperture facing inward (dorsally), in contrast to an everted aperture which faces outward (ventrally); development of the latter may result in a hooked or retroverted theca. Pronounced sigmoidal curvature of the thecal axis is usually associated with a definite **geniculum** (JAANUSSON, 1960), an angular bend in the free ventral wall, separating a supra- from an infragenicular portion. The geniculum may be the site of median or paired spines, or may be associated with median or paired flanges overhanging and restricting the aperture of the preceding theca.

Simple straight thecae (Fig. 40,*1*) characterize the uniserial stipes of the vast majority of dichograptids, many simpler monograptids, and the biserial stipes of certain diplograptids; accordingly, this has been appropriately termed the **dichograptid type.** Exceptionally, however, even dichograptids may show more advanced types (*Didymograptus leptograptoides* MONSEN and *Aulograptus*) and the sinograptids have quite highly elaborated thecae (*Holmograptus* and *Sinograptus*).

The first widespread thecal elaboration is a gentle sigmoidal curvature of the ventral wall, accompanied by elongation and reduction in the angle of inclination, which results in the so-called **leptograptid type** (Fig. 41,*1*). This finds its typical development in the Nemagraptidae, occurring also in some monograptids. Somewhat similar is the type seen in *Glyptograptus* (Fig. 41,*2*) among biserial forms.

The **dicellograptid type** is characterized by the development of a geniculum and by introversion, usually accompanied by some degree of isolation of the apertural region (Fig. 41,*3-5*). The supragenicular wall is usually convex, in some forms angularly so, and provided with a mesial spine.

A more pronounced expression of this sigmoidal curvature results in the sharply angular geniculation of many **climacograptid types** (Fig. 41,*7*); many diplograptids and some species of monograptids (e.g., *Monoclimacis*) show this. The supragenicular wall is usually straight; it may be parallel to the axis of the rhabdosome, or inclined distally inward or outward, rarely curved (*Pseudoclimacograptus,* Fig. 41,*6*) and with everted apertures (*Clinoclimacograptus*). It is short and inwardly inclined in *Lasiograptus* (Fig. 41,*8*) and *Gymnograptus,* and extremely short in the modified forms of this theca occurring in *Cryptograptus* and *Hallograptus*.

Thecal modifications are most conspicuous in the Monograptidae; dichograptid, leptograptid, and climacograptid types occur here, but the more extreme modifications all involve some degree of apertural "isolation." In these Silurian forms, however, the theca, as it becomes free distally, twists outward with excessive development of the dorsal lip (retroversion), in contrast to the introversion of the more extreme dicellograptids and dicranograptids of the Ordovician. Accompanying this isolation is a reduction or loss of the interthecal septum and an increase in the prothecal ratio (Fig. 40,*3,4*); thecal overlap can only be recognized in a restricted sense and ceases to have any precise descriptive or systematic value. ELLES (1922) recognized three main lines of this monograptid development (hooked, lobate, isolate); the term "triangulate" has also been applied, particularly to early stages in the isolate development. In recent years, a varied suite of modifications has been recognized which involve the development of conspicuous apertural flanges, and may be referred to collectively as auriculate.

In the **hooked type,** the isolate distal portion of each theca grows back upon itself in the form of an open hook, familiar in the widely distributed *Monograptus priodon* (Fig. 42,*1*). Typically no transverse widening of an apertural region occurs but lateral spines are commonly present. The thecae may overlap sufficiently to produce an appreciable interthecal septum, but in others provisionally included here (*M. clingani*) this may be lost.

The **lobate type** is really a very compact form of hook developed exclusively by the dorsal wall, which grows out and back over the thecal aperture like a cowl. Its highest development is seen in such species as

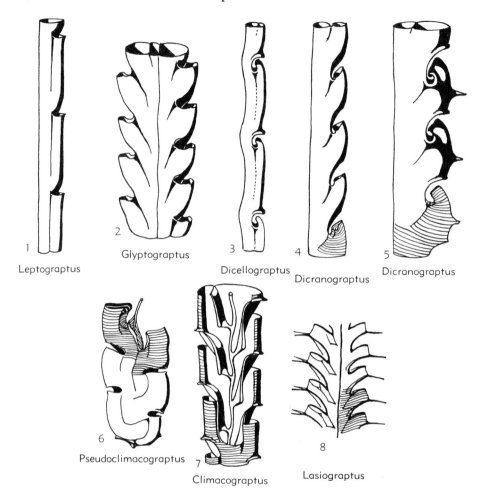

Fig. 41. Variations in graptolite thecae, mostly somewhat diagrammatic (21).

1. *Leptograptus*, ×15.
2. *Glyptograptus*, ×15.
3. *Dicellograptus geniculatus*, ×19.
4. *Dicranograptus pringlei*, ×11.
5. *Dicranograptus nicholsoni*, ×11.
6. *Pseudoclimacograptus scharenbergi*, ×22.
7. *Climacograptus typicalis*, ×19.
8. *Lasiograptus harknessi*, ×10.

Monograptus lobiferus (Fig. 42,2) and *M. becki*, where the thecal aperture is almost closed and the thecae appear in compressed material as a series of evenly spaced, rounded protuberances along the stipe. A related form is seen in *M.* sp. cf. *M. knockensis* (Fig. 42,3), where isolation is more strongly marked and the lobe less inflated.

In the **triangulate** and **isolate type** there is typically no trace of an interthecal septum, and the more extreme developments are so unlike normal Monograpti that they long ago received the separate generic name *Rastrites*. In less extreme development (*Monograptus spiralis* or *M. convolutus*, Fig. 42,5,6), the thecal segment is more or less triangular and the theca is distinctly hooked, but with an enrolled dorsal lip and transverse processes; the theca in profile view appears triangular, sometimes with a "flowing" apertural spine (one or other of the transverse processes), its apparent form depending on shearing). In *Rastrites* itself (Fig. 42,7) the straight slender thecal tubes terminate in a compact lobate aperture and are sharply differentiated from an extremely tenuous common canal, from which they extend at high angles as a row of uniformly spaced parallel tubes.

In addition to the above types, others are beginning to become known through the work of EISENACK, MÜNCH, BULMAN, and especially URBANEK on graptolites dissolved out of Silurian limestones; they are here provisionally referred to as the **auriculate group.** For the most part, these comprise long, slender thecae, with very little overlap and very short metathecal segments, of which a considerable portion of the apertural region is involved in large paired lateral lobes or auricles (cucullograptids) or a single lidlike shield developed from the dorsal margin (Fig. 42,8,9).

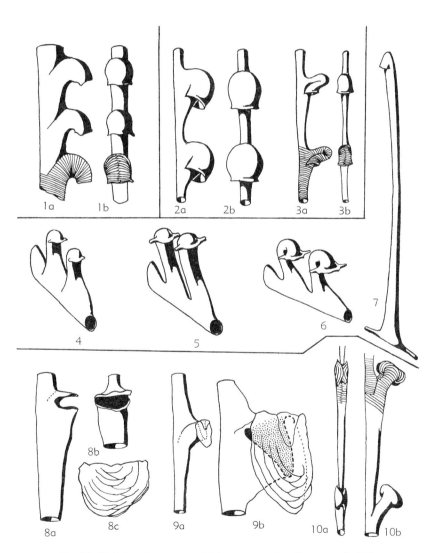

FIG. 42. Variations in monograptid thecae, somewhat diagrammatic (29).

1a,b. *Monograptus priodon.*
2a,b. *M. lobiferus.*
3a,b. *M. knockensis.*
4. *M. triangulatus.*
5. *M. convolutus.*
6. *M. spiralis.*
7. *Rastrites.*
8. *M.sp.* —— *8a,b*, ×35. —— *8c*, ×60 (Münch, 1938).
9. *M. huckei.* —— *9a*, ×25. —— *9b*, ×100 (Münch, 1938).
10a,b. *Cucullograptus (Lobograptus) scanicus.*

MONOGRAPTID TRENDS AND THEIR SIGNIFICANCE

The hooked, lobate, and isolate types of monograptid thecae appear to constitute definite trends (ELLES, 1922), each affecting large numbers of species. They were believed to operate during a comparatively short period of time and to be nonrecurrent, but some evidence indicates that hooked forms, at least, were either more persistent than originally was thought or actually reappeared after a considerable interval of time.

No graptoloid rhabdosome is composed of precisely similar thecae throughout; on the contrary, the thecae all undergo some change in form (commonly slight but in some very conspicuous) when traced along the stipe and this astogenetic succession is always remarkably regular. It is also one of the factors which complicate attempts to subdivide or define such genera as *Monograptus* on the basis of thecal form. ELLES (1922) recognized that new thecal types were introduced at the proximal end of the rhabdosome (Fig. 43), spreading distally in successive descendants, but exact data have not been available until recently and still regrettably few well-authenticated phylogenies can be cited. From the analysis of some triangulate monograptid lineages, SUDBURY (1958) was able to resolve the change into two processes, a distal spread of the new type (whereby more and more thecae acquired the new character) and a general change (whereby all thecae of the rhabdosome showed a slight and gradual trend toward the new type; Fig. 44). These processes were later referred to by URBANEK (1960) as *penetrance* and *expressivity*, respectively. In addition, SUDBURY presented some evidence that a new type might also be introduced distally. Convincing evidence of this distal introduction of new types has been provided by URBANEK (1966) in his precise work on the cucullograptids (Fig. 45). Moreover, he has attempted a biological interpretation of these evolutionary changes (URBANEK, 1960) on a morphophysiological gradient hypothesis. Briefly, he suggested that morphogenic substances, some acting as stimulators and some as inhibitors, were transmitted from the siculozooid in steadily decreasing quantities through the asexually budded succession of zooids, and that when these fell below a certain threshold level they no longer exerted any effect. Proximal introduction of a character results from increasing activity of a morphogenetic stimulator, together with a lowering of the threshold level; distal introduction results from diminishing activity of an inhibitor and a rise in threshold level. (See also section on "Cladia," p. V85.)

The few examples so far described are all monograptid, but comparable phenomena seem to be of general occurrence among Graptoloidea, especially dicellograptids and diplograptids.

APERTURAL PROCESSES, SPINES AND LOCALIZED THICKENING OF PERIDERM

Certain species of most graptolite genera show a development of apertural spines associated with some or all thecae of the rhabdosome, and spines may also be developed at other points, such as the ventral wall of a theca (mesial spines) of rarely

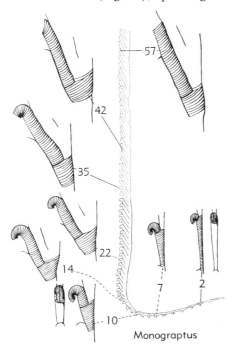

FIG. 43. Thecal changes along rhabdosome of *Monograptus argenteus* (NICHOLSON); outline of rhabdosome, ×2; enlarged thecae, ×10 (29).

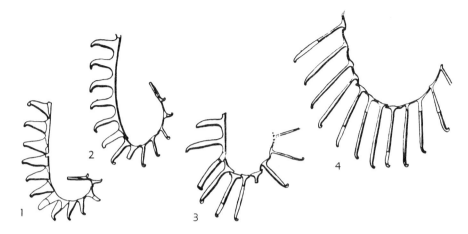

Fig. 44. Proximal introduction and distal spread of new thecal type in triangulate monograptids, all ×5 (230).——*1. Monograptus triangulatus separatus.*——*2. M. triangulatus triangulatus.*——*3. M. triangulatus extremus.*——*4. Rastrites longispinus.*

laterally (e.g., *Dicaulograptus*) or from the dorsal wall of a stipe (e.g., *Didymograptus nodosus*) or even along the edges of the septum (e.g., *Nymphograptus*, see Fig. 93,4).

Apertural spines are usually median in position and unpaired, and every gradation is found between a blunt denticle and slender aciculate spine; they are laid down in the manner of the sicular virgella (Fig. 39,2) and strengthened by deposition of secondary tissue. Such spines are particularly associated with the first two thecae of the rhabdosome (especially in dicellograptids and biserial genera) and in *Climacograptus bicornis*, where these spines are unusually large, the first two thecae may become completely involved, with their distal portions growing back along the spines. In other *Climacograptus* species, comparable spines are associated with flanges (*C. papilo*, see Fig. 70,11).

More rarely, apertural spines are paired structures (e.g., *Orthograptus quadrimucronatus* and various monograptids) and these are not produced from overlapping, alternating fuselli but arise as localized secretions of lateral fuselli (monofusellar tissue), as in *Saetograptus chimaera* (see Fig. 101,3). In some varieties of the *O. quadrimucronatus* groups, the spines of one particular thecal pair may be exaggerated in length.

Somewhat more complicated are the dorsal and lateral spines described by WHITTINGTON & RICKARDS (1969) as hollow structures associated with the microfusellar hood overhanging the apertures in *Glossograptus*.

The most remarkable spinous developments occur among the lasiograptids (see Fig. 93) where apertural and mesial spines of great length may break up distally to form an interlacing network (**lacinia**) outside the thecal apertures. Such structures are presumably of cortical tissue.

Spines usually have been regarded as protective in function and RUEDEMANN (1947) has observed that they are commonly placed at exposed portions of the rhabdosome; but the possible relationship of apertural processes to the lophophore of the graptolite zooid, both as supporting structures and hydrodynamic tunnels related to feeding, has been pointed out by URBANEK (1966).

LOCALIZED THICKENING. Certain graptolites show a marked thickening of parts of the periderm along structural lines, accompanied by attenuation or reduction of intervening areas of the test; the "shell" or box construction of the normal rhabdosome is replaced by a structural framework (**clathria**) of **lists** (strengthening rods) carrying only the most delicate cuticular "skin" or in extreme cases none at all. This is associated commonly with a profuse development of spines (apertural, mesial and lateral), in some breaking up into a filamentous network (**lacinia**) beyond the limits of the rhabdosome proper. Such features are developed to a varying extent in the Diplograptidae, Glossograptidae, and Lasiograptidae; and the recent discovery of *Dinemagraptus* shows that comparable features were developed even in the Didymograptina.

A thickened selvage on the apertural margin (apertural list) occurs generally; this may be extended laterally (pleural lists) and distally (mesial list) to form a bent ring, strengthening the aperture. Such is found in some climacograptids and more particularly in amplexograptids, and may be continued into mesial and apertural spines, as in the lasiograptids. The latter usually exhibit also at least the beginnings of parietal lists along edges of the interthecal septum, and an aboral list formed by thickening of the inner edge of the septum. Completing this framework (or clathria) a longitudinal list may occur along each lateral wall of the rhabdosome, and, except where thecal overlap is almost total, a longitudinal list also appears connecting the mesial with the apertural list along the mid-ventral line of each theca.

Among the Retiolitidae, the clathria supports a delicate skeletal network, the **reticulum;** in well-preserved specimens this seems

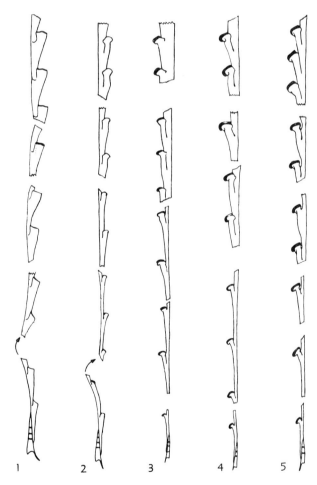

FIG. 45. Distal introduction and proximal spread of new thecal type in cucullograptids; all ×20 (253).——*1. Cucullograptus (Lobograptus) progenitor.*——*2. C. (L.) simplex.*——*3. C. (L.) scanicus parascanicus.*——*4. C. (L.) scanicus amphirostris.*——*5. C. (L.) scanicus scanicus.*

to be covered by an exceedingly thin, possibly structureless, film, but its relation to the normal graptolite periderm are obscure.

The three structures, clathria, reticulum and lacinia, need not be closely correlated; *Retiolites,* with perfect reticulum and clathria, has no lacinia; in *Nymphograptus* and most *Lasiograptus,* the clathria is associated with a well-developed lacinia; and in *Plegmatograptus nebula* the clathria is barely developed, the reticulum being distinct but irregular and the lacinia well developed.

NEMA

Among the Dichograptidae, with the exception of *Phyllograptus, Cardiograptus* and *Oncograptus,* a threadlike nema commonly extends from the apex of the prosicula and probably served (as suggested by LAPWORTH) for fixation of the rhabdosome, at least in juvenile stages; examples are known where it terminates distally in a somewhat irregular attachment disc. In the genera *Leptograptus* and *Dicellograptus,* the nema is usually so short that it is not obvious how it can have served, at a mature stage of astogeny, for attachment of the rhabdosome; and *Dicranograptus* appears to have lacked a nema. Among the Diplograptidae, a nema is invariably present as a central axis, then commonly referred to as the virgula, embedded in the median septum or in aseptate forms lying freely in the cavity of the common canal and in some species anchored by fusion to the bases of the interthecal septa (Fig. 46,1). Even where embedded in the median septum, additional strengthening lists may be developed (also in the median septum) connecting the nema with the lateral walls of the rhabdosome (Fig. 46,2). In monograptids, the nema is embedded in the dorsal wall of the stipe. Its relations in cladia-bearing monograptid genera are described elsewhere (p. V85).

The doubtful relations of the nema in retiolitids have been resolved largely by the work of EISENACK (1951) and it is now known that the "straight axis" in the lateral wall of *Retiolites* is the true nema. In other retiolitids it may be free and axial in position, and in still others it ends blindly within the rhabdosome, perhaps continued as an unsclerotized thread.

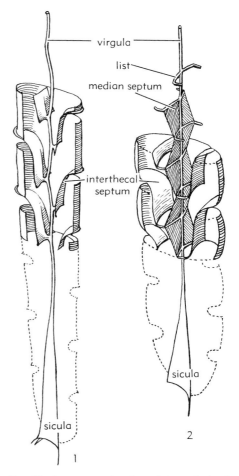

FIG. 46. Stereograms showing relations of virgula to median septum and interthecal septa in biserial graptolites (19).——*1.* Aseptate, *Climacograptus typicalis.*——*2.* Septate, *Pseudoclimacograptus scharenbergi.*

Several biserial graptolites possess what has been interpreted as a float at the distal end of the nema; this is discussed in the section on buoyancy, p. V93.

The nema of the neanic and adult rhabdosome replaces the embryonic *nema prosiculae* (Fig. 39,1), which is a hollow tubular projection from the apex of the prosicula. Lengthening of the *nema prosiculae* occurs during growth of the sicula; subsequent breakage and replacement of this by a regenerative nema has been described by URBANEK as "almost normal" in some monograptids, but in other graptolites the one appears to have developed from the other. Whether the nema remains hollow or becomes a solid rod is disputed, though the latter seems the more probable; and without doubt the nema can continue to lengthen during astogeny and this must be due to the presence of some covering layer of living tissue. Lengthening is convincingly demonstrated in synrhabdosomes comprising colonies in all stages of development, including young growth stages with very short nemata. Irregular flanges and thickenings may occur (e.g., *Climacograptus parvus*) on the outside of the nema in addition to the terminal "float" (see Fig. 69).

REGENERATION

Graptolites possessed considerable powers of regeneration of damaged skeletal tissue and examples are to be found in any large collection of fossils dissolved from the matrix (KRAFT, 1926; BULMAN, 1932; EISENACK, 1940; KOZŁOWSKI, 1949; URBANEK, 1958; etc.). Where the repair is to damage at the growing edge of the rhabdosome, it is effected by normal fuselli (though the growth lines are of course unconformable to those of the undamaged tissue). In other places on the rhabdosome, where it could no longer be repaired by that portion of the zooid concerned with the secretion of fusellar tissue, it is composed of a structureless film presumably comparable with cortical tissue and secreted by some part of the organism responsible for cortical tissue.

Prosiculae are particularly susceptible to injury in early stages of growth; indeed, a

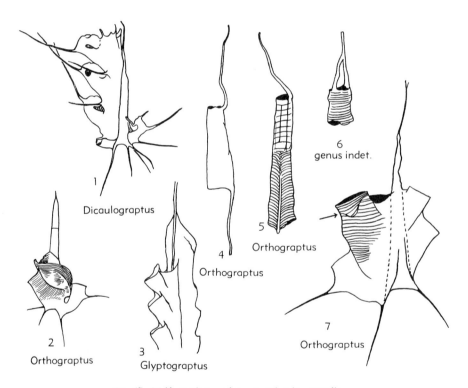

FIG. 47. Malformations and regeneration in graptolites.

1. Suppression of stipe above second-formed theca ($th1^2$) in biserial rhabdosome, *Dicaulograptus hystrix*, ×18 (19).
2. Malformation of proximal end, *Orthograptus gracilis*, ×13 (19).
3. Suppression of one stipe in biserial rhabdosome at distal end, *Glyptograptus dentatus*, ×10 (19).
4. Loss of prosicula, virgula attached to metasicula open at both ends, *O. gracilis*, ×30 (19).
5. Damage to prosicula open-ended distally, *O. gracilis*, ×40 (56).
6. Loss of prosicula, bifurcating virgula attached to metasicula, genus indet., ×30 (56).
7. Regeneration of damaged theca, *O. gracilis*, ×30 (126).

high proportion of diplograptids and monograptids possess regeneration nemata and Urbanek has concluded that in *Saetograptus chimaera* it is almost invariable. Damage to the graptolite rhabdosome in general is usually ascribed to some unknown predator, but may be due to wave action, and damage to the nema appears to have been due to insecure attachment during initial stages in the change from planktonic siculae (with nonfunctional nema prosiculae) to the epiplanktonic attachment of neanic stages.

The most remarkable instances of regeneration, involving the development of an entire colony (pseudocladium), have been described by Urbanek (1963) in the linograptids; this process is related to the development of cladia and is discussed further in that section (p. V89).

ABNORMALITIES IN DEVELOPMENT

Regularity in development of the graptolite rhabdosome is one of the most distinctive features of the group, but malformations due to some pathological cause and not to damage and regeneration are also known. Some examples concern single thecae only and generally affect the shape of the aperture or apertural processes; others affect several thecae before normal development is resumed. It is not clear whether the example illustrated in Figure 47,1 is truly pathological, or follows damage to $th2^1$; but the normally biserial rhabdosome has been converted to an exceptional uniserial colony.

Urbanek (1958) has described monograptid siculae with "twin pores," one lying on either side of the virgella, which may be due to some acceleration in the budding process. In another example, abnormality in *th1* (in *Saetograptus chimaera*) is associated with an abnormal budding process; *th2* is completely partitioned off from *th1* and a lateral resorption foramen in the first and another foramen in the second theca mark the points of exit and re-entrance of the stolon, which in its extrathecal course remained unprotected by any sclerotized skeleton.

Thecal occlusion, commonly recorded in dendroids and camaroids, is rare among the Graptoloidea.

DEVELOPMENT[1]

GENERAL DISCUSSION

Development commences with the secretion of the prosicula, of which three stages have been recorded (Kraft, 1926). The earliest of these (Fig. 48,1) consists of a delicate bottle-shaped object, usually 400 to 500 microns in length, open at the base and closed at the neck or *nema prosiculae*. It is faintly marked with a spiral thread *(Schraubenlinie)* which may be coiled indifferently right- or left-handedly; this spiral thread may be strengthening, or perhaps (according to Kraft) marks the line of fusion of a continuous spiral growth band. In the second stage (Fig. 48,2), a group of three or four longitudinal strengthening fibers is laid down from the *nema prosiculae* to the aperture of the prosicula. Finally (Fig. 48,3), secondary longitudinal fibers are secreted between the primaries, extending one-half to two-thirds the way from the aperture to the *nema prosiculae* and the growth of the prosicula then is complete. A ring of secondary tissue may (rarely) be developed at the margin of the prosicula, marking a pause in growth.

An entirely new structure, the metasicula, then begins to appear (Fig. 48,4,5). This contrasts sharply with the prosicula in its incremental mode of growth, being composed of normal peridermal fuselli laid down in alternating growth bands with typical zigzag suture along opposite (dorsal and ventral) sides. Rarely, the ventral and dorsal zigzag sutures, according to Kozłowski (1954), may be absent at the commencement of the metasicula. These growth bands tend to extend forward along the ventral side (Fig. 48,4), resulting in a rounded projection and later a short spine, which is finally reinforced to form the virgella spine. In nearly all dichograptids, a true virgella is absent and is represented only by a blunt or rounded ventral process. Kozłowski (1949) has suggested that the actual spine formation is deferred progressively in stratigraphically younger forms; Barrass (1954) has claimed that the late formation in *Climacograptus* contrasts with an early formation in *Diplograptus (s.l.)*. When growth of the metasicula is com-

---
[1] See footnote on page V32.

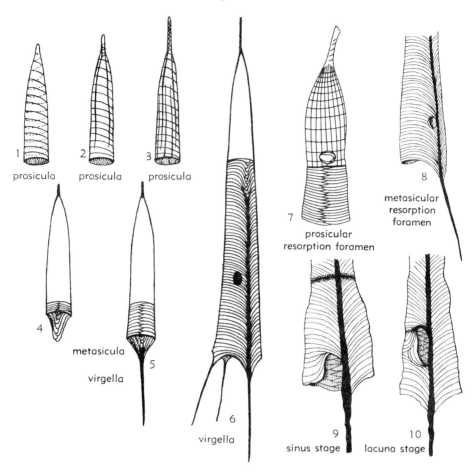

Fig. 48. Diagrams illustrating development of sicula and initial bud (*7-10* from 123, 126, and 249).

1-3. Prosicula.
4,5. Beginning of metasicula and formation of virgella.
6. Completed sicula with apertural spines, virgella and resorption foramen.
7. Prosicular resorption foramen, *Didymograptus* sp., ×65.
8. Metasicular resorption foramen, *Orthograptus gracilis*, ×45.
9. Sinus stage in monograptid, *Pristiograptus bohemicus*, ×65.
10. Lacuna stage in same, ×65.

plete, apertural spines, almost invariably paired, may develop on the dorsal side opposite the virgella (Fig. 48,6), but more elaborate apertural modifications occur rarely (Fig. 39,5). The presence of annuli among monograptids has already been mentioned (p. *V*59).

POSITION AND FORMATION OF PORUS

The **porus**, or pore through which the initial bud passes to the exterior, may be situated either in the prosicula or metasicula. In primitive and geologically early forms, such as many dichograptids and as in all known dendroids, it is prosicular (Fig. 14,*1,2;* 48,7) but in typical graptoloids it is metasicular (Fig. 46,8) in position, and seemingly the pore tends generally to arise progressively lower down on the metasicular wall until it comes to lie quite close to the aperture. Normally it is placed close to the virgella and on its (biologically) right-hand side; but in *Didymograptus formosus* (see Fig. 51) the prosicular pore lies on the side opposite to the virgella, which becomes incorporated in the ventral wall of the $th1^2$ and *D. rozkowskae* and *D. artus* appear to be other exceptions.

Two methods of pore formation are discriminated. Most commonly, and invariably when it is prosicular in position, a pore is produced by resorption, as clearly demonstrated by its relation to growth lines (or the spiral line on the prosicula). But in *Monograptus* it is contemporary with adjacent sicular tissue, being formed as a notch in the transient apertural margin, later closed by forward-bending growth bands. EISENACK (1942) termed these successive phases the **sinus** and **lacuna stages;** and where sicular annuli occur, the "budding ring" forms an apertural thickening to the proximal rim of the sinus. The formation of pores in dimorphograptids is not known in detail, but they do not appear to have acquired the sinus type of formation.

INITIAL BUD

The initial bud grows out from the porus and down the side of the sicula, except in monograptids and most dimorphograptids where it is erect. As in the dendroid stolotheca no true inner wall is present and the bud is a split tube fused at its edges to the sicula (see Fig. 50,2). Growth bands initially extend uninterruptedly from side to side, but later alternate to produce a median zigzag suture. All varying types of graptolite rhabdosome are derived from this single initial bud by different methods of branching.

In the development of a bilateral rhabdosome two buds must be borne by a single zooid, inhabiting the **dicalycal theca.** This may be the first-formed theca, $th1^1$, but it tends to be progressively deferred in the thecal succession; as a consequence of this delay, the earliest thecae acquire an alternating arrangement, growing across the sicula on the reverse side to open on the side opposite that of their origin. The proximal (prothecal) portions of such thecae constitute the **crossing canals.** If $th1^1$ is the dicalycal theca, there is one crossing canal, $th1^2$. If $th1^2$ is the dicalycal theca, there are technically two crossing canals, $th1^2$ and $th2^1$ (Fig. 49), though the second may be short and inconspicuous. When the $th2^1$ is the dicalycal theca, there are three crossing canals ($th1^2$, $th2^1$ and $th2^2$). With the ultimate stage of this process, no dicalycal theca is developed and all thecae either alternate to form an aseptate diplograptid rhabdosome or form a uniserial monograptid rhabdosome.

LATER DEVELOPMENT[1]

Recognition of four main types of development—dichograptid, leptograptid, diplograptid and monograptid—is due to ELLES (1922), but the structural detail now becoming revealed as a result of more modern techniques indicates a more complex range of rhabdosomal development. Stages in these main types recognized by BULMAN (e.g., 1955) retain some value as concise descriptive terms, but the concept of the dicalycal theca introduces a necessary discontinuity into what was formerly regarded as continuous gradual change.

The proximal end development is now known with full growth-line detail from transparencies in nearly 40 species of graptolites, of which the majority are biserial or monograptid. In the dichograptid type, where $th1^1$ is dicalycal, the first daughter theca, $th1^2$, may be either right- or left-handed, as in many dendroids; but a biologically right-handed origin is more usual, with $th1^2$ growing across the front of the sicula oriented with the virgella on the right (Fig. 49). In the isograptid type of development (*minutus, extensus* and *gibberulus* stages of Fig. 49) and presumably also in the leptograptid type, $th1^2$ is again right-handed in origin and the dicalycal theca is $th1^2$ (or possibly $th2^1$ in some leptograptids). In all the various examples of the diplograptid type so far described, $th1^2$ originates differently and the few known dicranograptids resemble the diplograptids in this respect. Here $th1^2$ arises behind $th1^1$ (left-handedly) and consequently grows around and across the front of $th1^1$ as well as of the sicula (Fig. 49). But as in isograptids and leptograptids, subsequent budding "fans out" in various ways on the reverse side, leaving the sicula largely free on the obverse side. All these developments have been collectively termed **platycalycal** (BULMAN, 1968). In contrast to them, the biserial rhabdosomes of *Cryp-*

---

[1] See footnote on page V32.

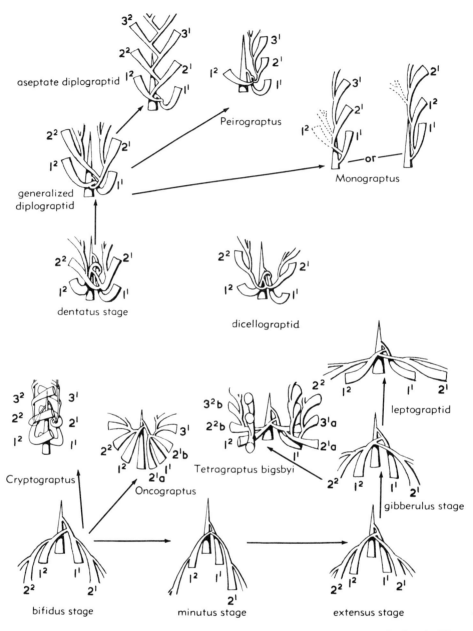

Fig. 49. Thecal diagrams illustrating progressive trends in development of the proximal end. Thecae are numbered according to convention in manner indicating their order of budding in different series, as $th1^1$ (first-formed theca developed from sicula, $th1^2$ (next-formed theca, budded from $th1^1$), $th2^1$ (third-formed theca, budded from $th1^1$ or $th1^2$) and so on. Bifidus stage relates to dichograptid type of development; isograptid type includes *minutus*, *extensus*, and *gibberulus* stages (Bulman, n).

tograptus and *Glossograptus* develop from a dicalycal $th1^1$ but with $th1^2$ originating left-handedly and growing behind the sicula (Fig. 49); during the course of development the sicula becomes largely enclosed and concealed on both sides, and in this **pericalycal** type of proximal end the terms obverse and reverse have little meaning.

DICHOGRAPTID TYPE

A type of graptolite development characterized by a single crossing canal was originally defined by ELLES (1922) as dichograptid type, for it was believed to characterize the bulk of the Dichograptidae. ELLES recognized the occurrence of a second crossing canal in *Didymograptus (Isograptus) gibberulus*, however, so that this species subsequently was separated along with other comparable species as the isograptid type (BULMAN, 1932).

The dichograptid type, now defined as a platycalycal mode of development with $th1^1$ as the dicalycal theca and a single crossing canal, is in fact relatively rare. The *minutus* stage (BULMAN, 1955) must be transferred to the isograptid type, leaving the dichograptid type with content of the *bifidus* stage alone. Details are known in only two examples, *Didymograptus rozkowskae* (Fig. 50,*1-3*) and *D. artus* (SKWARKO, 1967), but a number of pendent didymograptids, including *D. bifidus*, undoubtedly developed on this plan. An unusual modification is seen in *Parazygograptus* (Fig. 50,*4*), where $th1^1$, after producing a right-handed bud $th1^2$, undergoes no further development and the crossing canal is not associated with presence of a dicalycal theca.

ISOGRAPTID TYPE

In the isograptid type, the third theca $(th2^1)$ lacks any connection with the first $(th1^1)$ but develops from $th1^2$, which is the dicalycal theca originating right-handedly from $th1^1$. This type, with two crossing canals, is widespread and full structural details are known for *Didymograptus formosus* (Fig. 51), *D. minutus* (Fig. 52), *Aulograptus cucullus*, *Isograptus geniculatus* and *Tetragraptus* sp. cf. *T. bigsbyi* (Fig. 53); numerous well-preserved though not transparent proximal ends further extend the range of its occurrence.

Perhaps little purpose justifies attempting to distinguish between the *minutus*, *extensus*, *gibberulus* and *hirundo* stages (Fig. 49) which were based mainly on the precise levels of origin of $th2^1$ and $th2^2$; it is possible, if not indeed probable, that the dichograptid and isograptid types of development originated independently and stand in parallel rather than serial relationship to one another. The relationship of both to the anisograptid (dendroid) proximal end is obscure and critical evidence from such forms as *Kiaerograptus* is not yet available.

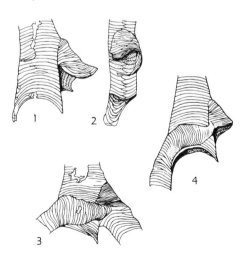

FIG. 50. Dichograptid type of development (*bifidus* stage) (116).——*1,2*. *Didymograptus rozkowskae*, initial bud and right-handed origin of $th1^2$.——*3*. *D. rozkowskae*, single crossing canal.——*4*. *Parazygograptus erraticus*, modified *bifidus* stage with abortive $th1^1$. All ×35.

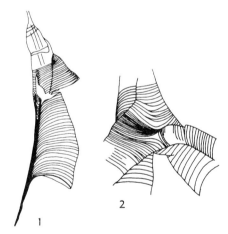

FIG. 51. Isograptid type of development (*extensus* stage), *Didymograptus formosus* (215).——*1*. Prosicular foramen and initial bud on anti-virgella side of sicula, ×43.——*2*. Crossing canal of $th1^2$ (right-handed) and $th2^1$, ×47.

LEPTOGRAPTID TYPE

Least well-known of all types of proximal end development in graptolites is that named leptograptid type. It was originally

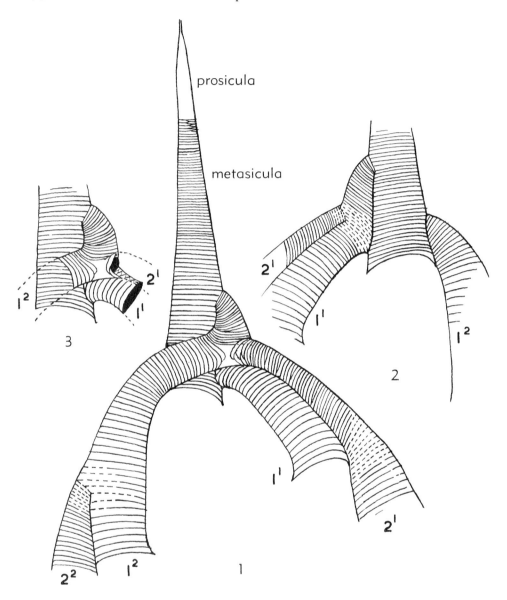

FIG. 52. Isograptid type of development (*minutus* stage), *Didymograptus minutus*.——*1*. Reverse aspect with single crossing canal.——*2*. Obverse aspect, ×40 (91).——*3*. Hypothetical growth stage. [Growth lines where doubtful shown by broken lines.]

defined by the presence of two crossing canals and the horizontal instead of downward direction of growth of even the earliest thecae as reported by ELLES (1922), but her accompanying diagram (Fig. 39) could be interpreted only in terms of three crossing canals. The leptograptid type was later accepted as possessing two or three crossing canals, with a horizontal or even slightly reclined direction of growth of the earliest thecae (BULMAN, 1955) and taken to include certain *Dicellograptus* species. These latter are here excluded and transferred to the diplograptid-dicranograptid type on the basis of the mode of origin of $th1^2$.

If species of *Leptograptus* should prove to exhibit a righthanded origin of the dicalycal $th1^2$ (Fig. 49), then the development would appear to represent a simple modification of the geologically earlier isograptid type. But if the mode of origin of the dicalycal theca proves to be left-handed, as in *Dicellograptus* and *Dicranograptus*, subsequently growing across $th1^1$ and the sicula, it would hardly be possible to separate this type from the diplograptid-dicranograptid type described below.

DIPLOGRAPTID TYPE

The compact proximal end of a biserial diplograptid rhabdosome includes a wide variety of forms and some 15 species are known in complete detail, with many more in rather less perfect preservation. In all cases some upward component in the growth direction of at least the distal portion of each proximal theca is generally well marked, and the prothecal portions of at least the first three thecae constitute crossing canals. Thus the dicalycal theca is $th2^1$ or some later theca. The origin of $th1^2$ is left-handed, the prothecal portion forming a hood which grows across the parent $th1^1$, as well as the sicula, and the development is platycalycal (Fig. 54).

In species of *Dicellograptus* the scandent element in thecal growth is less extreme and the sicula may lie exposed in the axil

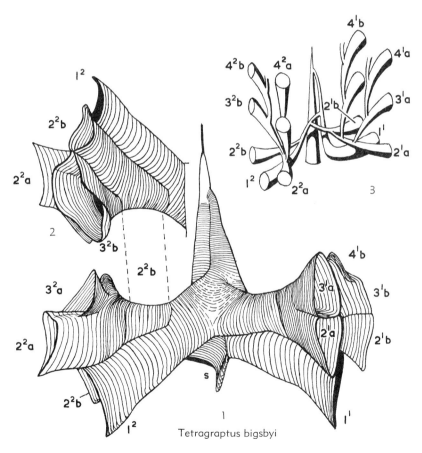

FIG. 53. Late isograptid or leptograptid type of development (shown by *Tetragraptus* sp. cf. *T. bigsbyi*, ×30) (29).——*1*. Reverse aspect with 2 crossing canals.——*2*. Obverse aspect of left side to demonstrate relations of $th2^2a$, $th2^2b$, and $th1^2$.——*3*. Thecal diagram illustrating relations of early thecae and mode of branching. [Growth lines where doubtful shown by broken lines. Sicula, *s*.]

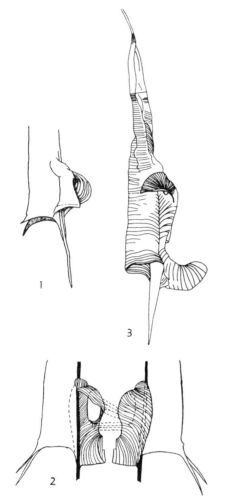

Fig. 54. Early stages of diplograptid development showing left-handed origin of $th1^2$ and growth of crossing canal across $th1^1$ and sicula.——*1.* *Orthograptus gracilis*, initial bud and partly formed foramen of $th1^2$ (cf. Fig. 50), ×30 (19).——*2.* *Amplexograptus cf. A. maxwelli*, reverse and obverse views of completed foramen of $th1^2$, ×40 (255).——*3.* *Glyptograptus austrodentatus oelandicus*, reverse view of slightly later stage, ×35 (215).

of the rhabdosome, though in many forms it is incorporated in the dorsal wall of one stipe (Fig. 55). *Dicranograptus* is typically diplograptid and theoretically at least the dicalycal theca could be some theca later than $th2^1$ (Fig. 56).

The diplograptid development may be divided into two groups, **streptoblastic** and **prosoblastic** (BULMAN, 1963). In the former, $th1^2$ is recumbent S-shaped, with an initially upward direction of growth, followed by a pronounced downward direction of the middle portion before the theca finally turns upward again at its distal end (Fig. 56). Associated with this, the initial portion of $th2^1$ has a pronounced downward direction of growth. This peculiar twisted configuration of the second and third thecae was originally described as the "*dentatus* stage," but later it was recognized in numerous genera including *Dicellograptus, Dicranograptus, Glyptograptus, Pseudoclimacograptus, Dicaulograptus,* and *Gymnograptus*. The prosoblastic type appears to develop from this by a gradual straightening-out of $th1^2$ and $th2^1$; $th2^1$ comes to grow entirely upward (Fig. 57) and ultimately (e.g., *Climacograptus brevis*) even $th1^2$ may grow upward from its origin. The process appears to occur independently in many different lineages and within a single genus.

The dominantly upward growth of the early thecae, coupled with some delay in separation of $th1^2$ and later thecae, leads to a condition in which the crossing canals pass across the nema rather than the sicula, which latter tends to be far more completely exposed even on the reverse side (e.g., *Cephalograptus*, see Fig. 91,*8*).

Independent of this straightening-out of the initial parts of early thecae, the separation of two linear series of thecae by a median septum may be progressively delayed as more and more of the proximal thecae alternate and the number of crossing canals steadily increases (Fig. 58); the dicalycal theca shifts progressively distally and the rhabdosome eventually becomes aseptate.

The development of the retiolitids remains imperfectly understood. One group, the Archiretiolitinae, appear to develop on lines generally similar to the diplograptids; the sicula is fully sclerotized, but it is difficult to trace the relations of later thecae owing to reduction of the periderm (see Fig. 95, 96). The Retiolitinae and Plectograptinae present a different appearance. The sicula is unsclerotized or at most represented by the prosicula (Fig. 59,*10*) and the familiar early growth stages are replaced by the **ancora** and **corona stages** (Fig. 59,*1-6*).

MONOGRAPTID TYPE

In the monograptid type (Fig. 60) a downward direction of growth no longer affects even the initial bud, which grows upward from its first appearance, following a nonresorption type of porus. The origin of this development presents another unsolved problem. To convert a biserial to a uniserial scandent rhabdosome requires the reduction and loss, or reorientation, of $th1^2$ in addition to the disappearance of a dicalycal theca (see p. V108). Most dimorphograptids appear to be aseptate (or cryptoseptate) and thus to lack a dicalycal theca, which could give rise by its suppression to a monograptid rhabdosome (Fig. 61).

PERICALYCAL TYPE

The development of *Cryptograptus* and *Glossograptus* is strikingly different from that of the biserial diplograptids. As in the dichograptid type, the first theca $th1^1$ is dicalycal and there is but a single crossing canal (Fig. 62). Theca $th1^2$ appears always to originate left-handedly (unlike the dichograptids) and instead of crossing over $th1^1$ and the sicula (as in the diplograptids) it grows down the back of the sicula on what corresponds to the obverse side, and the bases of the two monopleural stipes enclose the sicula in front and behind. Hence this type of development has been called **pericalycal**. A left-handed origin of $th1^2$ is not in itself sufficient to produce pericalycal budding and a monopleural rhabdosome, but the few known examples of monopleural astogeny are all left-handed. In *Cryptograptus* $th1^1$ and $th1^2$ are curved distally slightly upward and in fact curve right around the sicula (Fig. 62,*1-3*); but in

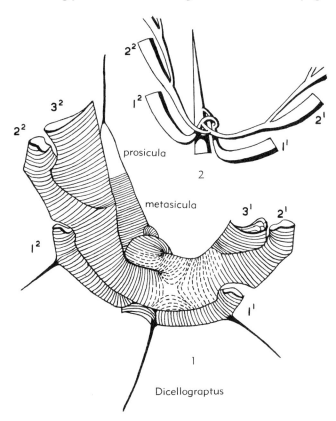

FIG. 55. Diplograptid type of development (illustrated by *Dicellograptus* sp., ×40) (29).——*1*. Reverse aspect showing 3 crossing canals.——*2*. Thecal diagram. [Growth lines where doubtful shown by broken lines.]

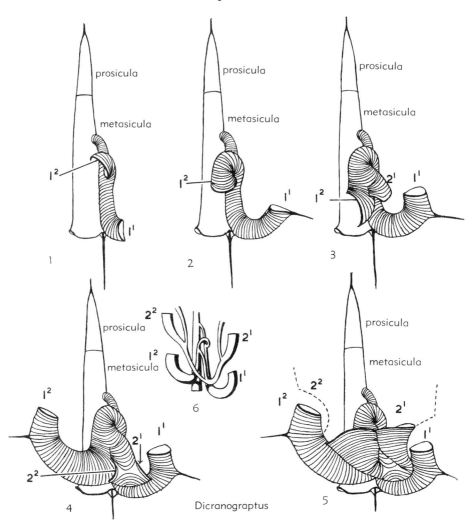

Fig. 56. Diplograptid type of development.——*1-6.* Series of growth stages of *Dicranograptus nicholsoni* illustrating *dentatus* stage, growth lines slightly schematic, ca. ×30 (29).

Glossograptus the first two thecae (in *G. sinicus* perhaps the first four thecae) appear to be straight, opening proximally on either side of the sicula (Fig. 62,4); and in *Skiagraptus* (Fig. 62,5), which has been interpreted on a similar plan, all the thecae are nearly straight.

In contrast to the platycalycal type of astogeny, where collectively the formation of the dicalycal theca is progressively delayed and ultimately eliminated, the evidence suggests that here the dicalycal theca cannot be other than $th1^1$. In this connection the development of *Isograptus manubriatus* is of some interest; the proximal thecae show an unusual curvature reminiscent of a *Glossograptus* and even encroach the sicula to some extent on the obverse side; but the astogeny is clearly based on the isograptid type, with $th1^2$ originating righthandedly and the development does not constitute a true transition between platycalycal and pericalycal types (Fig. 63).

Oncograptus and *Cardiograptus* are biserial forms with dichograptid affinities

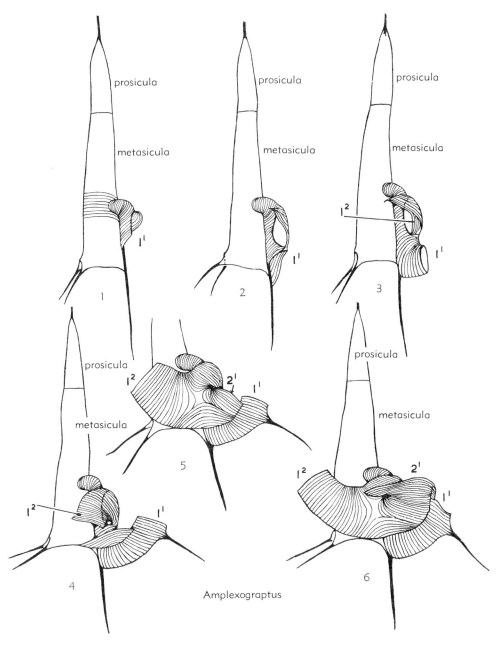

Fig. 57. Diplograptid type of development.——*1-6*. Series of growth stages of *Amplexograptus* sp. *cf. A. maxwelli*, slightly schematic, ×40 (255). [*3* shows disconformity between $th1^2$ and $th1^1$; *5* shows disconformity between $th2^1$ and $th1^2$.]

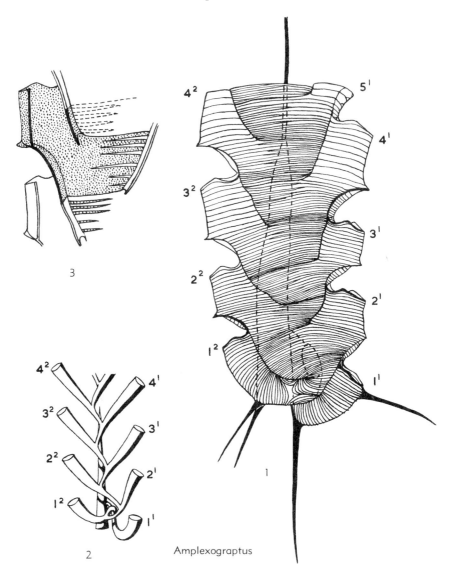

Fig. 58. Diplograptid type of development illustrated by *Amplexograptus* sp. cf. *A. maxwelli*, ×40 (255).——*1.* Proximal end of rhabdosome.——*2.* Thecal diagram.——*3.* Diagram showing growth relations of adjacent thecae and formation of interthecal septum.

and possibly dichograptid or isograptid development (like the scandent *Phyllograptus*), but few details are available.

BRANCHING OF RHABDOSOME

Branching in the graptolite rhabdosome may be either dichotomous or lateral; in the former, the two branches diverge symmetrically, whereas in the latter one branch continues the original direction of growth and the other is thrown off laterally. Lateral branching (e.g., *Trichograptus, Nemagraptus, Pleurograptus*) is less common than dichotomous; the two types may occur in the same rhabdosome (e.g., *Schizograptus*) and may (as in *Goniograptus*) be difficult to distinguish.

Reduction in number of branches is a general tendency in graptolite evolution. Indeed, it is a process carried to completion within the Dichograptidae and almost to completion within dendroid Anisograptidae (*see* also Phylogeny, p. *V*103). NICHOLSON & MARR (1895) suggested that this reduction was to insure a more adequate food supply to the zooids; whether or not this was a factor, symmetry and balance seem to have exerted a controlling influence throughout and a pronounced tendency leads toward regularity in the pattern of the rhabdosome (BULMAN, 1958).

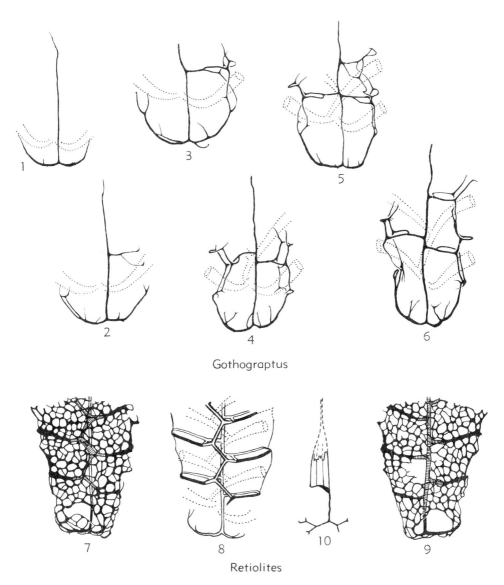

FIG. 59. Retiolitid development.——*1-6*. Series of growth stages of *Gothograptus tenuis* showing (1) ancora stage and (3) corona stage, approx. ×20 (57).——*7-9*. Proximal end of *Retiolites geinitzianus* showing corona, reticula and clathria, ×20 (Bulman, n).——*10*. Prosicula and ancora stage of *R. geinitzianus*, ×20 (Kühne, 1953).

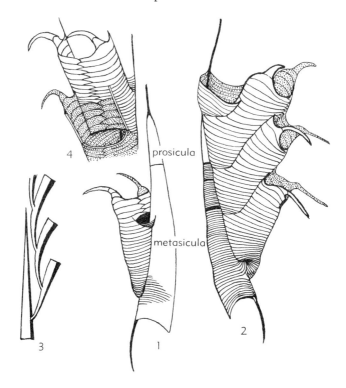

Fig. 60. Monograptid type of development illustrated by *Saetograptus chimaera* (256).——*1,2*. Early growth stages showing annular deposits in metasicula, primary notch and initial bud, and growth relations of successive thecae, ×30.——*3*. Thecal diagram.——*4*. Diagram illustrating formation of interthecal septum.

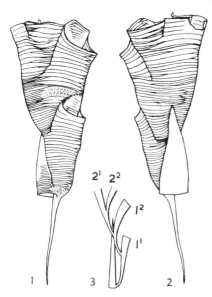

Fig. 61. Proximal end of *Dimorphograptus* sp. ——*1*. Reverse view, ×32.——*2*. Obverse view, ×32.——*3*. Thecal diagram. The order of succession of thecae is clearly shown, but diplograptid homologies in *3*, the thecal diagram, are conjectural (Bulman, n).

The actual process of branch division is very little known. In the Dendroidea, since a branching node consists of two stolothecae and an autotheca, it is due ultimately to the development of two autothecae in place of an autotheca and bitheca (Fig. 13). In the Graptoloidea, since only one type of theca is present, it must be due to the development of a dicalycal theca. *Tetragraptus* sp. cf. *T. bigsbyi* (Fig. 53) is the only graptolite rhabdosome so far known in which growth-line evidence indicates details of the branching division, and here the dichotomy appears as a replica of the isograptid proximal and development; the dicalycal theca constitutes the basal theca of one branch, while the predicalycal theca is carried over into the other at its basal theca, and a sort of crossing canal,

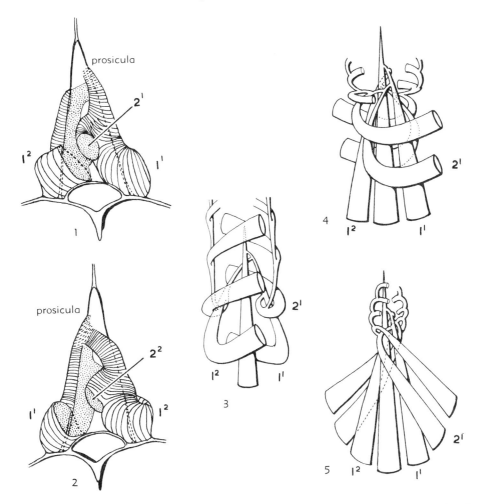

Fig. 62. Pericalycal type of development (23; 4 and 5 based on 261).——*1,2. Cryptograptus tricornis*, "obverse" and "reverse" aspects, ×20.——*3*. Thecal diagram of *C. tricornis*.——*4*. Thecal diagram of *Glossograptus holmi*.——*5*. Thecal diagram of *Skiagraptus* sp.

$th3^1b$ occurs. SKEVINGTON's dichograptid sp. *a* (1965) is closely comparable, as also is JAANUSSON's *Goniograptus* sp. (Fig. 64). What appears superficially here as lateral branching is seen to be dichotomous, and the main stipe is composed of dicalycal thecae alternating with the proximal ends of normal-type thecae which open on the lateral branches (Fig. 64,3). Those biserial graptolites which possess a complete septum may still be regarded as essentially two-stiped and the stipes may (pathologically) develop independently of one another (Fig. 47,3).

Details of lateral branching are virtually unknown. It occurs among Lower Ordovician dichograptids and again among the nemagraptids of the Upper Ordovician. All these examples may prove to be early instances of cladia production (see below) and it has even been suggested that *Janograptus* is a "pseudo-*Didymograptus*" with a sicular cladium.

CLADIA

The development of cladia in various monograptid genera has been investigated

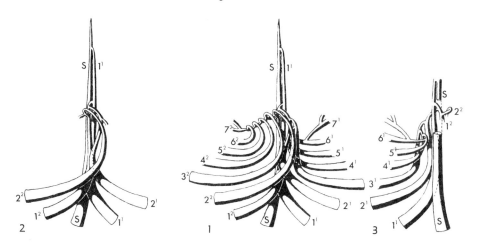

FIG. 63. *Isograptus manubriatus* (35).——*1.* Thecal diagram (reverse view).——*2.* Thecal diagram omitting $th3^1$ and subsequent thecae.——*3.* Thecal diagram (obverse view) omitting distal portions of $th1^2$ and later thecae of that series. [S, sicula.]

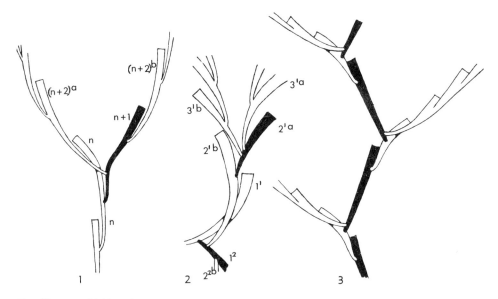

FIG. 64. Graptoloid branching.——*1.* Dichograptid sp. *A* of SKEVINGTON (215).——*2.* *Tetragraptus cf. T. bigsbyi* (Bulman, n).——*3.* *Goniograptus* sp. (104). [Dicalycal thecae shown solid black; n indicates an arbitrary number.]

in recent years by STRACHAN (1952), THORSTEINSSON (1955), TELLER (1962), and especially URBANEK (1963). The suggestion that the "bilateral" *Diversograptus* rhabdosome results from the production of a sicular cladium was confirmed by STRACHAN and the mode of development of thecal cladia in *Cyrtograptus* was described in detail by THORSTEINSSON; later work has centered mainly around *Neodiversograptus* and *Linograptus*.

Stages in the development of thecal

cladia in *Cyrtograptus* are well illustrated by THORSTEINSSON's data for *C. rigidus* var., shown in Figure 65. The first indication is commonly the elongation of one (obverse) of the lateral apertural spines of the mother theca, which is destined to become the pseudovirgula of the cladium. This is followed or accompanied by the appearance of the initial flange, and then the ventral hood, which by their ankylosis produce both the tubular initial portion of the cladium and a secondary aperture to the mother theca, the cladial activity retaining at the same time unrestricted communication with the cavity of the parent theca. The cladium appears, therefore, to be developed from an asexually produced bud on one of the mainstipe thecae.

For any given species, the thecal number of the mother theca (counted from the proximal end) is nearly constant and a more or less constant number of thecae (some three or four) are added to the distal end before the beginnings of cladial generation become manifest. Furthermore, the process of development outlined above occupies a time represented by the formation of several more thecae, so that by the time

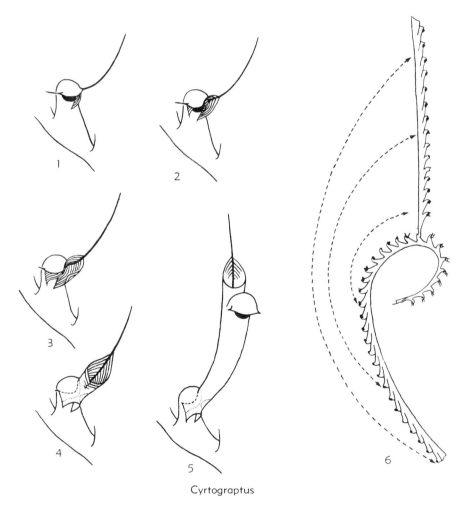

Cyrtograptus

FIG. 65. Cladial generation in *Cyrtograptus rigidus* var. (236).——*1-5*. Successive stages, somewhat schematic.——*6*. Diagram of mature rhabdosome, showing thecal relations of cladium and main stipe, ×2.

the first cladial theca is completed, the position of the mother theca has become seventh or eighth from the distal end. Progressive change in thecal form is usual in cyrtograptids and the characters of the first cladial theca correspond closely to those of the contemporary theca on the main stipe; subsequent growth of the cladium keeps in step with the main stipe and any further changes in thecal characters on the main stipe are paralleled by the cladial thecae (Fig. 65,6). Thus, at any time, thecae of the same size and shape are developing at both the free ends of the rhabdosome. The same principle seems to apply to more complicated cyrtograptids and to those with several "orders."

Sicular cladia are comparable in that each initial thecal tubule is based on an elongate apertural spine which becomes the pseudovirgula of the cladium. Full details are not yet available for *Diversograptus*, but in *Neodiversograptus* the pseudovirgula is either a symmetrical dorsal apertural spine of the sicula (Fig. 66) or an asymmetrical spine produced from one of a pair of rounded lappets; it is never the virgella. Stages are not known in quite the same detail as for sicular cladia. In *Linograptus* (exemplified by *L. posthumus*), four or more sicular cladia are produced; the first and second are based on apertural sicular spines, the third and fourth on spines arising from the adapertural plate (a thickened basal expansion of the preceding cladial tubule). The internal cavities of the first thecae of all cladia communicate with the cavity of the sicula, not with one another, and must have been budded independently and directly from the siculozooid (Fig. 67,5).

Unlike the thecal cladia of *Cyrtograptus*, the sicular cladia of *Linograptus* are produced in rapid succession. The number of thecae on the main stipe (procladium of URBANEK) and first sicular cladium is equally balanced by the time their second thecae are fully developed; the second sicular cladium has originated by *th4-th5* and the third by *th6-th7*. This preserves an approximate balance about the virgellarium as the assumed organ of buoyancy (Fig. 67,1).

In *Diversograptus* a few, and in *Sinodiversograptus* many thecal cladia are pro-

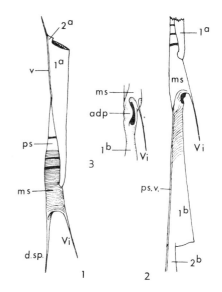

FIG. 66. Production of sicular cladium in *Neodiversograptus beklemischevi*, ×25 (252).——*1*. Sicula and first 2 thecae.——*2*. Initial thecae of cladium (1^b, 2^b).——*3*. Same, ventral view, showing opening of sicula and adapical plate. [1^a, 2^a, thecae of main stipe (procladium); *adp*, adapical plate; 1^b, 2^b, thecae of sicular cladium (metacladium); *d.sp.*, dorsal sicular spine; *ms*, metasicula; *ps*, prosicula; *ps.v.*, pseudovirgula; *v*, virgula; *Vi*, virgella.]

duced in addition to the sicular cladium, but the most extreme complication is represented by *Abiesgraptus* (see Fig. 104). The latter has four principal stipes (three of which appear to be sicular cladia) at right angles, and on one sicular cladium and the "main stipe" (procladium) paired thecal cladia arise at regular intervals.

URBANEK (1963) has interpreted these as representing stages in monograptid astogeny. Some analogy can be inferred between the primary bud of monograptids, which does not emerge from a resorption foramen but arises as an apertural bud (*via* sinus and lacuna stages) and the apertural budding of a cladium, either sicular or thecal, which justifies the use of the term **procladium** for the main stipe. This represents a stable astogenetic phase; elaborations are subsequently introduced by the development of **metacladia** (thecal or sicular), resulting in the series: 1), monograptid stage, with procladium or main stipe only; 2), cyrtograptid stage, with thecal cladia only; 3), diversograptid stage, with

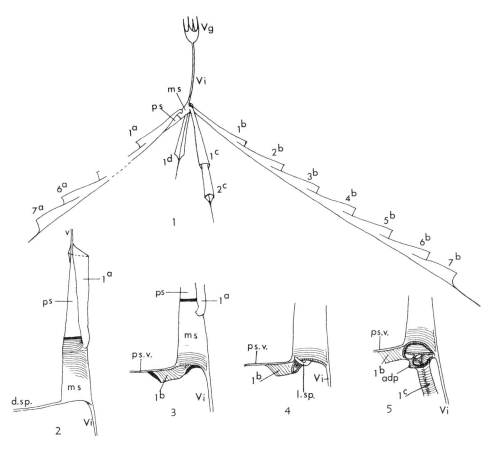

FIG. 67. *Linograptus posthumus* (252).——*1*. Restoration of early stage of development, with 3 sicular cladia.——*2-5*. Stages in development of main stipe (procladium) and first 2 sicular cladia (metacladia). [1^a-7^a, thecae of main stipe; 1^b-7^b, thecae of first sicular cladium; 1^c, 2^c, thecae of second sicular cladium; 1^d, first theca of third sicular cladium; *l.sp.*, lateral spine which becomes pseudovirgula; *Vg*, virgellarium. Other letters as in Fig. 66.]

one sicular cladium; 4), linograptid stage, with numerous sicular cladia; and 5), abiesgraptid stage, with numerous sicular and thecal cladia.

Finally, reference may be made to bipolar monograptid colonies which lack a sicula, investigated by BOUČEK and PŘIBYL (1953), JAEGER (1960) and URBANEK (1963). Various examples have been described, including *Cucullograptus (Lobograptus) scanicus,* but interpretations of such colonies are as yet hypothetical; it seems probable that they represent examples of regeneration involving the formation of **pseudocladia** following serious injury. On this interpretation (illustrated diagrammatically in Fig. 68), the regenerated pseudocladium, removed from the morphogenetic influence of the siculozooid, will exhibit thecal characters of the distal portion of the original stipe; the degree of contrast between the original stipe and the pseudocladium at their junction will depend mainly upon the level at which amputation occurred.

SYNRHABDOSOMES

In 1865, HALL figured a stellate group of *Lasiograptus* [*Retiograptus*] *eucharis* united by their virgulae, and in 1895, RUEDEMANN described a large series of such associations

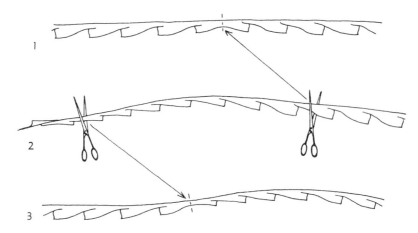

Fig. 68. Astogenetic succession of thecae in normal monograptid colony *(2)* and relations of thecae in pseudocladia regenerated after breakage at 2 different points on the rhabdosome *(1,3)* (252).

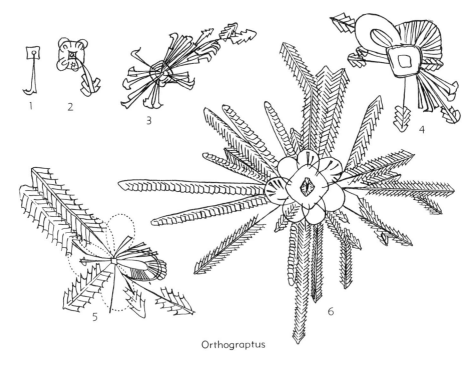

Fig. 69. Development of synrhabdosome in *Orthograptus* sp. (200).——*1-4.* Successive stages, ×3.——*5.* Partially developed synrhabdosome, ×4.——*6.* Fully developed synrhabdosome, ×1.

belonging to various biserial graptolites and attempted to relate them to the life cycle. This paper of Ruedemann's still provides the best and most completely illustrated description of these synrhabdosomes, as he called them; but a new interpretation has lately been suggested by Kozłowski.

The synrhabdosome consists essentially

of a stellate group of rhabdosomes in various stages of development, surrounding a central, almost square disc; and in better-preserved specimens this squarish disc partially overlies and is surrounded by several oval or circular discs which apparently contain bundles of siculae. In addition to this more or less adult arrangement, however, a series of immature stages was discribed: a single sicula or a single rhabdosome attached by its virgula to the squarish disc (Fig. 69,*1*); groups of siculae or very young rhabdosomes so attached (Fig. 69,*2-4*); and finally groups of rhabdosomes in all stages of development attached to the disc, with the subsidiary oval discs and their included siculae (Fig. 69,*6*). RUEDEMANN named the oval discs gonangia, and considered that the central disc was in fact a vesicle or float; but it might have been adhesive (a disc of attachment).

According to KOZŁOWSKI (1949), a clue to the interpretation of these synrhabdosomes is to be sought in the formation of buds on the peduncle (in some on a sterile peduncle) in *Cephalodiscus*. Admittedly these buds, though they may be so numerous as to form a circlet, do not remain attached to the maternal zooid; but in *Cephalodiscus* the organism itself does not form true colonies. The suggestion then is that synrhabdosomes result from a comparable process of budding from the apical portion of the original sicula, the buds here remaining associated. If this is true, the siculae of synrhabdosomes are not strictly comparable with the sexually produced siculae of normal rhabdosomes and should lack a differentiated prosicula portion; this point has not yet been verified. Such asexually produced siculae were termed *pseudosiculae* by KOZŁOWSKI.

Synrhabdosomes are comparatively rare, and are known only in a few species of biserial graptolites.

PALEOECOLOGY

The view that the true graptolites or Graptoloidea were sessile organisms, living erect with the sicula embedded in the mud of the sea bottom, rests upon a total misconception of the nature of the proximal end and long has been abandoned; the current view, that they were floating organisms, was first expressed by HALL (1865) and later developed by LAPWORTH (1897). In his classic paper, LAPWORTH draws a comparison with the Recent Sargasso Sea, picturing the graptolite rhabdosomes attached distally by their nemata to masses of floating weed, the periodic foundering of which supplied both carbonaceous matter and graptolite remains to the slowly accumulating, fine-grained black shales. This theory explains the significance of the nema and accounts for the wide geographical distribution of the graptolites (one of their most distinctive features), for their relation to the enclosing sediments, and to some extent for the lithology of the rocks in which they most commonly occur.

SIGNIFICANCE OF GRAPTOLITIC FACIES

Graptoloidea are not exclusively associated ecologically with any particular type of sedimentary environment, but their remains may occur in almost any kind of sediment, shallow or deep, to which they have sunk or drifted. Nevertheless, their most distinctive association is with black muds devoid of almost any other fossils, and this occurrence constitutes the "graptolitic facies."

These euxinic black shales represent a type of sediment that is not confined to the Lower Paleozoic but occurs at various horizons through the geological column, and the problem of its origin is a general one to which considerable attention now is being given. But of modern black mud conditions catalogued by DUNHAM (1961), few are in any way comparable. While they owe much of their sooty black color to the presence of carbon, such shales also may have a high iron-sulphide content (which imparts a black color) and must have accumulated under anaerobic conditions. Few analyses are available, but as much as 11 percent carbon and 7 percent sulphur has been recorded. Some more recent analyses indicate 7 to 8 percent carbon, but others show only 3 to 3½ percent; even this is some 15 times the amount present in normal shales. Pyrite infilling graptolite rhabdosomes in full relief indicates that the sulphide was syngenetic.

The essential condition is complete lack

of bottom circulation so that dissolved oxygen, soon exhausted, cannot be replenished; while a high proportion of decaying organic matter may be contributed by animal and plant remains falling from the superficial aerated layers. Depth of water in itself has little or no controlling effect. At the present time, comparable conditions

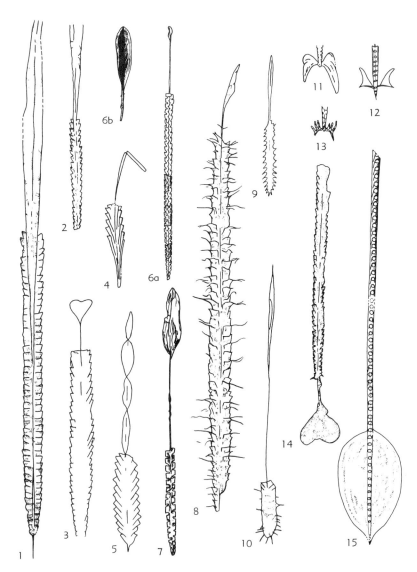

Fig. 70. Distal and proximal structures associated with buoyancy of graptolite rhabdosomes; all figures, except 6b, ×1.5 (34).

1. *Cystograptus vesiculosus.*
2. *Glyptograptus dentatus appendiculatus.*
3. *Diplograptus decoratus.*
4. *Cephalograptus tubulariformis.*
5. *Petalograptus speciosus.*
6a,b. *Pseudoclimacograptus scharenbergi* (6a, rhabdosome; 6b, "float," ×5.5).
7. *Climacograptus parvus.*
8. *Hallograptus mucronatus nobilis.*
9. *Cryptograptus tricornis schaeferi.*
10. *Glossograptus ciliatus.*
11. *Climacograptus papilio.*
12. *C. ensiformis.*
13. *C. venustus.*
14. *C. antiquus bursifer.*
15. *Monograptus pala.*

occur in the Black Sea (HUNDT, 1938), but also in narrow landlocked embayments such as the Norwegian fjords (STØRMER, 1938; STRØM, 1936) and even in coastal lagoons (SCUPIN, 1921; GRABAU, 1929).

It is certainly probable that such embayments and deltaic lagoons existed in the past, and reasonable to assume that graptolite remains might have drifted in from the open seas and accumulated there; but the extensive literature describing parallel grouping and current orientation of rhabdosomes (HUNDT, 1933-38; KLÄHN, 1930; KRAFT, 1926) and even ripple marks, rain prints and sun cracks in graptolite shales (FREBOLD, 1928; ÖPIK, 1929) probably gives an exaggerated impression of their importance. More or less uniform, haphazard distribution of graptolites in shales occurring over large areas is considered far more characteristic by RUEDEMANN (1935), SCHMIDT (1935), STRØM (1936), RAYMOND (1942), and many others. That suitable conditions might develop in large seas and even oceans, especially in a warm or temperate climate and with submarine barriers to restrict bottom circulation, is admitted by STRØM, and in at least one instance (WILLIAMS & BULMAN, 1931) the existence of such barriers is considered probable on wholly independent grounds.

It is true, however, that graptolite remains often are confined to thin layers separated by considerably greater thicknesses of unfossiliferous strata generally of different lithology. Many of these successions are now recognized to be turbidites, which introduces the possibility of current transport of rhabdosomes as well as or instead of gravity settling.

RUEDEMANN (1925b) has pointed out also that another class of graptolite shale exists, representing conditions which, though unfavorable, were by no means lethal to bottom life. Such shales as the Utica (in contrast to the Hartfell or Normanskill) contain some benthonic organisms and a variety of animals other than graptolites. Even these, however, contrast with nongraptolitic (e.g., Lorraine) shales in the nature of other organisms, seaweeds and cephalopods with small arthropods and horny brachiopods predominating in the one, but polyzoans, brachiopods, and benthonic mollusks in the other. Such "mixed" shales probably represent depositions in quiet waters beyond the littoral zone, where muds probably were carried out by strong undertow.

It is likely enough that no general explanation will account for all occurrences of these types of graptolite shale, however, but that each deposit must be assessed individually on its lithological and faunal characteristics and in relation to its general stratigraphical background. But it is a safe generalization that typical graptolite shales represent conditions more or less inimical to bottom life, and that graptolites themselves owe their distinctively wide geographical distribution to their superficial drifting mode of life.

BUOYANCY MECHANISM

Until recently, it has been assumed that graptolites were epiplanktonic organisms, living attached by their nemata (or virgulae) to floating seaweed comparable in magnitude and buoyancy to the modern *Sargassum*. On this analogy, the almost universal distribution of graptolites would necessitate an almost universal spread of *Sargassum*-type weed throughout the Lower Paleozoic. In several graptolites, however, no nema has even been recorded (e.g., *Dicranograptus, Phyllograptus*), at least in the adult, and in other examples the size and weight of the rhabdosome or length of the nema which would be required (e.g., most *Dicellograptus* species) or the mechanics of nematic support in relation to center of gravity and other factors (e.g., *Monograptus turriculatus,* see Fig. 72,3, *Cyrtograptus*) makes an epiplanktonic mode of life somewhat dubious (BULMAN, 1964). It is recognized, moreover, that in a wide range of biserial graptolites the distal prolongation of the virgula into a so-called "float" suggests that a truly planktonic mode of life was quite common.

The existence of living tissue external to the skeleton, which has to be postulated on other grounds, may have played some part in the buoyancy of graptolite rhabdosomes, more plausibly, perhaps, through the occurrence of gas bubbles in this tissue than a development of fat bodies. In all cases, the so-called floats, originally believed to be bladderlike, are now known to consist of two, or more usually three, vanes attached to the distal end of the nema, but these may well have supported vesicular tissue (Fig. 70) and comparable masses may have been related to the web structures of certain dichograptids (Fig. 71). The well-developed tufts of scleroproteic fibers commonly interpreted as "root fibers" in such forms as *Dictyonema flabelliforme* might equally well be related to an aggregation of buoyant tissue. The occurrence of thin-walled plates or vanes, commonly associated with the virgella or with prominent apertural spines at the proximal end of a rhabdosome (Fig. 70,*11-15*), suggests that, unless they acted as stabilizers, such species had a reversed orientation and floated "upside down" (see also the virgellarium of linograptids, Fig. 67,*1*).

In other instances, the peculiar configuration of the rhabdosome (Fig. 72) suggests possible gyratory movement in response to slight eddies in the water which could have been a significant factor in flotation.

Convincing attachment discs have been figured by RUEDEMANN, mainly in immature rhabdosomes (Fig. 73), and bearing in mind that immature dicranograptids (in the early biserial stage) may possess a short nema, it does seem possible that epiplanktonic attachment in juvenile stages may have been quite frequent. Such minute rhabdosomes, however, would not have required masses of buoyant weed comparable to *Sargassum,* but could have availed themselves of much smaller and more widely distributed planktonic algae. In synrhabdosomes, likewise, it is the juvenile stages, comprising numerous siculae and very early growth stages, which show the most convincing discs of attachment

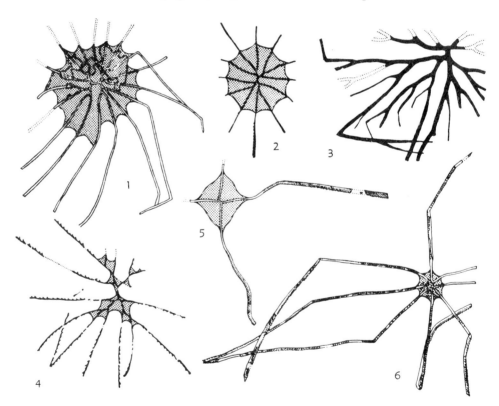

FIG. 71. Proximal web structures in various graptolites (34).——*1. Loganograptus logani,* ×0.75.——*2. L. kjerulfi,* ×0.5.——*3. Clonograptus callavei* with flanges extending along sides of stipes, ×1.——*4. Goniograptus palmatus,* ×0.75.——*5. Tetragraptus headi,* ×0.33.——*6. Dichograptus octobrachiatus,* ×0.33.

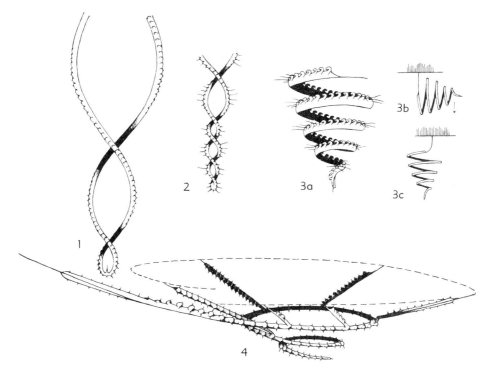

FIG. 72. Rhabdosome forms suggestive of gyratory motion (34).——*1. Dicellograptus caduceus,* ×2.——*2. Dicranograptus furcatus bispiralis,* ×2.——*3. Monograptus turriculatus; 3a,* rhabdosome, ×3.5.——*3b,c,* diagrams to show how attachment would distort nema *(3c)* so that growing rhabdosome could not attain regular helical form, unless oriented in defiance of gravity *(3b)*, diagram.——*4.* Immature *Cyrtograptus* rhabdosome, based on *C. solaris,* ×3.

(Fig. 69,*1-4*); most adult examples show nothing centrally other than a tangle of slender fibers and it is unlikely that the synrhabdosome association constitutes primarily a buoyancy mechanism.

GEOGRAPHIC DISTRIBUTION

Since the distribution of the Graptoloidea during life was essentially dependent upon current drifting, the distribution of their fossil remains may be practically worldwide, almost coextensive with that of the rocks of a particular age. Nearly all families, the majority of genera, and some species are almost cosmopolitan; *Nemagraptus gracilis* and *Monograptus turriculatus* are examples of such species. It does not follow, however, that all graptolites are more or less universal in their occurrence and to judge by present records, many have a decidedly restricted distribution.

Partly, no doubt, this may be attributed to imperfect collecting and recording; but evidence of the existence of faunal provinces is found and on a much smaller scale, of geographical races, though the erratic distribution of some graptolites is not read-

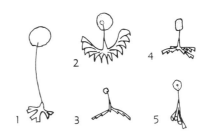

FIG. 73. Rhabdosome suspension; discs of attachment in immature rhabdosomes; all figures ×2 (201).——*1. Dictyonema flabelliforme.*——*2. Tetragraptus similis.*——*3. Adelograptus lapworthi.*——*4. Staurograptus dichotomous.*——*5. Tetragraptus fruticosus.*

ily accounted for in any of these ways. The genus *Rastrites* remains virtually unknown in North America. *Pleurograptus*, again, is peculiarly localized; it occurs characteristically in Scotland but not in England and Wales, or indeed elsewhere in Europe, though it has been recorded from Australia (Victoria) and rather doubtfully from North America. On the other hand, *Pterograptus* is a somewhat rare genus with a wide distribution; species are known from North and South America, northwest Europe, China, and Australasia.

Clear evidence of the existence of faunal provinces among the Graptoloidea can be cited, though they remain to be more precisely defined, a process complicated by a mass of misidentifications among existing records. Two examples are here indicated broadly. First, the *Oncograptus-Cardiograptus* fauna, originally described from Victoria, Australia, and now recognized to some extent in North America, Texas, and China, does not reach eastern North America or Europe, apart from a few specimens of *Oncograptus* recorded from western Ireland, whereas the late pendent didymograptid fauna so characteristic of the Llanvirn of Europe and South America is unrecorded in the contemporaneous beds of Australia and New Zealand. Second, on a somewhat smaller scale, the Tremadoc *Anisograptus-Triograptus* fauna, originally described from eastern Canada and later recognized in Taimyr, Norway, and South America, contrasts with the contemporary *Clonograptus-Adelograptus* fauna of Sweden and Britain, now also known from the Sahara. The lower Tremadoc fauna of Australasia (and China) has a superficially different composition and perhaps even origin; considerable doubt attaches to Asiatic records of *Dictyonema flabelliforme* and probably to the records of *Clonograptus tenellus* from Australasia, while the reported New Zealand *Triograptus* is now known to be a misidentification.

The distribution of various distinctive species may provide clues to marine connections between various regions in the Early Paleozoic. Certain Australian species are now being recorded from northwestern Europe, and reference may be made to *Aulograptus* [formerly *Didymograptus*] *climacograptoides,* which was originally described from South America (where it occurs in Peru, Bolivia and Argentina) and which is now known from the English Lake District, southern Sweden, and Belgium. Again, distinctively Bohemian monograptid species are now being recognized in British Silurian assemblages.

At a low taxonomic level, the various subspecies of *Glyptograptus austrodentatus* and other species described from different countries suggest geographical variation, and most graptolithologists will agree that local differences are often detectable in a widely distributed species.

STRATIGRAPHIC DISTRIBUTION

The transition from Dendroidea to Graptoloidea involves the loss of bithecae and a consequent simplification of branch structure, but the process appears to have been a gradual one in several respects. It has probably occurred independently in more than one line of descent, and instances are known where bithecae are not uniformly present (i.e., they are present distally but absent proximally in *Kiaerograptus*). The relatively poor preservation of many of these early graptolites adds to uncertainty as to their nature; thus, it is not known whether early Arenig species of *Clonograptus* and *Bryograptus* possess bithecae. For these reasons, it is not yet possible to indicate the earliest graptoloid with any precision.

Several species of *Didymograptus* and some of *Tetragraptus* have been described from rocks of Tremadoc age in various parts of the world, but correctness of the graptoloid attribution of at least some of these records is quite doubtful.

Undoubtedly *Didymograptus* species occur in the basal Arenig associated in most areas with such multiramous forms as *Clonograptus* and *Bryograptus* which may or may not possess graptoloid branch structure. From such beginnings, the Graptoloidea quickly established their position as one of the most important components of marine fauna of the lower Paleozoic. On present records, they became extinct in the early Devonian (Siegenian or perhaps as late as Emsian).

The range of individual genera (Fig. 74) is usually short, few extending through more than part of a single geological period; that of individual species is variable, some being confined to a single zone, others extending through five or six zones. The zones themselves represent variable time spans. Radioactivity figures suggest that the Silurian Period endured some 40 million years, and in Britain this embraces some 20 graptolite zones; in Central and Eastern Europe, another eight or 10 have been claimed for the upper Ludlow. On this basis the duration of a single zone (or the length of life of a short-ranged species) would be somewhat less than two million years. The time value of other zones may be even shorter; in Australia HARRIS & THOMAS recognized 11 zones in the series La3 to Ya2, approximately equivalent to the English Arenig and presumably representing not more than 10 or 12 million years.

In most parts of the world, the shaly facies of the lower Paleozoic has now been zoned by means of graptolites, which are of exceptional value for long-range correlation; but while a general similarity in the succession obtains, local differences occur and it would be out of place here to attempt a world-wide correlation of graptolite zones. It is possible, however, to indicate a sequence of graptolite faunas capable of fairly general application. Faunal provinces are discernible in the earlier portion of the sequence, while the upper portion is more uniform and cosmopolitan. It should be emphasized that the terms employed are descriptive of general aspect and dominant composition of a fauna rather than definitive of its precise upper and lower limits. The full stratigraphical range of the families Dichograptidae and Diplograptidae, for example, is not coincident with the upper and lower boundaries of the faunas bearing their name.

Tabulations of the generally recognized British and Australian graptolite zones, representing the European and Pacific provinces in the Lower Ordovician, are given at the end of this section on "Stratigraphic Distribution."

ANISOGRAPTID FAUNA

The anisograptid fauna characterizes Tremadoc beds and their equivalents. Strictly it is not a graptoloid fauna at all, since it comprises various pelagic species of *Dictyonema* (such as *D. flabelliforme*) and their pendent and horizontal anisograptid descendants; the extent of any graptoloid (dichograptid) component is at present indefinite.

In northwestern Europe, North Africa, eastern North America and South America, *Dictyonema flabelliforme* and its subspecies are widespread, though in Quebec *D. canadense* and other species occur and *D. flabelliforme* has not yet been recorded. Probably no great disparity in age marks these species, since the associated anisograptid fauna is closely related to that succeeding the *D. flabelliforme* Zone of Norway and South America. The *Clonograptus-Adelograptus* fauna and the *Anisograptus-Triograptus* fauna, both of which succeed the *D. flabelliforme* Zone, if not mutually exclusive, dominate particular regions, the former northwestern Europe (except Norway) and North Africa, the latter eastern North America and South America. It is not yet decided whether the *Anisograptus-Triograptus* occurrence associated with *Clonograptus* in Texas is lower or upper Tremadoc.

Dictyonema flabelliforme (or varieties ascribed to it) have been described from China and Korea, but the determinations and hence the inferred horizons are questionable and some could even be Arenig species. In Australia, a staurograptid and two small siculate species *(D. scitulum* and *D. campanulatum)* presumably represent the lower Tremadoc fauna.

Upper Tremadoc graptolites are extremely rare. The best-known fauna is that described by MONSEN (1925) from the *Ceratopyge* Shale near Oslo, comprising *Kiaerograptus* and *Didymograptus?* with fragmentary *Clonograptus* and *Bryograptus;* and a similar fauna immediately underlies the *Ceratopyge* Limestone. The La2 fauna from Victoria and New Zealand is believed to be upper Tremadoc. The presence of any significant graptoloid element in even upper Tremadoc begins to appear unlikely.

DICHOGRAPTID FAUNA

The dichograptid fauna as now defined

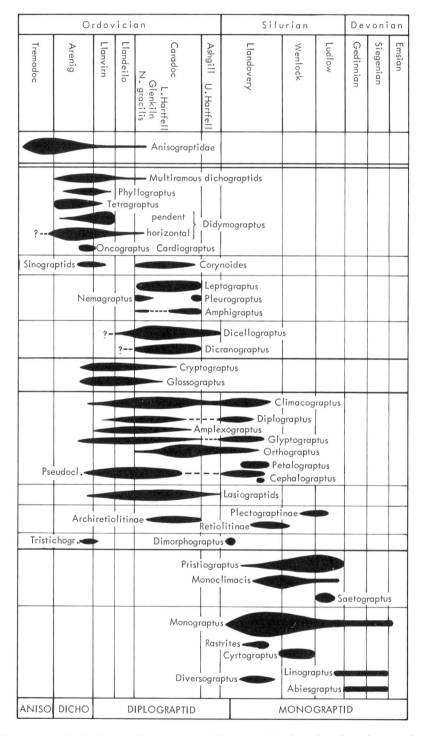

Fig. 74. Stratigraphic distribution of certain graptoloid genera. Number of species only approximately indicated; no attempt made to show relative importance of genera (Bulman, n). [Pseudocl.=*Pseudoclimacograptus;* Tristichogr.=*Tristichograptus.*]

characterizes Arenig beds and their equivalents and comprises a wide range of multiramous and pauciramous dichograptids.

In most countries, the base is marked by appearance of *Tetragraptus approximatus* associated with *Clonograptus* and with extensiform and pendent didymograptids and tetragraptids. Declined didymograptids and phyllograptids are well represented in the middle portion and tuning-fork didymograptids begin to be prominent in the upper portion, where also *Oncograptus, Cardiograptus*, and isograptids are distinctive, particularly in areas around the Pacific. Biserial graptolites (both monopleural and dipleural) make their appearance in the uppermost portion, but are not a numerically significant constituent of this fauna.

DIPLOGRAPTID FAUNA

Biserial graptolites became a distinctive element of the fauna by Llanvirn times and although in some areas they tended to be obscured by an extraordinary profusion of tuning-fork graptolites, it is at or just below this level that they underwent their major generic differentiation. They characterize the whole of the Ordovician graptolite fauna and are the only graptolites found in the basal part of the Silurian until the advent of *Monograptus*[1] and the dimorphograptids in the *Cystograptus vesiculosus* zone. The suggested subdivisions of this large time span are provisional, but are descriptive of general faunal characteristics of most areas. The boundary between the second and third subdivisions is indistinct, and zonal correlation at about this level (between the *Nemagraptus gracilis* and *Dicranograptus clingani* zones) is notoriously difficult.

GLYPTOGRAPTUS-AMPLEXOGRAPTUS SUBFAUNA, LLANVIRN AND LLANDEILO

The two genera named in the title are characteristic, but *Diplograptus, Hallograptus, Pseudoclimacograptus* and *Climacograptus* also occur, along with *Cryptograptus, Glossograptus*, and *Tristichograptus*. In addition to tuning-fork didymograptids and isograptids, late dichograptids include a number of multiramous genera, of which *Pterograptus* is particularly distinctive. The earliest *Dicellograptus* occurs in the *Glyptograptus teretiusculus* zone.

NEMAGRAPTUS-DICELLOGRAPTUS SUBFAUNA, BASAL CARADOC

The incoming of *Nemagraptus gracilis* produces an easily recognizable base to this subfauna, and *Leptograptus, Dicellograptus* and *Dicranograptus* rapidly assume prominence. Diplograptids are abundant. The last stragglers of the Dichograptidae (*Didymograptus superstes*, etc.) persisted into the basal levels.

ORTHOGRAPTUS-DICELLOGRAPTUS SUBFAUNA, CARADOC AND ASHGILL

Various species of *Orthograptus*, especially the *O. truncatus* and *O. calcaratus* groups, are the dominant diplograptids in most parts of the world, beginning at a level somewhat below the *Dicranograptus clingani* zone. *Leptograptus* and *Dicranograptus* disappear below the top of the subfauna, but *Dicellograptus* persists in association with *Orthograptus* and *Climacograptus* to the end of the Ordovician. The fauna of the upper portion, like that of the succeeding subfauna, has some of the characters of an impoverished fauna.

ORTHOGRAPTUS-CLIMACOGRAPTUS SUBFAUNA, BASAL SILURIAN

This subfauna is linked to those above and below, with both of which it has several species in common, but it is composed of biserial graptolites (dipleural); *Monograptus*[2] and the dimorphograptids have not yet appeared. Dwarfed forms of *Orthograptus* of the *O. truncatus* group persist from the underlying levels. Species referred to *Glyptograptus* and *Diplograptus* are present, though of a somewhat different aspect from the Ordovician forms and for this reason were not utilized in the subfaunal title. The most widespread species are *Climacograptus* commonly assigned to the *C. scalaris* group, and *Cephalograptus* [?=*Akidograptus*] *acuminatus* characterizes the upper portion.

MONOGRAPTID FAUNA

Like the diplograptid fauna, the monograptid fauna represents a large strati-

---
[1] See p. *V*100.

[2] See footnote on p. *V*100.

graphic and time interval, from the incoming of *Monograptus* slightly above the base of the Silurian[1] to its extinction in Early Devonian times; the latest representatives seem at present to be *M. yukonensis, M. atops* (in Bohemia) or *M. thomasi* (in Australia), possibly Emsian in age. In terms of the graptolite succession, the Silurian-Devonian boundary is now taken to lie at the base of the *M. uniformis* zone.

The Llandovery succession begins with a subfauna of dominantly simple thecal type, though it is becoming evident that many apparently simple forms reveal unexpected apertural modifications when adequately preserved. Dimorphograptids occur in association with the monograptids but do not appear to antedate the earliest ones. Monograptids with triangulate thecae give a distinctive aspect to the middle part of the Llandovery, and hooked and lobate forms to the upper part. In addition to these typical monograptid rhabdosomes (=procladia of URBANEK), the diversograptid stage of astogenetic development is represented by *Diversograptus* as recorded from all three of these subfaunas.

The Wenlock is distinguished particularly by *Cyrtograptus,* with its conspicuous thecal metacladia, occurring in association with hooked monograptids.

The Ludlovian has yielded a diversity of monograptid forms about which is difficult to generalize. Forms with simple thecae are abundant, but others with quite extreme apertural modifications occur and include examples of lateral asymmetry (e.g., *Cucullograptus*), very rare among graptolites. The upper Ludlow fauna also includes species with modified apertures in addition to pristiograptids and it is at these high levels that greatest astogenetic complexity is reached in such forms as *Abiesgraptus*.

The Lower Devonian as presently known yields a sparse but widely distributed graptolite fauna which includes *Monograptus uniformis, M. hercynicus, M. thomasi,* and *M. yukonensis*. Most species possess hooked proximal thecae, passing distally into thecae with straight supragenicular walls and hoodlike structures overhanging the apertures. *Abiesgraptus* and *Neodiversograptus* occur in the lower part of this succession.

Almost to the last, therefore, the graptolites appear as an actively evolving group of organisms and no satisfactory explanation of their extinction has been suggested. Conditions obtaining in a local geosyncline may be quite irrelevant to the problem of graptolite extinction, especially if the main centers of evolution and distribution of these organisms were oceanic (e.g., Pacific). Biological as well as physical factors may be involved, acting through food supply or the appearance of a more efficient group of predators, but the cause of graptolite disappearance remains entirely speculative (JAEGER, 1959).

BRITISH AND AUSTRALIAN GRAPTOLITE ZONES

Tabulations of the graptolite zones thus far distinguished and generally recognized in Britain and Australia are introduced here. Graptolites are unrepresented in Devonian and upper Ludlow rocks of Britain, but occur elsewhere in Europe, mainly in Thuringia, Czechoslovakia and Poland. A standard zonal succession has not yet been agreed, but the following tentative scheme gives some indication of what is missing at the top of the sequence in Britain.

Devonian graptolites are represented in Australia by *Monograptus aequabilis* and *M. thomasi,* assigned to the Gedinnian (and perhaps early Siegenian) by JAEGER (1966). A large number of graptolites, including many well-known European species, attest the presence of lower Llandovery, Wenlock and lower Ludlow strata, but these have not yet been formally zoned.

Unpublished work by R. A. COOPER suggests that in New Zealand it may not be possible to distinguish all the finer subdivisions of the Bendigonian and Chewtonian.

PRINCIPLES OF CLASSIFICATION

At generic level, the Graptoloidea present a somewhat confused picture of taxa based on a variety of criteria. A century ago, the now obsolete catch-all genus *Graptolithus* began to be subdivided on

---

[1] Since writing this, a *Monograptus* species has been discovered in the basal Silurian zone (Rickards & Hutt, 1970).

Graptolite Zones Distinguished in Britain and Other European Countries

Series or Stage	Zone
	Devonian
Siegenian	{ *Monograptus hercynicus* *Monograptus praehercynicus*
Gedinnian	*Monograptus uniformis*
	Silurian
Upper Ludlow	{ *Monograptus angustidens* *Pristiograptus transgrediens* *Monograptus perneri* *Monograptus bouceki* *Saetograptus lochkovensis* *Pristiograptus ultimus* *Pristiograptus fecundus* *Saetograptus fritschi linearis*
Lower Ludlow	{ *Saetograptus leintwardinensis* *Pristiograptus tumescens* *Cucullograptus (Lobograptus) scanicus* *Neodiversograptus nilssoni*
Wenlock	{ *Pristiograptus ludensis* (=*Monograptus vulgaris*) *Cyrtograptus lundgreni* *Cyrtograptus ellesae* *Cyrtograptus linnarssoni* *Cyrtograptus rigidus* *Monograptus riccartonensis* *Cyrtograptus murchisoni* *Cyrtograptus centrifugus*
Llandovery	{ *Monoclimacis crenulata* *Monoclimacis griestoniensis* *Monograptus crispus* *Monograptus turriculatus* *Rastrites maximus* *Monograptus sedgwicki* *Monograptus convolutus* *Monograptus gregarius* { *Monograptus leptotheca* / *Diplograptus magnus* / *Monograptus triangulatus* *Monograptus cyphus* (lower part sometimes distinguished as *Monograptus acinaces*) *Cystograptus vesiculosus* (=*Monograptus atavus*) "*Akidograptus*" *acuminatus* *Glyptograptus persculptus*
	Ordovician
Ashgill	{ *Dicellograptus anceps* *Dicellograptus complanatus*
Caradoc	{ *Pleurograptus linearis* *Dicranograptus clingani* *Diplograptus multidens* & *Climacograptus peltifer* *Nemagraptus gracilis*
Llandeilo	*Glyptograptus teretiusculus*
Llanvirn	{ *Didymograptus murchisoni* *Didymograptus bifidus*
Arenig	{ *Didymograptus hirundo* *Didymogratus extensus* { *Isograptus gibberulus* / *Didymograptus nitidus* / *Didymograptus deflexus* ....
Tremadoc	{ .... *Clonograptus tenellus* & *Bryograptus hunnebergensis* *Dictyonema flabelliforme* { *Dictyonema flabelliforme flabelliforme* / *Dictyonema flabelliforme sociale*

Graptolite Zones Distinguished in Australia

Stage	Zone	
ORDOVICIAN		
Bolindian	*Dicellograptus complanatus*	
	Pleurograptus linearis	
Eastonian	*Dicranograptus hians*	
	Climacograptus baragwanathi	
Gisbornian	*Climacograptus peltifer* & *Diplograptus multidens*	
	Nemagraptus gracilis	
Darriwilian	*Glyptograptus teretiusculus*	
	Diplograptus decoratus	
	Glyptograptus intersitus	
	Glyptograptus austrodentatus	
Yapeenian	Ya 2	*Oncograptus* & *Cardiograptus*
	Ya 1	*Oncograptus*
Castlemainian	Ca 3	*Isograptus caduceus maximodivergens*
	Ca 2	*Isograptus caduceus victoriae*
	Ca 1	*Isograptus caduceus lunata*
Chewtonian	Ch 3	*Didymograptus balticus*
	Ch 2	*Didymograptus protobifidus*
	Ch 1	*Didymograptus protobifidus* & *Tetragraptus fruticosus*
Bendigonian	Be 4	*Tetragraptus fruticosus* (3-br)
	Be 3	*Tetragraptus fruticosus* (3-br and 4-br)
	Be 2	*Tetragraptus fruticosus* (4-br)
	Be 1	*Tetragraptus fruticosus* & *Tetragraptus approximatus*
Lancefieldian	La 3	*Tetragraptus approximatus*
	La 2	*Bryograptus* & *Clonograptus*
	La 1	*Staurograptus* & *Dictyonema*

the basis of easily recognizable features of gross morphology such as distinctive general form *(Phyllograptus, Dicranograptus)*, number of branches *(Tetragraptus, Dichograptus, Monograptus)* or biseriality *(Diplograptus)*; occasionally some more minute feature, such as reticulate periderm *(Retiolites)* was utilized, but only rarely were thecal characters employed *(Rastrites, Climacograptus)*. At the other extreme are various recently described genera and subgenera *(Cucullograptus, Lobograptus)* defined on an accurate knowledge of the details of thecal form, and many of these at least approximate to phyletic entities. Between the extremes lies a whole range of genera based on somewhat more refined rhabdosomal characters, or on rather less exact and more contentious thecal characters.

Most of the characters which determine gross morphology of the rhabdosome seem to result from parallel evolution; often they represent the grades of biological improvement which constitute **anagenesis** (HUXLEY, 1958). In consequence, a high proportion of graptolite genera are polyphyletic. Thecal characters, which are largely used in specific diagnosis, are believed to represent **cladogenesis** and to provide a more reliable clue to genetic affinity; when such characters are used for generic diagnosis, they may define something approaching "natural genera."

A "natural classification" (i.e., a purely phyletic system) would classify products of cladogenesis (or genetic divergence) and ignore those of anagenesis (or grades of general biological improvement). Thus it would unite in a single taxon genetically related species of *Didymograptus, Tetragraptus*, and multiramous dichograptids (and ultimately of anisograptid dendroids as well as diplograptids and monograptids) while ignoring the existence of taxa named *Didymograptus, Tetragraptus*, and the like. It would trace a phyletic line using thecal similarity as its principal guide; but it would fail to recognize certain conspicuous features and stages, even though these have an obvious practical value. Any workably useful classification is a compromise and as concerns the Graptoloidea such compromise is determined by recognizing small, approximately phyletic units (genera) within larger grades of long-established "form genera."

However, the statement that thecal characters provide a reliable clue to genetic affinity requires qualification, for some evidence now shows that even "thecally based" genera are not necessarily monophyletic. The process of thecal differentiation within the grade represented by *Diplograptus (s. lat.)* appears to be essentially cladogenetic, not anagenetic, and the rapid diversification of the ancestral biserial forms with glyptograptid thecae results in the appearance of *Climacograptus, Diplograptus (s. str.), Amplexograptus,* and *Orthograptus*. But these genera seem also to include later gradations from one to another and it is possible that such transitions may occur in both directions. Thus the original transition from *Glyptograptus* to *Climacograptus* which occurs at the top of the Arenig may be followed by another in the lower Caradoc (*G. siccatus* to *C. brevis*) and yet another may be found of more uncertain direction in the Lower Silurian.

Ultimately, no doubt, such genera as *Climacograptus* will be subdivided into smaller and more "natural" units; already *Pseudoclimacograptus* has been discriminated, first as a subgenus and now as a genus divided into three subgenera. But the validity of such subdivision depends upon the accuracy of morphological diagnosis and any partitions based on full growth-line evidence of thecal structure and ontogeny are rare. All too often adequately detailed information is lacking and taxa defined without this information may be actually misleading. For this reason, many "technically valid' genera are not accepted here. It may also be noted that rare preservation of structural details may suggest desirable bases of classification which nevertheless cannot be applied to normally occurring and hence imperfectly preserved material. EISENACK's (1951) work on retiolitids illustrates this dilemma.

Finally, reference must be made to the taxonomic implications of penetrance in the introduction of new thecal types and the consequent occurrence of "bi-form" rhabdosomes combining different thecal types in the same colony (see also p. V66). These "penetrance intermediates" are of more importance than "expressivity intermediates" in the monograptids and it is regrettable that the detailed work now beginning along these lines no longer retains an undivided genus *Monograptus* as its *basis operandi*. Instead, a score or so of generic and subgeneric names are recognized, some founded on mere silhouette preservation and many so broadly conceived as to constitute form genera themselves. (See Addendum, p. V149.)

PHYLOGENY

The order Graptoloidea is divisible into four suborders, Didymograptina, Glossograptina, Diplograptina and Monograptina, which reflect three main events in graptoloid evolution, namely, origin of the Didymograptina from the order Dendroidea, development of scandent biserial rhabdosomes (monopleural and dipleural), and development of scandent uniserial rhabdosomes (Monograptina). As described below, the origin of the Graptoloidea results from a change which appears to be gradual and transitional forms occur. The other two changes are abrupt (mega-revolutionary) and no intermediates are recognized; they express major changes in rhabdosome construction resulting from the orientation and relationships of the earliest proximal thecae. The Didymograptina, Diplograptina, and Monograptina constitute a phyletic sequence, whereas the Glossograptina represent a relatively shortlived offshoot from the Didymograptina which died out without descendants.

DIDYMOGRAPTINA

DICHOGRAPTIDAE

The earliest of all graptoloid families is the Dichograptidae and its origin is synonymous with that of the Graptoloidea. It involves essentially the loss of bithecae and of sclerotized stolons (with consequent simplification of branch structure) and it follows upon the adoption of a pelagic mode of life. The change was gradual and the Anisograptidae is a family so completely transitional between the Graptoloidea and Dendroidea that it could be included in either, for the mode of life appears to have been graptoloid, whereas the branch structure remained characteristically dendroid. Bithecae have not yet been demonstrated in Arenigian species of *Clonograptus* and *Bryograptus,* though present in Tremadocian species of the same genera; but the inclusion of some forms with bithecae in a dendroid family (and, provisionally at least, in bitheca-bearing genera) involves no greater inconsistency than the converse. One effect of classing the Anisograptidae with the Dendroidea is to emphasize the polyphyletic origin of the Graptoloidea—and of the Dichograptidae; but this could only be avoided, if at all, by extending the scope of the Dichograptidae to include *Dictyonema flabelliforme (s. lat.).*

It was through their work on certain dichograptids that NICHOLSON & MARR (1895) first recognized the possibility that graptolite genera might be polyphyletic.[1] To these authors, the number of branches in a graptolite rhabdosome (though forming the basis of so many earlier generic definitions) was a feature of less importance than thecal characters, and using the criterion of thecal similarity and (to a less extent) angle of divergence, they recognized nine groups of *Didymograptus, Tetragraptus, Bryograptus,* and *Dichograptus* species as establishing the principle of stipe reduction. They stated: "It is comparatively easy to explain the more or less simultaneous existence of forms possessing the same number of stipes, but otherwise only distantly related, if we imagine them to be the result of variation of a number of different ancestral types along similar lines" (p. 537).

The principle of progressive stipe reduction thus propounded, with the generic series extended to become *Clonograptus-Loganograptus-Dichograptus-Tetragraptus - Didymograptus* (for horizontal forms), dominated discussions of dichograptid phylogeny for 40 years. Some difficulties were indicated by DIXON (1931) and it was more specifically challenged by HARRIS & THOMAS (1940a), who questioned the rigid application of this one criterion and advocated the exclusion of *Loganograptus* from any such series. It has been subsequently emphasized that no stratigraphic evidence supports strict chronological sequence of the genera involved in this series; indeed, species of *Tetragraptus* and *Didymograptus* appear to be the earliest true dichograptids. HARRIS & THOMAS indicated the probability that *Tetragraptus* and even *Didymograptus* also may have arisen from other multiramous ancestors (e.g., from *Schizograptus* or *Trochograptus,* by way of *Mimograptus*). And anisograptids may themselves achieve reduction to two stipes *(Kiaerograptus)* while still retaining typical dendroid branch structure. In effect, stipe reduction is now seen as a general feature of the evolution of the anisograptids as well as the dichograptids, and the phylogeny of the latter has assumed the character of a complicated network or plexus. The accompanying diagram (Fig. 75), doubtless inaccurate in detail, nevertheless gives some indication of possible lines of *Tetragraptus* and *Didymograptus* ancestry.

That reclined tetragraptids have given rise to the scandent quadriserial *Phyllograptus* can scarcely be doubted, though an appreciable time gap separates one possible morphological intermediate *Tetragraptus phyllograptoides* from the earliest species of *Phyllograptus.* Scandent biserial genera also exist which must be regarded as dichograptid. The relationship of *Oncograptus* and *Cardiograptus* to *Isograptus* is ambiguous; stratigraphical relations suggest analogy with the *Phyllograptus-Tetragraptus* series, but the only *Oncograptus* investigated in detail possesses a proximal end more "primitive" than any known isograptid. *Skiagraptus* is another biserial form which occupies a somewhat anomalous position, but which is here retained in the Dichograptidae.

No satisfactory subdivision of the large and varied family Dichograptidae on a formal subfamilial basis is yet possible and the arbitrary grouping into multiramous and pauciramous genera, further tentatively divided into arbitrary "sections," is here retained. Moreover, the recognition of such genera as *Pendeograptus* and *Ex-*

---

[1] WIMAN (1895) seems independently to have postulated a polyphyletic origin for *Monograptus,* regarding some of the different thecal types as implying a distinct ancestry, but he gave no details.

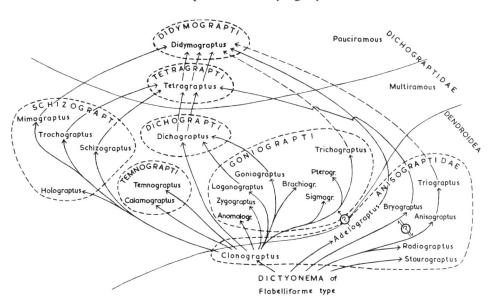

Fig. 75. Phylogeny of *Tetragraptus* and *Didymograptus* (tentative), with suggested grouping of the principal multiramous dichograptid genera and possible relations to the Anisograptidae (29).

tensograptus does not materially help to resolve the complicated phylogeny of *Tetragraptus* and *Didymograptus*. The unity of the genera comprising the Isograptidae of HARRIS is not yet convincingly demonstrated, but two compact families of highly specialized dichograptids have been recognized, the sinograptids and abrograptids.

SINOGRAPTIDAE AND ABROGRAPTIDAE

Characterized by progressive development of prothecal and metathecal folding, the Sinograptidae appear to represent a specialized offshoot from the Dichograptidae, though at what point is obscure, and some authorities assign the family an independent origin in the Anisograptidae. A stipe-reduction trend within the assemblage receives some support from the occurrence of the eight- and four-stiped *Pseudodichograptus* and *Allograptus* in the *Didymograptus hirundo* zone, while the thecally more extreme two-stiped genera *Tylograptus* and *Sinograptus* occur in the overlying *Amplexograptus confertus* zone of the Ningkuo Shale. *Nicholsonograptus* appears to represent the extreme of stipe reduction.

The highly specialized graptolites included in the Abrograptidae show a reduction of the periderm to a few sclerotized threads bearing apertural rings, but as in the retiolitid *Archiretiolites*, the sicula is normal. Their dichograptid origin is indicated by the presence of only a single crossing canal.

CORYNOIDIDAE

The affinities of the Corynoididae remain conjectural and some authorities assign it ordinal or subordinal status. Since the entire rhabdosome comprises not more than four individuals (including the sicula) it clearly represents arrested development. The prosicula appears to be devoid of the normal graptoloid longitudinal rods and it further resembles the dendroid prosicula in the apical position of the $th1$ resorption foramen, but these features may occur in other (primitive) dichograptids. The second theca was believed (BULMAN, 1947) to arise by resorption, but this was probably a misinterpretation of the delayed growth of $th2$, and KOZŁOWSKI (1953) has demonstrated that in *Corynites* the second theca (here the "microtheca") originates through a "primary notch." The mode of budding is thus probably normal and the alternating origin of the thecae is the equivalent of isograptid development. The extreme elongation of the metasicula and adnate thecae also recalls highly developed isograptids, for which reason the family is placed in its present position here; but a perceptible time gap sets off the last isograptids from the earliest *Corynoides* species. It is possible also that these forms represent or have arisen from giant larvae such as occur in modern plankton.

NEMAGRAPTIDAE AND DICRANOGRAPTIDAE

The leptograptid type of theca is foreshadowed in several species of *Didymograptus* and the superficial resemblance between *Leptograptus* and some slender *Dicellograptus* species has been considered to imply a phyletic relationship between these genera; but the presence of prothecal folds in several species of *Dicellograptus* is believed by other authorities to necessitate an independent origin from dichograptid or sinograptid stock (JAANUSSON, 1965; but also BULMAN, 1969). Lack of detailed information regarding the mode of development of the rhabdosome in *Leptograptus* is a source of uncertainty in discussions both of leptograptid ancestry and of *Leptograptus-Dicellograptus* relationships (see p. V76), for it is not known at what point the distinctive diplograptid (streptoblastic) development supersedes the dichograptid or isograptid type.

The view that *Dicranograptus* represents an intermediate phyletic stage between *Dicellograptus* and *Diplograptus* is no longer tenable and has been abandoned. Not only do various diplograptids long antedate the earliest known *Dicranograptus* (or *Dicellograptus*), but the distinctive dicranograptid theca is too specialized to be ancestral to that of any primitive diplograptid, and no progressive increase in length affects the biserial portion of the rhabdosome either in species time distribution, or in the range of a single *Dicranograptus* species. Rather do individual species give the impression of relatively stable semiscandent mutations, commonly ranging through several graptolite zones with negligible change. Rare examples of irregularity in rhabdosome construction have been figured by RUEDEMANN (1947) and have been named *Diceratograptus* by MU (1963). The mode of development of all known species is decidedly diplograptid (streptoblastic) and the possibility of evolution *from* a diplograptid ancestor is perhaps not altogether fanciful.

The precise significance of the branched nemagraptids is also unknown. Branching, where it occurs, is always lateral and in some paired (e.g., *Amphigraptus, Syndyograptus*) in a manner somewhat suggestive of thecal cladia, while the centribrachiate rhabdosomes of *Leptograptus* similarly suggest sicular cladia production rather than normal proximal end branching. That the branched condition of *Pleurograptus* is in some way secondary and not primitive is rather suggested by its high stratigraphic position and even lower Caradocian branched nemagraptids are separated from any multiramous dichograptid by a time gap.

It is not considered probable that *Pseudozygograptus* MU, LEE, & GEH represents a "leptograptid *Azygograptus*"; the dicalycal theca of a leptograptid, $th2^1$, is sufficient to prevent any simple derivation of the *Azygograptus* condition and the type of theca appears to be an inexact homeomorph.

GLOSSOGRAPTINA

In most areas of the world, the evolution of monopleural and dipleural biserial graptoloids is approximately contemporaneous, but their differences in structure and development are so great as to indicate a separate origin. That of the Glossograptina must lie in some unknown, presumably dichograptid, stock in which rhabdosome development was of a "primitive" type, with $th1^1$ the dicalycal theca and a single crossing canal. In this respect, the isograptids are already too advanced to be ancestral (*cf.* MU & ZHAN, 1966), and though evidence is insufficient as to the mode of development of *Oncograptus* and *Cardiograptus*, the rhabdosomes of these genera are not monopleural.

The Glossograptidae and Cryptograptidae make their first appearance in association in so many regions that it is not yet possible to assign priority to either; but they must have diverged rapidly assuming the existence of a common ancestor (based on their monopleural rhabdosomes and closely comparable mode of development). *Paraglossograptus* was erected for glossograptids with a well-developed lacinia (though its morphology is imperfectly known) and *Lonchograptus* also is clearly a derivative of *Glossograptus*. The affinities of *Nanograptus*, however, are less certain; it is an uncommon genus which combines some of the characters of both *Glossograptus* and *Cryptograptus,* though for stratigraphical reasons it can scarcely be a primary intermediate between them. The lack of conspicuous spines gives its rhabdosome a cryptograptid appearance (extremely fine apertural spines are definitely present in *N. phylloides*); but the characters of the thecae more closely resemble those of the Glossograptidae in which family it is provisionally included.

The relative lack of diversification and comparatively short stratigraphical range of this suborder indicates that for some reason it was the less efficient version of the scandent biserial rhabdosome. The dicalycal $th1^1$ compels an almost static proximal end.

DIPLOGRAPTINA

The Diplograptina occupy a dominant position among Ordovician graptolites and persisted until the Late Silurian (early Ludlow). Like the Glossograptina, the suborder must have arisen from dichograptid stock, but with more "advanced" proximal end, for the dicalycal theca is $th2^1$ (or some later theca) and three or more crossing canals are present. The earliest representative is a *Glyptograptus* of latest *Didymograptus extensus* or *Didymograptus hirundo* Zone age, but diversification was rapid and by early Llanvirn times (*Didymograptus bifidus* Zone) the genera *Diplograptus, Amplexograptus, Climacograptus* and *Pseudoclimacograptus* were present, together with representatives of the Lasiograptidae.

DIPLOGRAPTIDAE

A streptoblastic developmental plan of diplograptids is common but can no longer be claimed to be universal among these early representatives and its significance is obscure. Running through the family is also a general tendency for progressive delay in the siting of the dicalycal theca, resulting in a progressively incomplete septum and ultimately an aseptate rhabdosome. This should be applied as a phyletic criterion only with caution. Similar changes seem to affect the Lasiograptidae and the Retiolitidae.

Most genera are of long standing and were based mainly on thecal characters, originally determined in flattened material. This was a consequence of the very stable rhabdosome form. More precise studies on three-dimensional material have led to the establishment of further genera and subgenera and have served to indicate the complexity of diplograptid phylogeny without as yet providing sufficient evidence to offer a solution to the problem. It is probable that most genera are polyphyletic and they appear not only to define the results of original diversification but to include later gradations from one genus to another. This is strongly suspected in the case of *Glyptograptus* and *Climacograptus,* where three possible transitions are already known (p. *V*103). It is also possible that Silurian representatives of *Glyptograptus* and *Diplograptus* are homeomorphs, rather than descendants of Ordovician species of these genera. On current interpretation, therefore, these older genera are essentially "form genera" and their relationships have been likened to a bundle of rods with *Glyptograptus* forming a central core and with transitional connections from one to another at various levels. For all these reasons the family is no longer divided into subfamilies as in the first edition of this *Treatise.*

Two genera, both at present monotypic, have each been assigned to separate families: the *Dicaulograptidae* and the *Peiragraptidae.* The former may be an aberrant lasiograptid, but both have disconcertingly dicranograptid features; on the assumption that these are due to homeomorphy, both are placed in the Diplograptina.

LASIOGRAPTIDAE

Derivation of the Lasiograptidae from diplograptid stock seems certain; JAANUSSON even reduced the assemblage to a subfamily of his Diplograptidae. The thecal type in *Lasiograptus* suggests simply a more extreme development of the general climacograptid type of theca, but the supragenicular wall in *Gymnograptus* and *Hallograptus* is exceedingly short and the infragenicular portion straight; the analogy with *Cryptograptus* (and its relation to *Glossograptus*) suggests the possibility of an independent origin from a straight orthograptid type in these genera. The whole group is retained here with the rank of a family.

RETIOLITIDAE

More detail has come to light in recent years concerning the Retiolitidae, but it is still impossible to recognize any phyletic grouping of genera and therefore they are retained in three arbitrary subfamilies reflecting an approximate increase in specialization. The Archiretiolitinae are the least modified, but represent a marked advance on any normal diplograptid. Almost the whole rhabdosome is reticular, but normal fusellar tissue persists in the sicula and to a varying degree in the initial parts of early thecae. Two genera have been described from fragmentary (?immature) proximal ends and two others are not known three-dimensionally; but the limited evidence of thecal characters and mode of development suggest that the subfamily comprises various stages in retiolitid specialization affecting several lines derived from more than one diplograptid ancestor.

The Retiolitinae are more specialized; no conclusive evidence indicates any continuous periderm except in the prosicula, and the recognizably diplograptid development of the Archiretiolitinae has been replaced by the ancora and corona stages (Fig. 59). Whether independently evolved or related to the preceding family is uncertain. The Plectograptinae include the latest and most highly modified genera of all, with the skeleton commonly reduced to little more than an open clathria.

DIMORPHOGRAPTIDAE

Regarded from the viewpoint of adult rhabdosomes, the dimorphograptids occupy a morphologically intermediate position between Diplograptidae and Monograptidae (and between Diplograptina and Monograptina), but as with the dicranograptids they were probably not phyletically intermediate. They do not represent an essential intermediate in the astogenetic changes involved; no time significance is seen in length of the uniserial portion of the rhabdosome; and stratigraphically the species are later than the earliest monograptids.[1] They are here included in the Diplograptina not only because they appear to represent diplograptids that have failed to become monograptid, but because the mode of development is more diplograptid in the downward direction of growth of the initial part of $th1^1$ in several forms and because of the lack of sinus and lacuna type of porus formation.

The disappearance of the dicalycal theca is not in itself sufficient to convert a diplograptid into a monograptid rhabdosome; *Peiragraptus* illustrates (Fig. 76,2) what results merely from this step in the process, and shows that the real problem is the reorientation (or elimination) of $th1^2$.

---
[1] See p. V100.

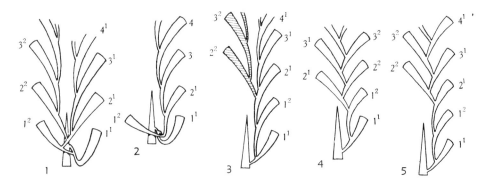

Fig. 76. Diagrams illustrating derivation of monograptid from diplograptid rhabdosome (Bulman, n).——*1*. Biserial rhabdosome with dicalycal $th2^1$.——*2*. *Peiragraptus*, a uniserial rhabdosome in which $th1^2$ retains its diplograptid orientation.——*3*. Hypothetical diplograptid with reoriented $th1^2$ and dicalycal $th2^1$; loss of $th2^2$ and descendant thecae (shaded) would convert to monograptid.——*4,5*. Possible dimorphograptid rhabdosomes with aseptate thecae.

Once this has been accomplished, the disappearance of the dicalycal theca, if it were $th2^1$, could result in the immediate production of a uniserial rhabdosome (Fig. 76,3) from a septate diplograptid. No such form as this has yet been recognized, possibly because it was extremely short-lived; most dimorphograptid species are aseptate (Fig. 76,4,5) or possess a partial septum, and from such forms the production of *Monograptus* would be a long and complicated process.

MONOGRAPTINA

Thecal elaboration affects the monograptids to an extent exceeding anything recognized among Ordovician graptolites. Unfortunately, preservation of early monograptids is generally very unsatisfactory and little detail is as yet available. Work in progress on the structure of many seemingly "simple" species, particularly of ELLES & WOOD's (1901-18) Group II, indicates unexpected thecal elaboration even at this level, and the possibility of polyphyletic origin for *Monograptus* cannot be excluded. At present the ancestry of such distinctive groups as the triangulate and hooked monograptids is quite unknown, and generic and subgeneric names for such groups are not here adopted. *Pristiograptus* and *Monoclimacis* are accepted in this edition, but these are long-ranging and may well prove to comprise unrelated species.

The long series of rhabdosomal changes, beginning with pendent or horizontal dichograptids, culminates little more than halfway through the geological history of the Graptoloidea in the scandent uniserial monograptids of the Silurian, and the development of cladia-bearing rhabdosomes represents the only further change possible. Reference has already been made (p. $V88$) to URBANEK's recognition of a series of astogenetic stages (monograptid to abiesgraptid) based on this cladia-production, which he compared with the developmental stages of other orders, especially the Diplograptina. Astogenetic stages in the Diplograptina have never been proposed as a basis for classification, but in the Monograptina they have been so used and to a large extent form the basis for the classification provisionally retained here. In this, the Monograptidae (without cladia) are separated from the Cyrtograptidae (with cladia) and the latter are subdivided into Cyrtograptinae (with thecal cladia only) and Linograptinae (with sicular cladia, with or without thecal cladia). We do not know to what cause cladia-production is a response and it may be that the same species may occur in more than one form. "*Monograptus*" *runcinatus* commonly appears to be a normal monograptid, but (as described by STRACHAN) it may develop a sicular cladium; the species then becomes recognizably diversograptid and would be assigned to a separate genus, here placed in a separate subfamily. *Neodiversograptus* doubtless provides comparable examples. Whether it is more reasonable to accept the diversograptid potentiality as the classificatory criterion or to accept the more common astogenetic condition, is clearly contentious, but the number of species involved seems at present to be small. Bipolar rhabdosomes lacking a sicula, which represent a regeneration process (p. $V89$) are not, of course, given any taxonomic rank.

SYSTEMATIC DESCRIPTIONS

Suborder DIDYMOGRAPTINA Lapworth, 1880, emend. Bulman, herein

[*nom. correct.* JAANUSSON, 1960, p. 309, *ex* Didymograpta LAPWORTH, 1880, p. 192] [=Didymograpta+Dicellograpta LAPWORTH, 1880; Dichograptina+Leptograptina OBUT, 1957; Didymograptina+Corynoidina *sensu* JAANUSSON, 1960]

Uniserial, pendent to reclined, rarely biserial (dipleural) or quadriserial graptoloids without virgula; development platycalycal, $th1^1$, $th1^2$ or $th2^1$ being the dicalycal theca. *Ord.*

Family DICHOGRAPTIDAE Lapworth, 1873

[Dichograptidae LAPWORTH, 1873, table 1 facing p. 555]

Rhabdosome bilaterally symmetrical, branching dichotomous or lateral, central disc present in some; pendent to scandent, usually declined or horizontal; stipes uniserial, rarely biserial or quadriserial; thecae typically simple, straight or with slight ventral curvature, denticulate, overlapping about one-half their length; development of dichograptid or isograptid type. *L.Ord.*

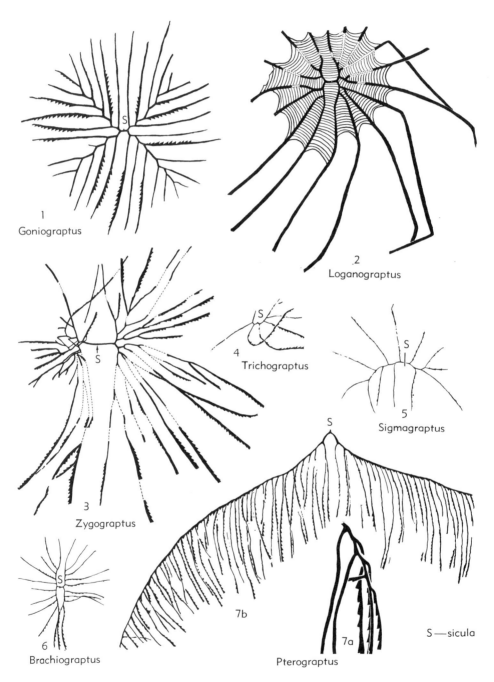

Fig. 77. Dichograptidae (Goniograpti) (p. *V*111-*V*112).

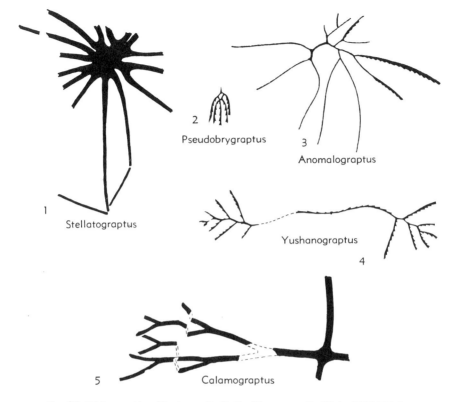

Fig. 78. Dichograptidae (Goniograpti) *(1-4)*; (Temnograpti) *(5)* (p. *V*111-*V*114).

MULTIRAMOUS FORMS

Pendent to horizontal, rarely reclined; branching dichotomously to produce at least third-order branches (first-order branches constitute the "funicle" of HALL), or laterally from one or both sides of two or four main stipes. *L.Ord.*

Section GONIOGRAPTI

Based on didymograptid or tetragraptid foundation, with compact regular branching either dichotomous or lateral. *Ord.*

Goniograptus M'COY, 1876; p. 130 [*Didymograpsus thureaui* M'COY, 1876, p. 129; M]. With 4 zigzag main stipes, from angles of which undivided lateral stipes are produced with great regularity, so that form suggests regularly alternating dichotomy (FIG. 64,3) and in one subspecies dichotomous division occurs in some quadrants; thecae with low inclination and slight overlap. *L.Ord.*, N.Am.(Deepkill-?Normanskill)-N.Z.-Australia(Bendigon.-Castlemain.)-?NW. Eu.——FIG. 77,1. *G. thureaui* (M'COY), Australia; ×1 (138).

Anomalograptus CLARK, 1924, p. 63 [*A. reliquus;* OD]. Late aberrant clonograptid with asymmetrical and irregular dichotomies up to 6th order. *L.Ord. (Glyptograptus dentatus* Z.*)*, Que.——FIG. 78,3. *A. reliquus;* ×1.5 (46).

Brachiograptus HARRIS & KEBLE, 1932, p. 43 [*B. etaformis;* M]. Small, composed of 4 main branches forming with funicle a letter H, from outer sides of which are produced close-set undivided lateral branches; thecae slender, with low inclination and slight overlap. *L.Ord.(Llanvirn),* N.Am.-S.Am.-Australia-?China.——FIG. 77,6. *B. etaformis,* Australia (Darriwil); ×1 (84).

Loganograptus HALL, 1868, p. 237 [*Graptolithus logani* HALL, 1858, p. 142; M]. Typically 16 to 8 branches, rarely exceeding 4th order, produced by proximally concentrated dichotomy; undivided terminal stipes mostly long and flexuous; central disc commonly present, enclosing proximal dichotomies; thecae moderately inclined with overlap of about one-half. *L.Ord.(low.Arenig-Llanvirn-?Normanskill),* NW. Eu.-N.Am.-Asia-Australia-N.Z.——FIG. 77,2. *L. logani* (HALL), Levis Sh., Que.; ×1 (77).

?**Oslograptus** JAANUSSON, 1965, p. 427 [*O. peculiaris*; OD]. Similar in rhabdosome form to *Pseudobryograptus*, but with only second order branches; stipes with pronounced dorsal folds at level of thecal apertures; development ?isograptid. *L.Ord.(L. Didymograptus Sh.)*, Eu.(Nor.).

Pseudobryograptus MU, 1957, p. 421 [*P. parallelus*; OD]. Rhabdosome small, pendent; branching dichotomous, up to third order; thecae dichograptid. *L.Ord.(up.Arenig-low.Llanvirn)*, China (Ningkuo Sh.)-Australia (Darriwil)-N. Am.-(Glenogle Sh.).——FIG. 78,2. *P. parallelus*, China; ×1.5 (149).

Pterograptus HOLM, 1881, p. 74 [*P. elegans* (=*Graptolithus gracilis* KJERULF, 1865, p. 4; *non* HALL, 1848), p. 274; M]. Pendent or declined, consisting of 2 primary stipes, each giving rise to undivided lateral branches alternately to right and left, forming a somewhat flabelliform rhabdosome; thecae denticulate, inclined at moderate angles. *L.Ord.(up.Arenig-Llanvirn)*, NW. Eu.-S.Am.-Australia-China.——FIG. 77,7a. *P. elegans*, U.*Didymograptus* Sh., S. Sweden; ×4 (Hadding, 1911).——FIG. 77,7b. *P. scanicus* MOBERG, U.*Didymograptus* Sh., S. Sweden; ×1 (Hadding, 1911).

Sigmagraptus RUEDEMANN, 1904, p. 701 [*S. praecursor*; OD]. With 2 slender main branches from which slender undivided lateral branches originate alternately on both sides (genus is essentially a 2-stiped *Goniograptus*); thecae extremely slender, inclined at low angles and with slight overlap. *L.Ord.*, N.Am.(Deepkill)-Australia-N.Z. (Bendigo).——FIG. 77,5. *S. praecursor*, Deepkill, N.Y.; ×1 (201).

Stellatograptus ERDTMANN, 1967, p. 343 [*S. stellatus*; M]. Like *Loganograptus* but with thick central web and tapering lateral alae to more distal branches. *L.Ord.(?up. Arenig, Levis Sh.)*, N.Am.(Que.).——FIG. 78,1. *S. stellatus*; ×1.5 (61).

Triaenograptus T. S. HALL, 1914, p. 115 [*T. neglectus*; M] [=*Tridensigraptus* ZHAO, 1964, p. 640 (type, *T. zhejiangensis*)]. Rhabdosome large, horizontal, composed of 4 main stipes, each with paired lateral branches some of which may bear paired (4th or higher order) branches. *L.Ord. (Isograptus* and *Didymograptus hirundo* Z.); Australia(Victoria)-China.

Trichograptus NICHOLSON, 1876, p. 248 [*Dichograptus fragilis* NICHOLSON, 1869; OD]. With 2 slender primary stipes, straight or flexuous, originating at about 180 degrees from sicula, with slender undivided lateral branches regularly produced from one side only; thecae elongate with low inclination and very slight overlap. *L.Ord. (Arenig-Llanvirn)*, NW. Eu.-S. Am.-Australia.——FIG. 77,4. *T. fragilis* (NICHOLSON), Skiddaw Sl., N.Eng.; ×1 (59).

Yushanograptus CHEN, SUN, & HAN, 1964, p. 239 [*Y. separatus*; OD]. Rhabdosome of 2 long, declined stipes with goniograptid branching distally. *L.Ord.(up. Arenig, Ningkuo Sh.)*, China.——FIG. 78,4. *Y. separatus*; ×1 (44).

Zygograptus HARRIS & THOMAS, 1941, p. 308 [*Graptolithus abnormis* HALL, 1857; OD]. With 2 long first-order stipes forming an exaggerated funicle, followed by repeated dichotomies at close intervals to 5th or higher order; thecae with moderate to low inclination and slight overlap.

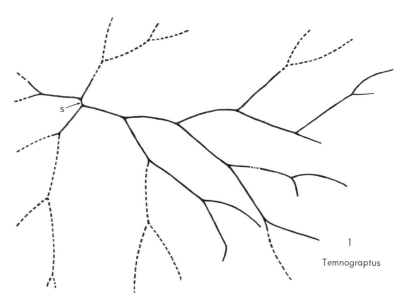

FIG. 79. Dichograptidae (Temnograpti) [*s*, sicula] (p. *V*113).

Graptoloidea—Didymograptina—Multiramous Forms

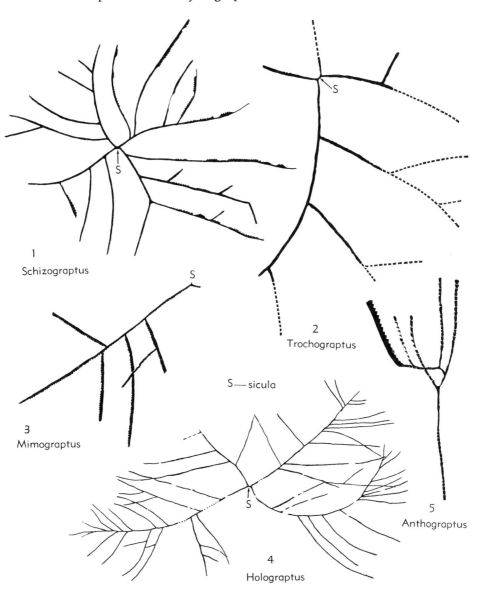

Fig. 80. Dichograptidae (Schizograpti) (p. V114).

L.Ord.(Arenig-?Llanvirn), Australia-N.Am.-?N.Z.——Fig. 77,3. *Z. abnormis (HALL), Levis Sh., Que.; ×1 (88).

Section TEMNOGRAPTI

Widely and evenly spaced dichotomous branching based on a tetragraptid foundation; rhabdosome usually of large size. L.Ord.

Temnograptus NICHOLSON, 1876, p. 248 [*Dichograpsus multiplex NICHOLSON, 1868, p. 129; OD]. Like Clonograptus, produced by regular dichotomous division but more consistently divergent, with very short funicle and long 2nd-order stipes, successive later orders being approximately equal in length to 2nd; thecae denticulate with moderate inclination and one-half to two-thirds overlap. L.Ord., NW.Eu.(Arenig)-N.Am.(?Athens).——FIG. 79,1. *T. multiplex (NICHOLSON), Didymograptus Sh., S.Sweden; ×0.7 [s, sicula] (241).

Calamograptus CLARK, 1924, p. 61 [*C. porrectus; OD]. Like Temnograptus but with branches of

Graptolithina

2nd order very long, higher orders slightly decreasing in length. *L.Ord.(Levis)*, N.Am.(Que.).——FIG. 78,*5*. **C. porrectus;* ×0.5 (46).

Section SCHIZOGRAPTI

Usually of large size, based on either didymograptid or tetragraptid foundation, with laterally produced secondary branches. *L.Ord.*

Schizograptus NICHOLSON, 1876, p. 248 [**Dichograpsus reticulatus* NICHOLSON, 1868, p. 143; OD]. Rhabdosome based on 4 main stipes produced by dichotomous division from short funicle; lateral branches on one side only of main stipe; tertiary lateral branches rarely developed. *L.Ord. (Arenig)*, NW.Eu.-N.Am.-Australia-N.Z.-?S.Am.——FIG. 80,*1*. *S. rotans* TÖRNQUIST, *Didymograptus* Sh., S.Sweden; ×0.7 (241).

?Anthograptus TÖRNQUIST, 1904, p. 22 [**A. nidus;* M]. Proximal end unknown; ?2nd-order stipes of great length, at distal end with lateral branches and stipes of higher order produced by irregular dichotomy. *L.Ord.(L.Didymograptus Sh.)*, S. Sweden.——FIG. 80,*5*. **A. nidus;* ×1 (241).

Holograptus HOLM, 1881, p. 45 [**H. expansus;* M] [=*Rouvilligraptus* BARROIS, 1893]. Like *Schizograptus* but lateral branches produced somewhat irregularly from both sides of 4 main stipes, particularly distally. *L.Ord.(Arenig)*, NW.Eu.-Boh.——FIG. 80, *4*. *H. deani* ELLES & WOOD, Skiddaw Sl., N.Eng.; ×0.13 (59).

Mimograptus HARRIS & THOMAS, 1940, p. 197 [**M. mutabilis;* M]. Robust, consisting of 2 main stipes diverging from sicula at less than 180 degrees, bearing lateral branches at irregular intervals which in turn may bear tertiary branches; forms with few or no lateral branches also occur. *L.Ord.(Chewton.)*, Australia.——FIG. 80,*3*. **M. mutabilis;* ×0.7 (87).

Trochograptus HOLM, 1881, p. 48 [**T. diffusus;* M]. Rhabdosome large, similar to *Schizograptus* but with more widely spaced lateral branches and tertiary branches common. *L.Ord.(Arenig)*, NW.Eu.-N.Am.-Australia. —— FIG. 80,*2*. **T. diffusus,* L. Didymograptus Sh., Oslo; ×0.7 (89).

Section DICHOGRAPTI

With eight or fewer stipes, dichotomously dividing to third order only; first two orders generally short, equal in length, third order long and usually flexuous; thecae denticulate, inclined at moderate angles and with considerable overlap, less commonly with low inclination, slight overlap, and (rarely) long apertural spines. *L.Ord.*

Dichograptus SALTER, 1863, p. 139 [*nom. correct*

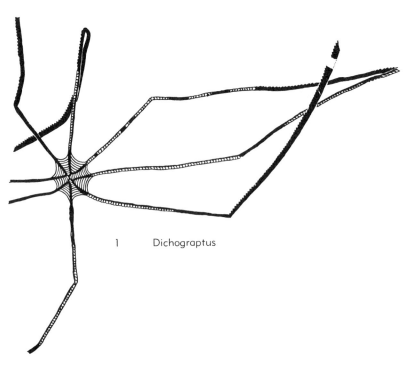

1 Dichograptus

FIG. 81. Dichograptidae (Dichograpti) (p. *V*114-*V*115).

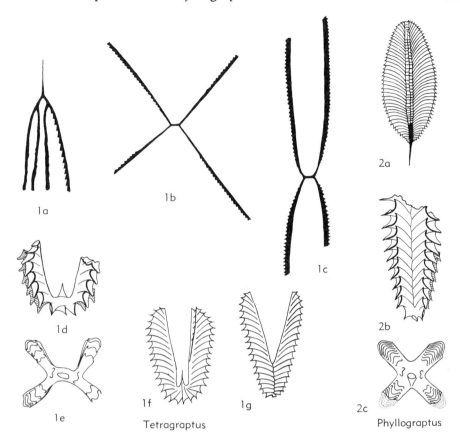

Fig. 82. Dichograptidae (Tetragrapti) (p. V115-V116).

HALL, 1865 (pro *Dichograpsus* SALTER, 1863), ICZN Opin. 650] [**Dichograpsus sedgwicki;* SD GURLEY, 1896, p. 64]. Characters of section; central disc in certain species. *L.Ord.(Arenig-Llanvirn)*, almost world-wide.——FIG. 81,*1. D. octobrachiatus* (HALL), Levis Sh., Que.; ×0.5 (77).

PAUCIRAMOUS FORMS

Pendent, deflexed, declined, horizontal, reflexed, reclined or scandent, wholly or in part; branching dichotomous to first or second order only; thecae simple, rarely with sigmoidal curvature or elaborated apertural modifications. *L.Ord.-U.Ord.*

Section TETRAGRAPTI

Rhabdosome pendent to scandent, composed of four stipes of second order, rarely two stipes of second order and one of first order; theca simple, denticulate. *L.Ord.*

Tetragraptus SALTER, 1863, p. 140 [*Fucoides serra* BRONGNIART, 1828, p. 71 (=*Graptolithus bryonoides* HALL, 1858, p. 150); OD] [*nom. correct.* HALL, 1865 (pro *Tetragrapsus* SALTER, 1863), ICZN, Opin. 650] [=*Etagraptus* RUEDEMANN, 1904, p. 644 (type, *Tetragraptus (Etagraptus) lentus* RUEDEMANN, 1904, p. 666); *Eotetragraptus* BOUČEK & PŘIBYL, 1951, p. 7 (type, *Graptolithus quadribrachiatus* HALL, 1858, p. 125); *Pendeograptus* BOUČEK & PŘIBYL, 1951, p. 12 (type, *Tetragraptus pendens* ELLES, 1898, p. 491); *Paratetragraptus* OBUT, 1957, p. 33, 38 (type, *Tetragraptus approximatus* NICHOLSON, 1873, p. 136); *Ramulograptus* ROSS & BERRY, 1963, p. 84 (type, *R. surcularis*)]. Bilaterally symmetrical, pendent to reclined; central disc in some horizontal species; funicle usually short, commonly bearing one theca only; development dichograptid or isograptid. *L.Ord.(Arenig-Llanvirn),* worldwide.——FIG. 82,*1a. T. fruticosus* (HALL), Levis Sh., Que.; ×1 (77).——FIG. 82,*1b. T. quadribrachiatus* (HALL), Levis Sh., Que.; ×1 (77).——FIG. 82,*1c. T. approximatus* NICHOLSON, L.*Didymograptus* Sh., S.Sweden; ×1 (241).——FIG. 82,

1d,e. T. bigsbyi (HALL), *Orthoceras* Ls.(Ontikan), Öland, Sweden; *1d,e*, lat. and ventral views of specimens dissolved from limestone, ×4 (91).——FIG. 82,*1f,g. T. phyllograptoides* LINNARSSON, L.*Didymograptus* Sh., S.Sweden; *1f,g*, ×2 (91).

Phyllograptus HALL, 1858, p. 137 [**P. typus;* OD]. Quadriserial, composed of 4 scandent 2nd-order stipes; nema unknown; thecae simple, slightly curved, with high inclination and large overlap; development where known isograptid. *L.Ord. (Arenig-Llanvirn),* worldwide.——FIG. 82,*2a. *P. typus,* Levis Sh., Que.; ×1 (77).——FIG. 82,*2b,c. P. angustifolius* HALL, *Orthoceras* Ls.(Ontikan), Öland, Sweden; *2b,c*, lat. and ventral views of specimens dissolved from limestone, ×4 (91).

Tristichograptus JACKSON & BULMAN, 1970 [**Graptolithus ensiformis* HALL, 1859, p. 133; OD] [=*Trigonograpsus* NICHOLSON, 1869, p. 231 (type, *T. lanceolatus*); *Pseudotrigonograptus* MU & LEE, 1958, p. 416 (type, *P. uniformis*)]. Rhabdosome scandent, triserial, without virgula, elongate fusiform in shape, triangular in cross section; thecae with slight ventral curvature; development elaborated on basis of dicalycal *th1²*. In compressed examples, rhabdosome appears biserial and apertural margins produce an even line. *L.Ord.* (*up.Arenig,* almost worldwide; *Llanvirn,* Pacific province).——FIG. 98 (see p. *V*132). **T. ensiformis* (HALL), compressed specimen, Skiddaw Sl., N.Eng.; ×2 (59).

Section DIDYMOGRAPTI

Pendent to scandent, composed of not more than two stipes. *L.Ord.-U.Ord.*

Didymograptus M'COY in SEDGWICK & M'COY, 1851, p. 9 [**Graptolithus murchisoni* BECK, in MURCHISON, 1839, p. 694; SD MILLER, 1889] [*nom. correct.* HALL, 1865 (*pro Didymograpsus* M'COY, 1851) ICZN Opin. 650] [=*Cladograpsus* GEINITZ, 1852, p. 29 (type, *Graptolithus murchisoni* BECK, 1839, p. 694; SD BULMAN, 1929, p. 169); *Expansograptus* BOUČEK & PŘIBYL, 1951, p. 13 (type, *Graptolithus extensus* HALL, 1858, p. 132); *Corymbograptus* OBUT & SOBOLEVSKAYA, 1964, p. 27 (type, *Didymograpsus v-fractus* SALTER, 1863, p. 137); *Cymatograptus* JAANUSSON, 1965, p. 423 (type, *Didymograptus undulatus* TÖRNQUIST, 1901, p. 10)]. Pendent to reclined; development of dichograptid or isograptid type; thecae typically simple, straight or with slight ventral curvature. *L.Ord.-U.Ord.(Nemagraptus gracilis Zone),* worldwide.——FIG. 83,*1a. D. extensus* (HALL), L.Ord. (Levis Sh.), Que.; ×1 (77).——FIG. 83,*1b. *D. murchisoni* (BECK), L.Ord.(Llanvirn), S.Wales; ×1 (59).——FIG. 83,*1c. D. nicholsoni* LAPWORTH, L.Ord.(Skiddaw Sl.), N.Eng.; ×1 (59).

Atopograptus HARRIS, 1926, p. 59 [**A. woodwardi;* OD]. Horizontal didymograptid with everted, reflexed thecal apertures; sicula unknown. *L.Ord.* (*Darriwil.*), Australia (Victoria)-China.——FIG. 83,*2. *A. woodwardi;* ×5 (81).

Aulograptus SKEVINGTON, 1965, p. 25 [**Didymograptus cucullus* BULMAN, 1932, p. 15; OD]. Pendent didymograptid with climacograptid thecae, with distally directed or slightly everted apertures; development isograptid. *L.Ord.(up. Arenig or low.Llanvirn),* NW.Eu.-S.Am.(Arg.-Peru)-?China.——FIG. 83,*3. *A. cucullus* (BULMAN), *Orthoceras* Ls. (Ontikan), Sweden(Öland); *3a,* proximal end; *3b,* diagram. long. sec. through thecae; ×10 (19).

Azygograptus NICHOLSON (*ex* LAPWORTH MS), 1875, p. 269 [**A. lapworthi;* OD] [=*Pseudazygograptus* MU, LEE, & GEH, 1960, p. 37 (type, *Azygograptus incurvus* EKSTRÖM, 1937, p. 33)]. Asymmetrical, unilateral, composed of a single stipe which may be pendent to reclined. *L.Ord.-U.Ord.(Glenogle Sh.),* Eu.-China-N.Am.-S.Am.——FIG. 83,*4. A. suecicus* MOBERG, L.Ord. (L. *Didymograptus* Sh.), S. Sweden; ×2 (144).

Cardiograptus HARRIS & KEBLE, 1916, p. 66 [**C. morsus;* M] [=*Paracardiograptus* MU & LEE, 1958, p. 419 (type, *P. hsüi*)]. Biserial, elongate-ovate, emarginate distally, resembling an *Oncograptus* in which distal uniserial stipes have failed to develop. *L.Ord.(up.Yapeen.-Darriwil.),* Australia-China-N.Am.——FIG. 83,*10. *C. morsus,* Australia(Victoria); ×1 (234).

Isograptus MOBERG, 1892, p. 345 [**Didymograptus gibberulus* NICHOLSON, 1875, p. 271 (?=*D. caduceus* SALTER, 1853, p. 87); M]. Reclined; thecae elongate with high inclination and large overlap, especially proximally; development isograptid, 1st few thecae growing entirely downward. *L. Ord.(Arenig-L.Llandeilo),* NW.Eu.-N.Am.-S.Am.-Australia-Asia.——FIG. 83,*7. *I. gibberulus* (NICHOLSON), L.*Didymograptus* Sh., S.Sweden; *7a,* rhabdosome, ×1 (144); *7b,* proximal end, ×5 (19).

Janograptus TULLBERG, 1880, p. 314 [**J. laxatus;* M]. Resembling an extensiform *Didymograptus* but without apparent sicula, possibly representing pro- and pseudocladia. *L.Ord.(U.Didymograptus Sh.-L.Dicellograptus Sh.),* Sweden-Norway-S.Am.-China.——FIG. 83,*6. *J. laxatus,* S.Sweden; ×2 (242).

Kinnegraptus SKOGLUND, 1961, p. 391 [**K. kinnekullensis;* OD]. Declined to horizontal didymograptids with 2 or more long, exceedingly slender stipes, thecae and sicula with prominent apertural processes; development dichograptid or isograptid. *L.Ord.(L. Didymograptus Sh.),* NW.Eu.——FIG. 83,*5. *K. kinnekullensis,* Sweden; *5a,* immature rhabdosome, ×4; *5b,* proximal end with sicula, ×33; *5c,* apertural region of theca, ×33 (216).

Maeandrograptus MOBERG, 1892, p. 344 [**M. schmalenseei;* M]. Reclined stipes of almost uniform width, composed of somewhat undulating elongate thecae with low inclination and large overlap; development of isograptid type, 1st theca of each stipe distally reclined. L.Ord.(*L.Didymograptus Sh.*), S.Sweden.——FIG. 83,*8*. **M. schmalenseei*, S.Sweden; *8a*, rhabdosome, ×2 (144); *8b*, proximal end, ×5 (19).

Oncograptus T. S. HALL, 1914, p. 109 [**O. upsilon*; OD]. Initially scandent biserial, later diverging; thecae long, slender, with high inclination and considerable overlap; development dichograptid, $th1^1$ dicalycal, first thecae short, downwardly directed, increasing in length and changing direction distally. L.Ord.(Yapeen.), Australia-N.Am.-S.Am.-W.Ire.-China-USSR(Taimyr). —— FIG. 83,*11*. **O. upsilon*, Australia; ×1 (234).

Parazygograptus KOZŁOWSKI, 1954, p. 129 [**P. erraticus*; OD]. Like *Azygograptus*, but with single stipe based on $th1^2$ produced from initial bud, $th1^1$ without metathecal portion. L.Ord. (glacial boulders), Eu.(Pol.).——FIG. 50,*4*. **P. erraticus*; illustrating development of proximal end; ×35 (116). [Also p. 439 (Polish text).]

Skiagraptus HARRIS, 1933, p. 108 [**Diplograptus gnomonicus* HARRIS & KEBLE, 1916, pl. 1, fig. 5,6; OD]. Rhabdosome biserial; thecae short, proximal thecae growing entirely downward, later thecae horizontal and then distally directed; development pericalycal. L.Ord.(Yapeen.), Australia-N.Am.——FIG. 83,*9*. **S. gnomonicus* (HARRIS & KEBLE), Australia; schematic, ×2 (after Harris, 1933).

Family SINOGRAPTIDAE Mu, 1957

[Sinograptidae Mu, 1957, p. 423]

Thecae with initial prothecal folds and typically with introverted apertural modi-

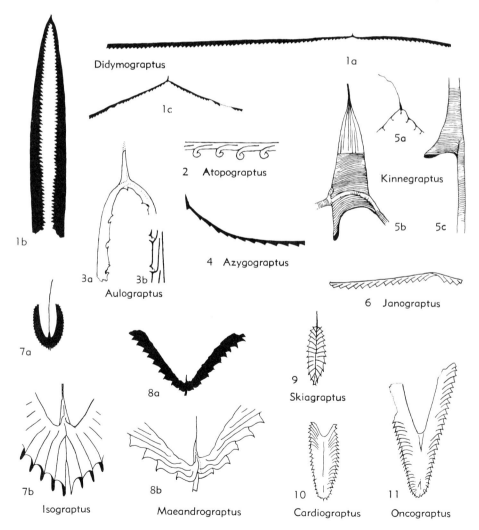

FIG. 83. Dichograptidae (Didymograpti) (p. *V*116-*V*117).

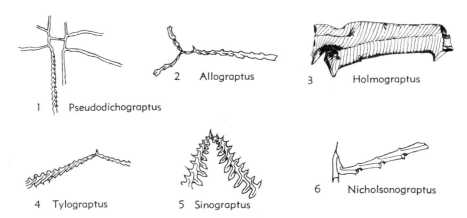

Fig. 84. Sinograptidae (p. V118).

fications; prothecal and metathecal nodes may bear spines; stipes typically showing pronounced increase in thecal overlap distally; development dichograptid. *L.Ord. (up.Arenig-low.Llanvirn)*.

Sinograptus MU, 1957, p. 434 [*S. typicalis;* OD]. Rhabdosome of 2 declined stipes; thecae with exaggerated prothecal and metathecal folds. *L. Ord.(low.Llanvirn)*, China *(Ningkuo Sh.)*-Yukon *(Road River F.)*.——FIG. 84,5. *S. typicalis, Amplexograptus confertus* Z., Changshan; ×3 (149).

Allograptus MU, 1957, p. 423 [*A. mirus;* OD]. Rhabdosome of 4 or 3 horizontal stipes; thecae with prothecal folds and relatively unmodified apertures. *L.Ord.(up.Arenig)*, China *(Ningkuo Sh.)*-Quebec *(Levis Sh.)*.——FIG. 84,2. *A. mirus, Didymograptus hirundo* Z., Changshan; ×3 (149).

Holmograptus KOZŁOWSKI, 1954, p. 126 [*Didymograptus callotheca* BULMAN, 1932, p. 16 (=?*D. lentus* TÖRNQUIST, 1911, p. 430); OD]. Rhabdosome of 2 declined stipes; thecae with prothecal folds accentuated by dorsal "notches" and introverted apertures with mesial spine and lateral lappets; an "apertural plate" on the succeeding metatheca further constricts the aperture. *L.Ord.(up.Arenig or low.Llanvirn)*, NW.Eu.——FIG. 84,3. *H. callotheca* (BULMAN), L. Ord. (glacial boulder), Pol.; ×35 (116). [Also p. 432 (Polish text).]

Nicholsonograptus BOUČEK & PŘIBYL, 1951, p. 14 [*Didymograpsus fasciculatus* NICHOLSON, 1869, p. 241; OD]. Rhabdosome of 1 reflexed stipe; thecae as in *Holmograptus. L.Ord.(low.Llanvirn)*, NW.Eu.-China-S.Am.(Peru).——FIG. 84,6. *N. fasciculatus* (NICHOLSON), L.Ord. (*Didymograptus bifidus* Z.), N.Eng.; ×7.5 (Skevington, 1966).

Pseudodichograptus CHU, 1965, p. 102 [*P. confertus;* OD]. Rhabdosome dichotomously dividing to 3rd order; thecae with prothecal folds and incipient apertural modifications. *L.Ord.(up. Arenig)*, China.——FIG. 84,*1*. *P. confertus, D. hirundo* Z., Chekiang; ×1.5 (45).

Tylograptus MU, 1957, p. 428 [*T. regularis;* OD] [=*Pardidymograptus* MU, GEH, & YIN, 1962, p. 73 (type, *P. acanthonotus*)]. Rhabdosome of 2 declined stipes, thecae with pronounced prothecal folding and weak to a strong apertural introversion; greatly increased thecal overlap distally. *L.Ord.(up.Arenig-low.Llanvirn)*, China-Australia. ——FIG. 84,*4*. *T. regularis, Amplexograptus confertus* Z., Changshan; ×2.25 (149).

Family ABROGRAPTIDAE MU, 1958

[Abrograptidae MU, 1958, p. 264)

Rhabdosome comprising two reclined stipes; sicula completely sclerotized, but stipe periderm reduced to one or two dorsal threads with complete or partial rings representing apertures; development dichograptid, with single crossing canal. *Ord. (?up.Arenig-Nemagraptus gracilis Z.)*

Abrograptus MU, 1958, p. 264 [*A. formosus;* OD] [=*Parabrograptus* MU & QIAO, 1962]. Stipes consisting of 2 dorsal threads united at intervals by apertural rings or half-rings. *Ord.*, China(*Glyptograptus teretiusculus* and *Nemagraptus gracilis* Z.); N.Am.(B.C.) (*Glenogle F.*), ——FIG. 85,*1*. *A. formosus;* diagram., ×5 (150).

Dinemagraptus KOZŁOWSKI, 1952, p. 87 [*D. warkae;* OD]. Stipes consisting of a single dorsal thread with complete apertural rings. *Ord.(?up. Arenig to Nemagraptus gracilis Z.)*, NW.Eu.-China.——FIG. 85,*2*. *D. warkae*, diagram.; ×4 (115). [Also p. 292 (Polish text).]

?**Jiangshanites** MU & QIAO, 1962, p. 7 [*J. ramo-

FIG. 85. Abrograptidae [*s*, sicula] (p. *V*118).

sus; OD]. A doubtful graptolite possibly representing a branched rhabdosome with comparable periderm reduction. U.Ord.(*Nemagraptus gracilis* Z.), China-N.Am.(B.C.) (156).

Family CORYNOIDIDAE Bulman, 1944

[Corynoididae BULMAN, 1944, p. 22] [*pro* Corynograptidae HOPKINSON & LAPWORTH, 1875, p. 633; Corynoideae RUEDEMANN, 1908, p. 233]

Rhabdosome consisting of a very long sicula, one or two pendent adnate thecae each bearing a broad, lamelliform apertural process, and one minute isolate theca; initial bud arises apically on prosicula; thecae alternating in origin. U.Ord.

Corynoides NICHOLSON, 1867, p. 108 [**C. calicularis*; M] [=*Corynograptus* HOPKINSON & LAPWORTH, 1875, p. 633]. Rhabdosome consisting of sicula with broad lamelliform virgella and 2 adnate thecae bearing broad apertural processes; where a 3rd theca occurs, it is small and isolate. U.Ord.(*Glenkiln* and *Hartfell Sh.*), NW.Eu.-N. Am.-Australia.——FIG. 86,*1*. **C. calicularis*, Ardwell Ser., S.Scot.; ×13 (23).

Corynites KOZŁOWSKI, 1956, p. 260 [**C. wyszogradensis*; OD]. Similiar to *Corynoides* but with only one adnate theca, the 2nd theca minute, coiled and distally directed; sicula curved and provided with elaborate apertural flanges. U.Ord. (glacial boulders), Eu.(Pol.).——FIG. 86,*2*. *C.*

divnoviensis KOZŁOWSKI, Pol.; ×13 (117).——FIG. 39,*5*. **C. wyszogradensis;* illustrating apertural modifications of sicula, ×35 (117).

Family NEMAGRAPTIDAE Lapworth (ex Hopkinson MS), 1873

[Nemagraptidae LAPWORTH (*ex* HOPKINSON MS, 1873, p. 556)] [=Leptograptidae LAPWORTH, 1879, p. 27]

Uniserial, bilaterally symmetrical, with two slender flexuous stipes having a primary angle of divergence of about 180 degrees; branches (if present) lateral, rarely paired, simple or compound; thecae elongate, typically inclined at low angles and with well marked sigmoid curvature (leptograptid type); development of leptograptid type. L.Ord.(*Glyptograptus teretiusculus* Z.), U.Ord.

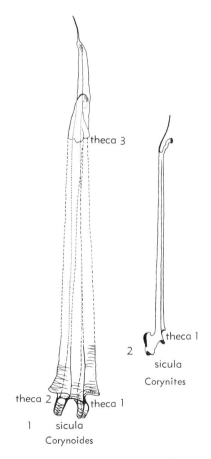

FIG. 86. Corynoididae (p. *V*119).

Nemagraptus EMMONS, 1855, p. 109 [*nom. correct.* HALL, 1859 (*pro Nemagrapsus* EMMONS, 1855), ICZN, Opin. 650] [**Graptolithus gracilis* HALL, 1848, p. 274 (=*Nemagrapsus elegans* EMMONS, 1855, p. 109; SD HALL, 1868, p. 211)] [=*Stephanograptus* GEINITZ, 1866, p. 124 (type, *G. gracilis* HALL, 1848); *Helicograpsus* NICHOLSON, 1868, p. 23 (type, *G. gracilis* HALL, 1848); *Geitonograptus* OBUT & ZUBTZOV, 1964, p. 320 (type, *G. suni*)]. Main stipes slender reclined or more usually curved to form letter S, with regularly produced lateral branches from convex side of each. *L.Ord.*

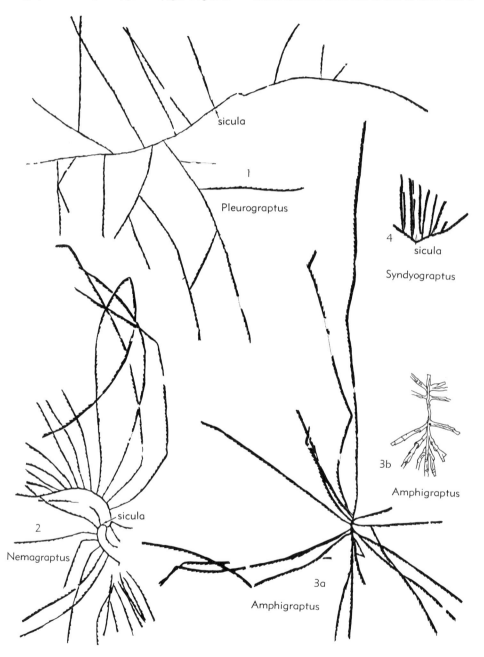

FIG. 87. Nemagraptidae (p. *V*120-*V*121).

(*Glyptograptus teretiusculus* Z.)-U.Ord.(*Glenkiln-Normanskill*), Eu.-N.Am.-S.Am.-Australia-Asia.——FIG. 87,2. **N. gracilis* (HALL), Glenkiln Sh., S.Scot.; ×1 (59).

Amphigraptus LAPWORTH, 1873, p. 559 [**Graptolithus divergens* HALL, 1859, p. 509; M] [=*Coenograptus* HALL, 1868, p. 179 (type, *Graptolithus divergens* HALL, 1859, p. 509; SD MILLER, 1889, p. 668); *Clematograptus* HOPKINSON, in HOPKINSON & LAPWORTH, 1875, p. 652 (type, *Graptolithus multifasciatus* HALL, 1859, p. 508; SD GURLEY, 1896, p. 93)]. Rhabdosome horizontal, composed of 2 straight main stipes with simple or compound, rigid lateral branches, typically produced in pairs. *U.Ord.(Normanskill-Hartfell)*, Eu.-N.Am.-China.——FIG. 87,3. **A. divergens* (HALL), 3a, rhabdosome from Hartfell Sh., S.Scot., ×1 (59); 3b, proximal end of specimen from Normanskill, N.Y., ×3 (201).

Leptograptus LAPWORTH, 1873, p. 558 [**Graptolithus flaccidus* HALL, 1865, p. 143; M]. Biramous, stipes slender, flexuous, slightly reclined, without secondary branches except in centribrachiate mutations. *?L.Ord.(Glyptograptus teretiusculus* Z.)-*U. Ord.(Bala-Normanskill-Utica-M. Dicellograptus* Sh.), Eu.-N.Am.-Australia.——FIG. 88,*1b,d*. **L. flaccidus flaccidus* (HALL), Ord.(Hartfell Sh.), S.Scot.; *1b*, centribrachiate form, ×1 (59); *1d*, proximal end, ×8 (21).——FIG. 88,*1a*. *L. flaccidus macilentus* ELLES & WOOD, Ord.(Hartfell Sh.), S.Scot.; ×1 (59).——FIG. 88,*1c*. *L. flaccidus trentonensis* RUEDEMANN, Ord.(Utica), N.Y.; ×3 (201).

Pleurograptus NICHOLSON, 1867, p. 257 [*nom. correct.* LAPWORTH, 1873 (*pro Pleurograpsus* NICHOLSON, 1867), ICZN Opin. 650] [**Cladograpsus linearis* CARRUTHERS, 1858, p. 467; OD] [=*Cladograpsus* CARRUTHERS, 1858, p. 467, *non* GEINITZ, 1852; *non* EMMONS, 1855) (type, *C. linearis*)]. Main stipes somewhat flexuous, from one or both sides of which simple or compound branches are given off rather irregularly. *U.Ord.* (*Hartfell-Utica*), NW.Eu.-N.Am.-Australia-?China.——FIG. 87,*1*. **P. linearis* (CARRUTHERS), Hartfell Sh., S.Scot.; ×1 (59).

Syndyograptus RUEDEMANN, 1908, p. 266 [**S. pecten*; OD] [=*Tangyagraptus* MU, 1963, p. 377 (type, *T. typicus*)]. Like *Amphigraptus* but with reclined main stipes and paired erect branches. *U.Ord.*, N.Am.-China.——FIG. 87,*4*. **S. pecten*, Normanskill Sh., N.Y.; ×1 (201).

Family DICRANOGRAPTIDAE Lapworth, 1873

[Dicranograptidae LAPWORTH, 1873, table 1 facing p. 555]

Uniserial or uni-biserial, reclined or initially scandent, without branches; thecae with conspicuous sigmoid curvature, some species elaborated; development of diplograptid type. *L.Ord.(Glyptograptus teretiusculus* Z.)-*U.Ord.*

Dicranograptus HALL, 1865, p. 112 [**Graptolithus ramosus* HALL, 1848, p. 270; OD] [=*Cladograpsus* EMMONS, 1855, p. 107 (type, *C. dissimilis*; SD BULMAN, 1929, p. 173); *Diceratograptus* MU, 1963, p. 377 (type, *D. mirus*)]. Proximally biserial, dividing distally to 2 uniserial reclined stipes. *L.Ord.-U.Ord.(Hartfell-Utica)*, Eu.-N.Am.-S.Am.-Australia-Asia.——FIG. 89,*2a*. *D. ramosus longicaulis* ELLES & WOOD, Hartfell Sh., S.Scot.; ×1 (59).——FIG. 89,*2b*. *D. nicholsoni* HOPKINSON, Balclatchie, S.Scot.; ×4 (23).

Dicellograptus HOPKINSON, 1871, p. 20 [*nom. correct.* LAPWORTH, 1873 (*pro Dicellograpsus* HOPKINSON), ICZN Opin. 650] [**Didymograpsus elegans* CARRUTHERS, 1868, p. 129; SD GURLEY, 1896, p. 70]. Rhabdosome of 2 reclined uniserial stipes, straight or curved. *L.Ord.-U.Ord.-(Glenkiln-Hartfell-Dicellograptus* Sh.), Eu.-N.Am.-S.Am.(Arg.)-Australia-Asia.——FIG. 89,*1a*. **D. elegans* (CARRUTHERS), M.Ord.(Hartfell Sh.), S.

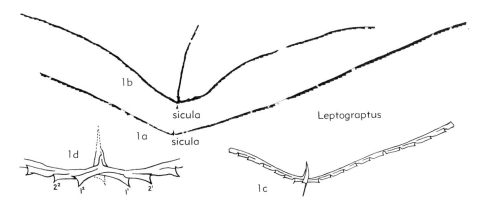

FIG. 88. Nemagraptidae (p. *V*121).

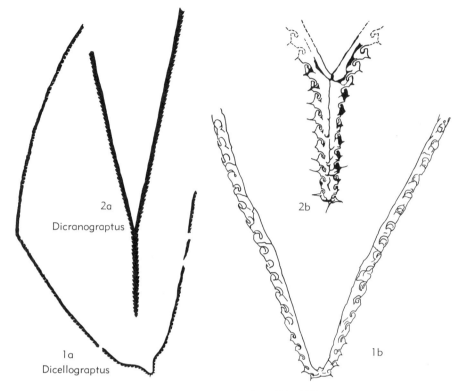

Fig. 89. Dicranograptidae (p. V121-V122).

Scot.; ×1 (59).——Fig. 89,1b. *D. morrisi*, M. Ord. (*Dicellograptus* Sh.), Sweden; ×4 (29).

Suborder GLOSSOGRAPTINA Jaanusson, 1960

[Glossograptina JAANUSSON, 1960, p. 319]

Biserial, monopleural, axonophorous graptoloids with pericalycal proximal end developed from dicalycal $th1^1$. *Ord.*

Family GLOSSOGRAPTIDAE Lapworth, 1873

[Glossograptidae LAPWORTH, 1873, table 1, facing p. 555]

Rhabdosome characteristically spined; thecae basically orthograptid but with apertural flanges in some species and commonly with secondary tissue at apertural margin. *Ord.*

Glossograptus EMMONS, 1855, p. 108 [*nom. correct.* HALL, 1865 (*pro Glossograpsus* EMMONS), ICZN Opin. 650] [**G. ciliatus*; SD LAPWORTH, 1873]. Rhabdosome with apertural, "dorsal" and lateral spines. *L.Ord.-U.Ord.*, almost worldwide.——Fig. 90,1a,b. *G. hincksi* (HOPKINSON); *1a,b,* biprofile and scalariform views; U.Ord.(Glenkiln Sh.), S. Scot., ×2 (59).——Fig. 90,1c. *G. holmi* BULMAN, Cow Head Gr., Newf., restor. of rhabdosome, ×4 (261). [*a.fl,* apertural flange; *as,* apertural spine; *ds,* "dorsal" spine; *il,* initial lacinia; *ls,* lateral spine; *s,* sicula; *v,* virgella.]

Lonchograptus TULLBERG, 1880, p. 313 [**L. ovatus;* M]. Like *Glossograptus* but with "dorsal" spines represented by a single pair of long, stout spines. *L.Ord.*, NW.Eu.——Fig. 90,3. **L. ovatus,* U. Didymograptus Sh., S.Sweden; *3a,* specimen showing thecal apertures; *3b,* outline of rhabdosome showing spines, ×2 (242).

Nanograptus HADDING, 1915, p. 328 [**N. lapworthi;* SD BULMAN, 1929, p. 179]. Rhabdosome minute; thecae denticulate or with very slender apertural spines; $th1^1$ and $th1^2$ opening downwards. *U.Ord.(Nemagraptus gracilis* Z.), Eu.(S.Sweden-Scot.).——Fig. 90,4. **N. lapworthi,* L. Dicellograptus Sh., S.Sweden; *4a,* rhabdosome, ×5; *4b,* early growth stage, ×5 (71).

Paraglossograptus HSÜ (*ex* MU MS), 1959, p. 187 [**P. latus;* SD BERRY, 1966, p. 431]. Like *Glos-*

sograptus but with well-developed lacinia; $th1^1$ and $th1^2$ with outwardly directed apertural region. L.Ord.(*up.Arenig-low.Llanvirn*), China-?Australia-N.Am.——FIG. 90,2. *P. typicalis* MU, Shihuigon Sh., China; ×1.5 (151).

Family CRYPTOGRAPTIDAE Hadding, 1915, emend. Bulman, herein

[Cryptograptidae HADDING, 1915, p. 332]

Characters of genus. *L.Ord.-U.Ord.*

Cryptograptus LAPWORTH, 1880, p. 174 [**Diplograpsus tricornis* CARRUTHERS, 1859, p. 25; OD]. Rhabdosome parallel-sided, without spines other than basal spines (sicular, $th1^1$ and th^2); distal portions of $th1^1$ and $th1^2$ curved to open outwardly and distally; subsequent thecae inclined at a high angle, with geniculum distally and very short, vertical supragenicular wall, somewhat thickened ventrally; ringlike apertural lists. *L. Ord.-U.Ord.*, almost worldwide.——FIG. 90,5. **C. tricornis*, U.Ord.(Hartfell Sh.), S.Scot.; *5a*, complete rhabdosome, ×2 (59); *5b*, restoration of proximal part, ×10 (23).

Suborder DIPLOGRAPTINA Lapworth, 1880, emend. Bulman, herein

[=*nom. correct.* OBUT, 1957, p. 17 (*ex* Diplograpta LAPWORTH, 1880, p. 191)] [=Diplograptina JAANUSSON, 1960, p. 321, excl. Monograptidae]

Biserial, dipleural, axonophorous graptoloids with platycalycal proximal end developed from dicalycal $th2^1$ or later theca. *L.Ord.-U.Sil.*

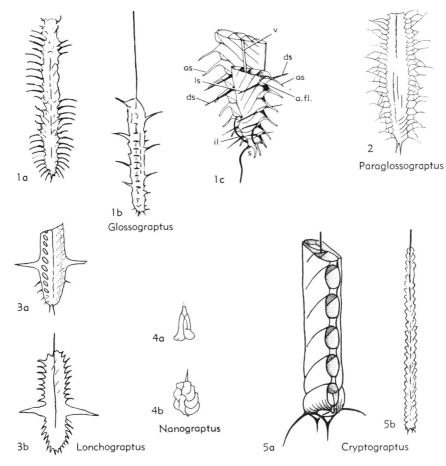

FIG. 90. Glossograptidae *(1-4)*; Cryptograptidae *(5)* (p. V122-V123).

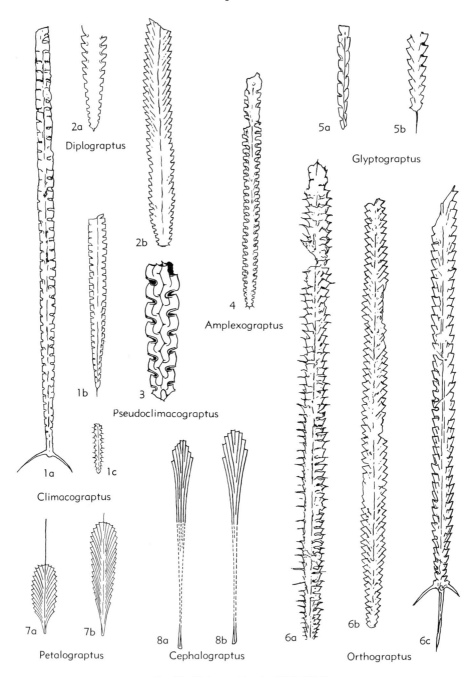

FIG. 91. Diplograptidae (p. V125-V126).

Family DIPLOGRAPTIDAE Lapworth, 1873

[Diplograptidae LAPWORTH, 1873, table 1 facing p. 555]

Rhabdosome biserial with or without median septum or with incomplete or partial septum; thecae straight (orthograptid) or with sigmoidal curvature (including glyptograptid) or with geniculum and variously-inclined supragenicular wall; usually unspined or with apertural or mesial spines restricted to base of rhabdosome, which is oval, circular or tabular in cross section; periderm continuous, rarely attenuated or supported by lists; development streptoblastic or prosoblastic. *L.Ord.-L.Sil.*

Diplograptus M'COY, 1850, p. 270 [*nom. correct.* HALL, 1865 (*pro Diplograpsis* M'COY, 1850) ICZN, Opin. 650] [**Prionotus pristis* HISINGER, 1837, p. 114; SD GURLEY, 1896, p. 78] [=*Mesograptus* ELLES & WOOD, 1907, p. 258 (type, *Graptolithus foliaceus* MURCHISON, 1839, p. 694)]. Basal thecae strongly sigmoidal with apertures in broad semicircular excavations (amplexograptid), becoming more gently sigmoid (glyptograptid) and almost straight (orthograptid) distally; periderm somewhat attenuated and with apertural lists proximally; cross section ovoid or nearly rectangular. *L.Ord.(Llanvirn)-L.Sil.*, almost worldwide.——FIG. 91,*2a*. **D. pristis* (HISINGER), U.Ord.(*Trinucleus* Sh.), Sweden; ×2. ——FIG. 91,*2b*. *D. foliaceus* (MURCHISON), L.Ord.(Meadowtown Ls.), Eng.; ×2 (Bulman, n).

Amplexograptus ELLES & WOOD, 1907, p. 258 [**Diplograptus perexcavatus* LAPWORTH, 1876, pl. 2, fig. 38; OD] [=?*Hedrograptus* OBUT, 1949, p. 13 (type, *H. janischewskyi*); *Comograptus* OBUT & SOBOLEVSKAYA in OBUT, SOBOLEVSKAYA, & MERKUREVA, 1968, p. 60 (type, *C. comatus*)]. Rhabdosome ovoid or subrectangular in cross section, with a tendency to reduction in thickness of periderm; thecae strongly geniculate, apertural excavations deep and long, generally with selvage round infragenicular wall, sometimes developed into genicular flange and sometimes confluent with apertural selvage; supragenicular wall typically slightly inclined outwards, rarely parallel to axis of rhabdosome. *L.Ord.(Llanvirn)-U.Ord.*, almost worldwide; *L.Sil.*(USSR).——FIG. 91,*4*. **A. perexcavatus* (LAPWORTH), U.Ord.(Glenkiln Sh.), S.Scot.; ×2 (59).

Cephalograptus HOPKINSON, 1869, p. 159 [*nom. correct.* LAPWORTH, 1873 (*pro Cephalograpsus* HOPKINSON, 1869), ICZN, Opin. 650] [**Diplograpsus cometa* GEINITZ, 1852, p. 26; OD]. An extreme development of *Petalograptus;* rhabdosome more or less triangular, with very elongate thecae and exposed sicula. *L.Sil.*, Eu.-Asia(China-Malaya)-USSR(Taimyr)-N.Am.(Arctic).——FIG. 91,*8*. **C. cometa* (GEINITZ), *Rastrites* Sh., Sweden; *8a*, obverse, *8b*, reverse; ×2 (239).

Climacograptus HALL, 1865, p. 111 [**Graptolithus bicornis* HALL, 1848, p. 268; OD] [=*Paraclimacograptus* PŘIBYL, 1947, p. 5 (type, *Climacograptus innotatus* NICHOLSON, 1869, p. 238)]. Rhabdosome nearly circular in cross section, scalariform views consequently common; thecae strongly geniculate, with deep apertural excavations, supragenicular wall straight, parallel to axis of rhabdosome. *L.Ord.-L.Sil.*, worldwide.——FIG. 91,*1a*. **C. bicornis* (HALL), U.Ord.(Hartfell Sh.), S.Scot.; ×2 (59).——FIG. 91,*1b*. *C. rectangularis* (M'COY), L.Sil.(Birkhill Sh.), S.Scot.; ×2 (59).——FIG. 91,*1c*. *C. innotatus* NICHOLSON, Birkhill Sh., S.Scot.; ×2 (59).

Cystograptus HUNDT, 1942, p. 206, *emend.* JONES & RICKARDS, 1967, p. 181 [**Diplograpsus vesiculosus* NICHOLSON, 1869, p. 237 (=*Cystograptus speciosus* HUNDT, 1942); SD JONES & RICKARDS, 1967, p. 181]. Rhabdosome rectangular in cross section, thecae with double sigmoid (ogee) curvature, apertures somewhat everted; point of origin of median septum variable; sicula typically elongate; vane structure commonly present distally on virgula. *L.Sil.*, Eu.(incl. USSR)-Asia(Malaya)-N.Am.(Arctic).——FIG. 92,*1*. *C. penna* (HOPKINSON), *Monograptus acinaces* Z., central Wales; ×10 (110).

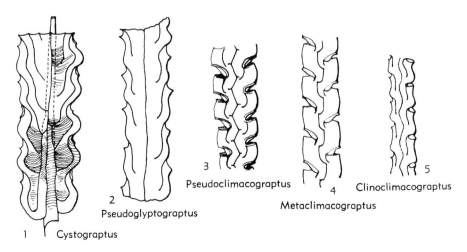

FIG. 92. Diplograptidae (p. *V*125-*V*126).

Glyptograptus Lapworth, 1873, table 1, facing p. 555 [*Diplograpsus tamariscus Nicholson, 1868, p. 526; OD]. Thecae with gentle sigmoidal curvature (glyptograptid); supragenicular wall almost straight, sloping outwards, or rarely with gentle double curvature and everted apertures; apertural margin commonly undulate. *L.Ord.(up. Arenig)-L.Sil.*, worldwide.
G. (Glyptograptus). Thecae with gentle sigmoidal curvature, apertural margins commonly undulate. *L.Ord.-L.Sil.*, worldwide.——Fig. 91,5a. *G. (G.) tamariscus* (Nicholson), L.Sil.(Birkhill Sh.), S.Scot.; ×4 (59).——Fig. 91,5b. *G. (G.) dentatus* (Brongniart), L.Ord.(*Orthoceras* Ls.), Öland, Sweden; ×4 (19).
G. (Pseudoglyptograptus) Bulman & Rickards, 1968, p. 13 [*G. (P.) vas; OD]. Supragenicular wall concavoconvex, with strongly everted aperture. *L.Sil.*, NW.Eu.——Fig. 92,2. *G. (P.) vas, Diplograptus magnus* Z., N.Eng.; ×10 (38).
Orthograptus Lapworth, 1873, table 1, facing p. 555 [*Graptolithus quadrimucronatus Hall, 1865, p. 144; OD] [=Glossograptus Ruedemann, 1947, partim (non Emmons, 1855); Rectograptus Přibyl, 1949, p. 25 (type, Diplograptus pristis var. truncatus Lapworth, 1876, pl. 1, fig. 28); Dittograptus Obut & Sobolevskaya, in Obut, Sobolevskaya, & Merkureva, 1968, p. 69 (type, D. fortuitus)]. Thecae straight or with very slight sigmoidal curvature; paired apertural spines in one group, large basal spines not uncommon; rhabdosome rectangular or ovoid in cross section. *U.Sil.-L.Sil.*, worldwide.——Fig. 91,6a. *O. quadrimucronatus* (Hall), U.Ord. (Hartfell Sh.), S.Scot.; ×2 (59).——Fig. 91,6b. *O. truncatus* (Lapworth), U.Ord.(Hartfell Sh.), S.Scot.; ×2 (59).——Fig. 91,6c. *O. calcaratus* (Lapworth), U.Ord.(Hartfell Sh.), S.Scot.; ×2 (59).
Petalograptus Suess, 1851, p. 100 [pro Diprion Barrande, 1850, and Petalolithus Suess, 1851 (ICZN pend.)]. [*Prionotus folium Hisinger, 1837, p. 114; SD Lapworth, 1873, table 1, facing p. 555]. Rhabdosome foliate, exaggeratedly rectangular in cross section; thecae long, straight or with gently ventral curvature, with large thecal overlap; $th1^1$ and $th1^2$ with pronounced upward direction of growth, leaving sicula largely exposed. *L.Sil.*, Eu.-Asia(USSR-China-Malaya)-Arctic Can.——Fig. 91,7. *P. folium* (Hisinger), *Rastrites* Sh., S.Sweden; 7a,b, obverse and reverse views, ×2 (Tullberg, 1881).
Pseudoclimacograptus Přibyl, 1947, p. 5 [*Climacograptus scharenbergi Lapworth, 1876, pl. 2, fig. 55; OD]. Like *Climacograptus* but with supragenicular walls convex, rarely nearly straight, or concavoconvex; median septum zigzag, angular and undulating in proximal region, sometimes becoming straighter distally; apertural excavations deep and short, often introverted. *L.Ord.(up.Arenig)-L.Sil.*, Eu.-Asia-N.Am.-?N.Afr.
P. (Pseudoclimacograptus). Supragenicular wall convex, apertural excavations short, deep and introverted; median septum mostly zigzag throughout. *L.Ord.* and basal *U.Ord.*, NW.Eu. (including USSR)-N.Am.-China.——Fig. 91,3; 92,3. *P. (P.) scharenbergi* (Lapworth), U.Ord. (Balclatchie beds), S. Scot.; 91,3, ×6 (23); 92,3, partly diagram.; ×10 (Bulman, 6).
P. (Clinoclimacograptus) Bulman & Rickards, 1968, p. 8 [*P. (C.) retroversus; OD]. Supragenicular wall convex proximally and concave distally; apertures strongly everted; median septum undulating proximally, straight distally. *L.Sil.*, NW.Eu.——Fig. 92,5. *P. (C.) retroversus*, Llandovery, Wales; partly diagram., ×10 (38).
P. (Metaclimacograptus) Bulman & Rickards, 1968, p. 3 [*Diplograpsus hughesi Nicholson, 1869, p. 235; OD]. Supragenicular wall gently convex or almost straight; apertural excavations short, deep, introverted and partly covered by flanges from geniculum of succeeding theca; median septum angular to undulating. *L.Sil.*, NW.Eu., ?China-?Malaya-?N.Afr.——Fig. 92,4. *P. (M.) undulatus* (Törnquist), Llandovery, Wales; partly diagram., ×17 (38).

Family LASIOGRAPTIDAE Lapworth, 1879

[Lasiograptidae Lapworth, 1879, p. 188]

Rhabdosome usually somewhat flattened, cryptoseptate or with complete or incomplete median septum; thecae geniculate, with short inwardly-inclined supragenicular wall (lasiograptid or gymnograptid); periderm commonly attenuated; more or less well-developed clathria and conspicuous development of genicular (and ?thecal) spines sometimes associated with a lacinia; development streptoblastic or prosoblastic. *Ord.*

Lasiograptus Lapworth, 1873, p. 559 [*L. costatus; OD] [=Thysanograptus Elles & Wood, 1908, p. 325 (type, Diplograptus Harknessi Nicholson, 1867, p. 262); Prolasiograptus Lee, 1963, p. 574 (type, Lasiograptus retusus Lapworth, 1880, p. 175)]. Thecae lasiograptid with somewhat inwardly-inclined supragenicular wall and inwardly-inclined (introverted) apertural margins, paired genicular spines associated with lacinia; clathria of apertural, pleural and weak parietal lists; development prosoblastic. *Ord.(up. Arenig-Caradoc)*, Eu.-N. Am.-S. Am.-Australia-China.——Fig. 93,1a. *L. costatus*, U.Ord. (*Climacograptus wilsoni* Z.), S.Scot.; ×2 (59).——Fig. 93,1b. *L. harknessi* (Nicholson), U.Ord. (Balclatchie beds), S.Scot.; somewhat schem.; ×15 (23).

Graptoloidea—Diplograptina

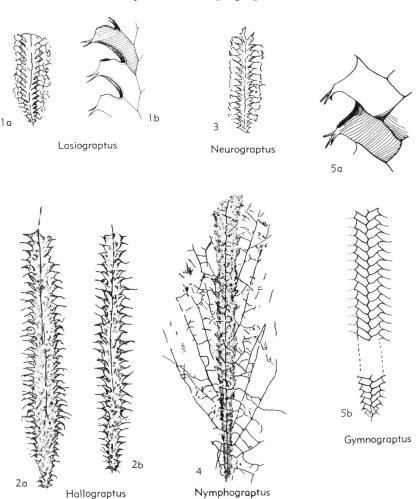

Fig. 93. Lasiograptidae (p. V126-V128).

Gymnograptus BULMAN (*ex* TULLBERG MS), 1953, p. 515 [*Diplograptus linnarssoni* MOBERG, 1896, p. 17; OD] [=?*Idiograptus* LAPWORTH, 1880, p. 169 (type, *I. aculeatus*)]. Rhabdosome somewhat flattened and more or less tabular in cross section; thecae gymnograptid, with very short supragenicular wall, everted (outwardly inclined) apertural margin accentuated by angular fuselli, and with paired genicular spines; median septum incomplete or cryptoseptate, with zigzag septal lists; development streptoblastic or prosoblastic. *L.Ord.-U.Ord.(Ogygiocaris Ser. & Ludibundus Ls.)*, NW. Eu.-China.——FIG. 93,5. *G. linnarssoni* (MOBERG); 5a, Pol.; enl. showing thecae with angular fuseli, glacial buolder, ×20; 5b, *Ogygiocaris* Ser., Norway; ×3 (250).

Hallograptus LAPWORTH (*ex* CARRUTHERS MS), 1876, p. 7 [*Diplograpsus bimucronatus* NICHOLSON, 1869, p. 236; M]. Thecae lasiograptid, with extremely short supragenicular wall and single or paired genicular spines; clathria weakly developed, lacinia absent; septal processes (scopulae) visible in scalariform view. *Ord.(Arenig-low.Caradoc)*, Eu.-N.Am.-Australia.——FIG. 93,2a. *H. bimucronatus* (NICHOLSON), U.Ord.(Glenkiln Sh.), S. Scot.; ×2 (59).——FIG. 93,2b. *H. mucronatus* (HALL), Glenkiln Sh., S.Scot.; ×2 (59).

Neurograptus ELLES & WOOD, 1908, p. 320 [=*Neurograptus* LAPWORTH, 1875, p. 641 *(nom. nud.)*] [*Lasiograptus margaritatus* LAPWORTH, 1876, pl. 2, fig. 60; SD BULMAN, 1929, p. 179]. Thecae as in *Hallograptus*; thecal spines breaking up distally into a highly developed lacinia; scopulate septal processes also well developed. *U.Ord.,* Eu.-N.Am.-Australia.——FIG. 93,3. **N. margaritatus* (LAPWORTH), Hartfell Sh., S.Scot.; ×2 (59).

Nymphograptus ELLES & WOOD (*ex* LAPWORTH MS), 1908, p. 320 [**N. velatus;* OD]. Thecae apparently as in *Hallograptus;* septal strands very strongly developed to form elaborate lacinia enveloping rhabdosome. *U.Ord.(Dicellograptus anceps* Z.-Easton.), Eu.-Australia.——FIG. 93,4. **N. velatus,* Hartfell Sh., S.Scot.; ×2 (59).

Family DICAULOGRAPTIDAE Bulman, n. fam.

Characters of genus. *L.Ord.*

Dicaulograptus RICKARDS & BULMAN, 1965, p. 278 [**Lasiograptus hystrix* BULMAN, 1932, p. 29; OD]. Rhabdosome minute; thecae 1^1 and 1^2 with isolate and introverted apertural region, mesial spine, and paired apertural spines; subsequent thecae almost dicranograptid, with angularly convex suprageniculate wall bearing elongate mesial spine, apertures introverted, with flattened lateral processes fused with rhabdosome wall to leave rounded lateral foramina; long slender spines at the base of the pleural lists; development streptoblastic. *L.Ord.,* Eu.(Sweden).——FIG. 94,*1a-c*. **D. hystrix* (BULMAN), Folkeslunda Ls. (*?Glyptograptus teretiusculus* Z.), Öland; *1a,* mature rhabdosome; ×6; *1b,* proximal end; ×14; *1c,* restoration showing thecal characters; ×14 (19).

Family PEIRAGRAPTIDAE Jaanusson, 1960

[*nom. transl.* BULMAN, 1963, *ex* Peiragraptinae JAANUSSON, 1960, p. 322]

Characters of genus. *U.Ord.*

Peiragraptus STRACHAN, 1954, p. 509 [**P. fallax;* OD]. Development of incomplete diplograptid type, with no dicalycal theca, $th2^1$ producing uniserial scandent stipe distal to $th1^1$, partially enclosing sicula; thecae geniculate, suprageniculate wall almost parallel to axis, apertural margins with rounded lateral lappets. *U.Ord.,* N.Am. ——FIG. 94,2. **P. fallax,* U.Ord.(?Vaureal F.), Anticosti Is.; ×7.5 (226).

Family RETIOLITIDAE Lapworth, 1873

[Retiolitidae LAPWORTH, 1873, table 1 facing p. 555]

Rhabdosome scandent, biserial, dipleural; periderm reduced to meshwork composed of reticulum or clathria or both, lacinia present in some forms. Thecae markedly alternate. *U.Ord.-U.Sil.*

This undoubtedly is a polyphyletic assemblage which may for convenience be provisionally divided into the following groups.

Subfamily RETIOLITINAE Lapworth, 1873

[*nom. transl.* BOUČEK & MÜNCH, 1952, p. 110 (*ex* Retiolitidae LAPWORTH, 1873)]

Well-developed reticulum supported on a distinct clathria, sicula unsclerotized or partially sclerotized (prosicula); development with partially developed ancora stage. *U.Ord.-M.Sil.*

Retiolites BARRANDE, 1850, p. 68 [*nom. conserv.* (ICZN Opin. 199)] [**Gladiolites geinitzianus* BARRANDE, 1850; M] [=*Gladiolites* BARRANDE, 1850, *nom. suppr.* ICZN Opin. 199; *Gladiograptus* LAPWORTH, 1875, p. 633 (type, *G. geinitzianus* BARRANDE, 1850); *Dimykterograptus* HABERFELNER, 1936, p. 92 (type, *D. bončevi*); *Pseudoretiolites* BOUČEK & MÜNCH, 1944, p. 22 (type, *Retiolites perlatus* NICHOLSON, 1868, p. 530)]. Reticulum on strongly developed clathria of parietal, pleural, apertural and aboral lists, with virgula rapidly incorporated on one side and dorsal list ("zigzag virgula") on other. *L.Sil.-M.Sil.,* almost worldwide.——FIG. 95,5. **R. geinitzianus* (BARRANDE), *5a,* rhabdosome from L.Sil., Boh., ×2 (13); *5b-d,* structural details of specimen from L.Sil., Dalarne, Sweden, ×12 (90).

Arachniograptus ROSS & BERRY, 1963, p. 159 [**A. laqueus;* M]. Like *Pseudoplegmatograptus,* but without lacinia. *U.Ord. (D. complanatus* Z.), N.Am.(Nev.).

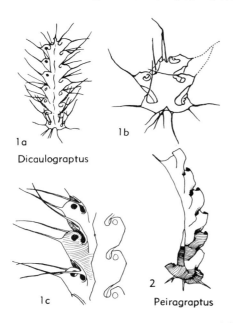

FIG. 94. Dicaulograptidae *(1);* Peiragraptidae *(2)* (p. *V*128).

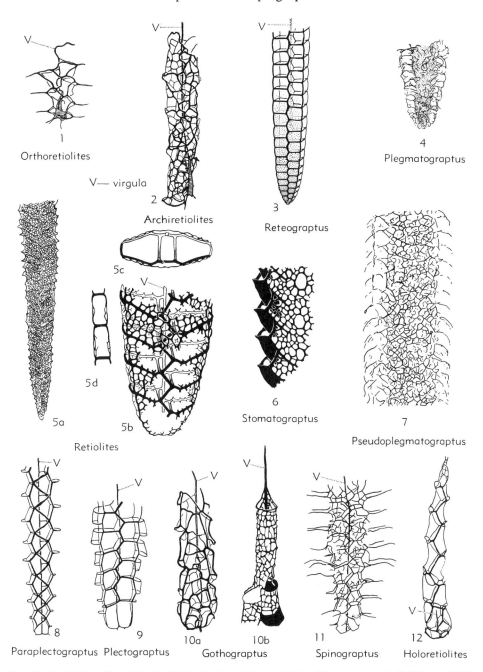

FIG. 95. Retiolitidae (Retiolitinae) *(5-7)*; (Archiretiolitinae) *(1-4)*; (Plectograptinae) *(8-12)* (p. V128-V131).

Pseudoplegmatograptus PŘIBYL, 1948, p. 22 [**Retiolites perlatus obesus* LAPWORTH, 1877, p. 137; OD]. Like *Retiolites* but with somewhat ill-defined clathria and well-developed lacinia. *L. Sil.*, Eu.-USSR(Kazakh.)-China.——FIG. 95,7. **P. obesus* (LAPWORTH), Gala, S.Scot.; ×4 (59).

Sinostomatograptus Huo Shih-Cheng, 1957, p. 521 [*S. mui; OD]. Like *Stomatograptus*, but with lacinia. *L.Sil.-M.Sil.*, China.

Stomatograptus Tullberg, 1883, p. 42 [*S. törnquisti* (=*Retiolites grandis* Suess, 1851, p. 99); M]. Like *Retiolites* but with solid interthecal septa, less overlapping thecae, and median row of large pores in reticulum. *L.Sil.-M.Sil.*, Eu.(incl. USSR)-Australia-Canad. Arctic.——Fig. 95,6. *S. grandis* (Suess), L.Sil., Dalarne, Sweden, ×12 (26).

Subfamily ARCHIRETIOLITINAE Bulman, 1955

[Archiretiolitinae Bulman, 1955, p. 88]

Sicula and initial portions of one or more proximal thecae sclerotized; development basically diplograptid. *U.Ord.*

Archiretiolites Eisenack, 1935, p. 74 [*A. regimontanus*; M]. Sicula and initial bud sclerotized; reticulum well developed, with irregular ill-defined clathria; thecae with ventral margin approximately parallel to axis of rhabdosome; virgula internal, with sporadic rodlike attachments to reticulum. *U.Ord.*, NW.Eu.——Fig. 95,2. *A. regimontanus*, glacial boulder, NW. Ger.; ×48 (54).

Orthoretiolites Whittington, 1954, p. 614 [*O. hami*; OD]. Sicula, initial bud and proximal portion of $th1^2$ sclerotized; clathria with traces of attenuated periderm but without reticulum; thecae orthograptid; virgula incorporated in obverse wall, zigzag "virgula" in reverse wall. *U.Ord.*, N.Am.——Fig. 95,1. *O. hami*, Viola Ls. (?*Nemagraptus gracilis* Z.), Okla.; ×8 (259).

Phormograptus Whittington, 1955, p. 846 [*P. sooneri*; OD]. Similar to *Archiretiolites*, but with reticulum extending below sicular aperture, supported on virgella and apertural spines, and with more horizontal direction of growth of $th1^2$. *U.Ord.*, N.Am.——Fig. 96,1. *P. sooneri*, Viola Ls. (?*Nemagraptus gracilis* Z.), Okla.; ×30 (260).

Pipiograptus Whittington, 1955, p. 839 [*P. hesperus*; OD]. Sicula, much of $th1^1$ and $th1^2$ and the initial part of $th2^1$ sclerotized; later thecae coarsely reticulate, clathria not clearly differentiated; thecal characters imperfectly known, but $th2^1$ with initial downward direction of growth. *U.Ord.*, N.Am.——Fig. 96,2. *P. hesperus*, Viola Ls. (?*Nemagraptus gracilis* Z.), Okla.; ×50 (260).

Plegmatograptus Elles & Wood, 1908, p. 340 [*P. nebula*; OD]. Reticulum with well-developed lacinia; ?membranous periderm and sclerotized sicula. Development unknown. *U.Ord.*, NW.Eu.-Australia-?N.Am.——Fig. 95,4. *P. nebula*, Hartfell Sh., S.Scot.; ×2 (59).

Reteograptus Hall, 1859, p. 518 [*R. geinitzianus*, p. 518; OD] [=*Retiograptus* Hall, 1865, p. 115

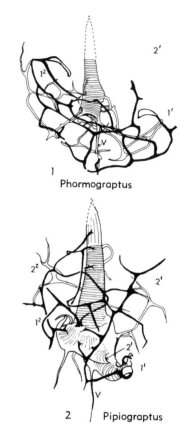

Fig. 96. Retiolitidae (Archiretiolitinae) [*V*, virgula] (p. *V*130).

(*nom. null.*); *Clathrograptus* Lapworth, 1873 (type, *C. cuneiformis*)]. Clathria only, supporting a membranous periderm at proximal end of rhabdosome; sicula ?sclerotized. *U.Ord.*, Eu.-N.Am.-China-?Australia. —— Fig. 95,3. *R. geinitzianus*, Normanskill, N.Y.; ×4 (201).

Subfamily PLECTOGRAPTINAE Bouček & Münch, 1952

[Plectograptinae Bouček & Münch, 1952, p. 110]

Clathria well developed, commonly without reticulum, lacinia absent; development with ancora stage; proximal end of rhabdosome usually somewhat inflated (corona), narrowing distally and in some genera terminating in a slender tubular "appendix." ?*L.Sil., M.Sil.-U.Sil.*

Plectograptus Moberg & Törnquist, 1909, p. 18 [*Retiolites macilentus* Törnquist, 1887, p. 491; M]. Rhabdosome rectangular in cross section, composed of open, subhexagonal meshes (clathria) with subordinate reticulum, open distally, with central (free) virgula. *M.Sil.-U.Sil.,* Eu.——FIG. 95,9. **P. macilentus* (TÖRNQUIST), low.Ludlow, Boh.; proximal portion of rhabdosome, ×4 (13).

Gothograptus FRECH, 1897, p. 670 [**Retiolites nassa* HOLM, 1890, p. 25; OD]. More or less circular in cross section, thecal apertures connected by ventral instead of pleural lists, reticulum usually fairly well developed; rhabdosome tapering distally and terminating in tubular appendix; virgula central in the corona, later incorporated in lateral wall. *U.Sil.,* Eu.-USSR(Taimyr)-Arctic Can.——FIG. 95,*10b*. **G. nassa* (HOLM), Baltic; ×12 (264).——FIG. 95,*10a*. *G. intermedius* BOUČEK & MÜNCH, Baltic; ×12 (264).

Holoretiolites EISENACK, 1951, p. 153 [**Retiolites mancki* MÜNCH, 1931, p. 1; OD] [=*Balticograptus* BOUČEK & MÜNCH, 1952, p. 117 (type, *Holoretiolites erraticus* EISENACK, 1951, p. 136)]. Tapering rhabdosome with inflated corona, usually with distal appendix, composed of clathria only; thecae climacograptid, their apertures connected by ventral lists; virgula central, confined to proximal end (corona). *U.Sil.,* Eu.——FIG. 95,*12*. **H. mancki* (MÜNCH), Baltic; ×10 (Münch, 1929).

Paraplectograptus PŘIBYL (*ex* BOUČEK & MÜNCH MS), 1948, p. 21 [**Retiolites eiseli* MANCK, 1917, p. 338; OD]. More or less square in cross section, with virgula embedded in one wall and pleural lists arranged in zigzag line in other; reticulum subordinate or absent. *?L.Sil., M.Sil.,* Eu.-?Australia.——FIG. 95,*8*. **R. eiseli* (MANCK), M.Sil., Boh.; ×4 (13).

Spinograptus BOUČEK & MÜNCH, 1952, p. 130 [**Retiolites spinosus* WOOD, 1900, p. 485; OD]. Like *Plectograptus* but, with better-developed reticulum and paired apertural spines. *U.Sil.,* Eu.-Arctic Can.——FIG. 95,*11*. **S. spinosus* (WOOD), low.Ludlow, Boh.; ×4 (13).

Family DIMORPHOGRAPTIDAE Elles & Wood, 1908

[Dimorphograptidae ELLES & WOOD, 1908, p. 347]

Proximal portion of rhabdosome uniserial, with loss or re-orientation of $th1^2$ and generally lacking further thecae of the secondary series, becoming biserial distally; biserial portion usually with partial septum (or aseptate); development of modified diplograptid type, or with initially upward-growing $th1^1$ but apparently lacking monograptid sinus and lacuna stages. *L.Sil.*

Dimorphograptus LAPWORTH, 1876, p. 545 [**D. elongatus;* SD BASSLER, 1915, p. 441] [=*Bulmanograptus* PŘIBYL, 1948, p. 46 (type, *Dimorphograptus confertus* NICHOLSON, 1868, p. 526);

Agetograptus OBUT & SOBOLEVSKAYA, in OBUT, SOBOLEVSKAYA, & MERKUREVA, 1968, p. 78 (type, *A. secundus*)]. Thecae orthograptid or glyptograptid, with a tendency in some species towards isolation of apertural region; uniserial portion of varying length; development with initial bud upwardly directed at origin. *L.Sil.,* Eu.-USSR-China-Malaya-Arctic Can.——FIG. 97,*2a*. *D. decussatus* ELLES & WOOD, Birkhill Sh., S.Scot.; ×2 (59).——FIG. 97,*2b*. **D. elongatus,* Birkhill Sh., S.Scot.; ×2 (59).

Akidograptus DAVIES, 1929, p. 9 [**A. ascensus;* OD]. Thecae climacograptid; proximal end obscure, without definite uniserial portion; initial bud downwardly directed at origin. *L.Sil.,* Eu.-China.——FIG. 97,*1*. **A. ascensus,* L.Birkhill Sh., S.Scot.; ×4.5 (47).

Rhaphidograptus BULMAN, 1936, p. 20 [**Climacograptus törnquisti* ELLES & WOOD, 1906, p. 190; OD] [=?*Metadimorphograptus* PŘIBYL, 1948, p. 46 (type, *Dimorphograptus extenuatus* ELLES & WOOD, 1908, p. 358)]. Thecae climacograptid; initial bud downwardly directed at origin. *L.Sil.,* Eu.-Malaya.——FIG. 97,*3a*. **R. toernquisti* (ELLES & WOOD), *Monograptus gregarius* Z.,

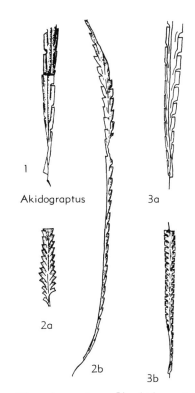

Akidograptus · Dimorphograptus · Rhaphidograptus

FIG. 97. Dimorphograptidae (p. *V*131-*V*132).

central Wales; ×3 (59).——FIG. 97,*3b*. *R. extenuatus* (ELLES & WOOD), Birkhill Sh., S.Scot.; ×2 (59).

FIG. 98. Dichograptidae (Tetragrapti) (p. *V*116).

Suborder MONOGRAPTINA Lapworth, 1880

[*nom. correct.* OBUT, 1957, p. 18 (*ex* Monograpta LAPWORTH, 1880, p. 191)]

Scandent uniserial graptoloids; development monograptid, with sinus method of pore formation and initially upward direction of growth of *th1*. L.Sil.-L.Dev.

Family MONOGRAPTIDAE Lapworth, 1873

[Monograptidae LAPWORTH, 1873, table 1 facing p. 555]

Scandent uniserial rhabdosomes without cladia. L.Sil.-L.Dev.

Monograptus GEINITZ, 1852, p. 32 [*pro Lomatoceras* BRONN, 1835 (*etiam Monoprion* BARRANDE, 1950) ICZN, Opin. 198, 1954] [**Lomatoceras priodon* BRONN, 1835, p. 56; SD BASSLER, 1915, p. 822] [=*Pomatograptus* JAEKEL, 1889, p. 677 (obj.) (type, *Lomatoceras priodon* (BRONN); SD BULMAN, 1929, p. 180)] [The following names, mostly proposed as subgenera, are technically valid, but are here included as subjective synonyms mainly owing to lack of adequate information on structural details. Reasons for placing these names in synonymy are discussed in the Addendum, p. *V*149. *Campograptus* OBUT, 1949, p. 24 (**Monograptus convolutus* var. *communis* LAPWORTH, 1876, p. 358; SD OBUT, 1964, p. 328); *Coronograptus* OBUT & SOBOLEVSKAYA in OBUT, SOBOLEVSKAYA, & MERKUREVA, 1968, p. 92 (**M. gregarius* LAPWORTH, 1876, p. 317; OD); *Demirastrites* EISEL, 1912, p. 27 (**Rastrites triangulatus* HARKNESS, 1851, p. 59; SD BULMAN, 1929, p. 175); *Globosograptus* PŘIBYL (*ex* BOUČEK & PŘIBYL MS), 1948, p. 37 (**Monograptus wimani* BOUČEK, 1932, p. 153; OD); *Lagarograptus* OBUT & SOBOLEVSKAYA, in OBUT, SOBOLEVSKAYA, & MERKUREVA, 1968, p. 90 (**L. inexpeditus;* OD); *Mediograptus* PŘIBYL (*ex* BOUČEK & PŘIBYL MS), 1948, p. 39 (**M. kolihai* BOUČEK, 1931, p. 300; OD); *Oktavites* LEVINA, 1928, p. 10 (**Graptolithus spiralis* GEINITZ, 1842, p. 700; SD OBUT, 1964, p. 328) (=*Obutograptus* MU, 1955, p. 10); *Pernerograptus* PŘIBYL, 1941, p. 9 (**Graptolithus argenteus* NICHOLSON, 1867, p. 239; OD); *Pribylograptus* OBUT & SOBOLEVSKAYA, 1966, p. 33 (**Monograptus incommodus* TÖRNQUIST, 1899, p. 11; OD); *Spirograptus* GÜRICH, 1908, p. 34 (**Graptolithus turriculatus* BARRANDE, 1850, p. 56; SD BULMAN, 1929, p. 182) (=*Tyrsograptus* OBUT, 1949, p. 24); *Streptograptus* YIN, 1937, p. 297 (**Monograptus nodifer* TÖRNQUIST, 1881, p. 436; OD); *Testograptus* PŘIBYL, 1967, p. 49 (**Graptolithus testis* BARRANDE, 1850, p. 53; OD)]. Thecae and shape of rhabdosome variable, comprising all Monograptidae other than the genera recognized below. L.Sil.(*Cystograptus vesiculosus* Z.)-L.Dev. (*Monograptus hercynicus* Z.), worldwide.——FIG. 99,*1a*. *M. cyphus* LAPWORTH, L.Sil.(L. Birkhill Sh.), S.Scot.; ×2 (59).——FIG. 99,*1b*. **M. priodon* (BRONN), L.Sil. (Gala), S.Scot.; proximal and distal ends of long rhabdosome, ×2 (59).——FIG. 99,*1c*. *M. convolutus* (HISINGER), L.Sil.(*Rastrites* Sh.), S.Sweden; ×2 (240).——FIG. 99,*1d*. *M. discus* TÖRNQUIST, L.Sil.(Tarannon), Wales; ×4 (59).——FIG. 99,*1e*. *M. turriculatus* (BARRANDE), L.Sil., Bohemia; ×2 (2).

Cucullograptus URBANEK, 1954, p. 78 [**C. pazdroi;* OD]. Thecae long, with elongate straight protheca and short metatheca; aperture round to slitlike, with lateral (monofusellar) apertural lappets or lobes, forming complex auriculate structures in extreme forms, symmetrical or asymmetrical. U.Sil.(*low.Ludlow*), chiefly *Lobograptus scanicus* Z.), NW.Eu.-Australia(Victoria)-?N.Am.

C. (Cucullograptus). Aperture slitlike, with asymmetrical lateral lobes, left lobe larger, right lobe always smaller or atrophied (L. cucullograptids). U.Sil.(*low.Ludlow*), between *Lobograptus scanicus* and *Saetograptus leintwardinensis* Z.), Eu. (Pol.-NW.Ger., boulders).——FIG. 100,*3*. *C. (C.) aversus rostratus*, ?*S. leintwardinensis* Z.,

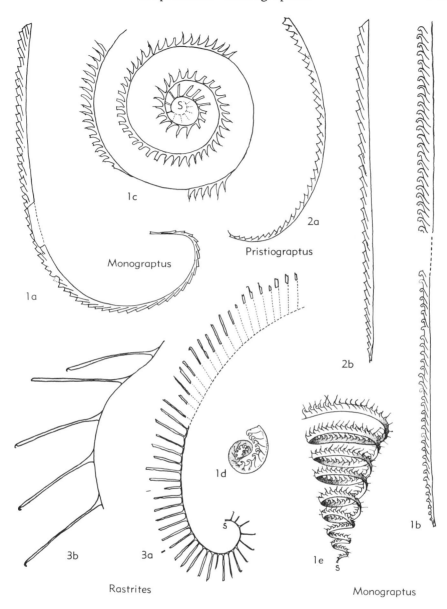

FIG. 99. Monograptidae [*S*, sicula] (p. *V*132, *V*134).

Pol.; *3a*, left side showing hypertrophied left lobe; *3b*, ventral view, ×35 (253).

C. (Lobograptus) URBANEK, 1958, p. 12 [*Monograptus scanicus* TULLBERG; OD]. Aperture rounded, with symmetrical, subsymmetrical or asymmetrical lateral lobes; where asymmetrical, the right lobe is larger (S and R cucullograptids). *U.Sil.(low.Ludlow, upper Neodiversograp-tus nilssoni* to basal *Saetograptus leintwardinensis* Z.), NW.Eu.-Australia(Victoria)-?N.Am.
——FIG. 100,*1*. *C. (L.) simplex* URBANEK, N. nilssoni Z., Pol.; *1a*, right side; *1b*, ventral aspect; ×35 (253).——FIG. 100,*2*. *C. (L.) scanicus parascanicus* (KÜHNE); low.Ludlow boulder, Pol.; *2a*, right side; *2b*, left side, showing smaller left lobe; ×35 (253).

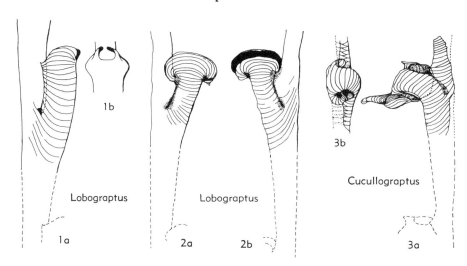

Fig. 100. Monograptidae (p. V132-V133).

Monoclimacis FRECH, 1897, p. 621 [*Graptolithus vomerinus* NICHOLSON, 1872, p. 53; OD]. Thecae geniculate, with straight suprageniculear wall approximately parallel to axis of rhabdosome; apertural margins somewhat everted; genicular flange of microfusellar tissue commonly present. *L.Sil.-U.Sil.*, almost worldwide.——FIG. 101,*1*. *M. micropoma* (JAEKEL), low.Ludlow (glacial boulder), Pol.; proximal end, approx. ×20 (249).

Pristiograptus JAEKEL, 1889, p. 667 [*P. frequens*; OD] [=*Bohemograptus* PŘIBYL, 1967, p. 134 (type, *Graptolithus bohemicus* BARRANDE, 1850, p. 40)]. Thecae simple, cylindrical, with straight or only slightly curved free ventral wall and without any distinctive apertural processes; rhabdosome straight or slightly curved ventrally. *L.Sil.-U.Sil.*, worldwide.——FIG. 99,*2a*. *P. bohemicus* (BARRANDE), low.Ludlow Sh., Wales; ×2 (59). ——FIG. 99,*2b*; 101,*2*. *P. dubius* (SUESS), low. Ludlow Sh., Eng.; *99,2b*, ×2 (59); 101,*2*, proximal end, somewhat schematic; ×10 (247).

Rastrites BARRANDE, 1850, p. 64 [*R. peregrinus*; SD HOPKINSON, 1869, p. 158] [=*Rastrograptus* HOPKINSON & LAPWORTH, 1875, p. 633 (*pro Rastrites* BARRANDE); *Corymbites* OBUT & SOBOLEVSKAYA, in OBUT, SOBOLEVSKAYA, & NIKOLAEV, 1967, p. 132 (type, *C. sigmoidalis*); *Stavrites* OBUT & SOBOLEVSKAYA, in OBUT, SOBOLEVSKAYA, & MERKUREVA, 1968, p. 111 (type, *S. rossicus*)]. Rhabdosome dorsally curved; thecae straight, isolate and tubular, with retroflexed (hooked) aperture and lateral spines in some, arising widely spaced from a threadlike "common canal" at high angles. *L.Sil.*(*Monograptus gregarius-M. turriculatus Z.*), worldwide except S.Am. and ?N.Am.——FIG. 99,*3a*. *R. longispinus* (PERNER), Birkhill Sh., S.Scot., ×2 (59).——FIG. 99,*3b*. *R. maximus* CARRUTHERS, U.Birkhill Sh., S.Scot.; ×2 (59).

Saetograptus PŘIBYL, 1942, p. 11 [*Graptolithus chimaera* BARRANDE, 1850, p. 52; OD] [=*Colonograptus* PŘIBYL, 1942, p. 2 (type, *Graptolithus colonus* BARRANDE, 1850, p. 42)]. Thecae straight, cylindrical, with lateral apertural processes (lappets or spines) of monofusellar tissue on proximal thecae or throughout rhabdosome. *U.Sil.*(*low. Ludlow, Neodiversograptus nilssoni-Lobograptus scanicus Z.*), almost worldwide.——FIG. 101,*3a*, *3c*. *S. chimaera* (BARRANDE), glacial boulder, Pol.; *3a*, proximal end showing long thecal spines, ×10; *3c*, almost complete rhabdosome; ×5 (249). ——FIG. 101,*3b*. *S. colonus* (BARRANDE); proximal end, somewhat schematic; ×10 (Bulman, n).

Family CYRTOGRAPTIDAE Bouček, 1933

[Cyrtograptidae BOUČEK, 1933, p. 1]

Scandent uniserial rhabdosomes with thecal or sicular cladia or both. *L.Sil.-L. Dev.*

Subfamily CYRTOGRAPTINAE Bouček, 1933

[*nom. transl.* YIN, 1937, p. 296 (*ex* Cyrtograptidae BOUČEK, 1933, p. 1)]

Main stipe (procladium) generally spirally coiled, helicoidally at proximal end, with one or more thecal cladia, sometimes bearing second- or higher-order cladia; production of cladia typically regular. *M.Sil. (Wenlock).*

Cyrtograptus CARRUTHERS, 1867, p. 540 [*nom. correct.* LAPWORTH, 1873 (*pro Cyrtograpsus* CARRUTHERS, 1867), ICZN, Opin. 650, 1963) [*Cyrtograpsus murchisoni*; OD] [=?*Damosiograptus* OBUT, 1950, p. 269 (type, *Cyrtograptus spiralis*

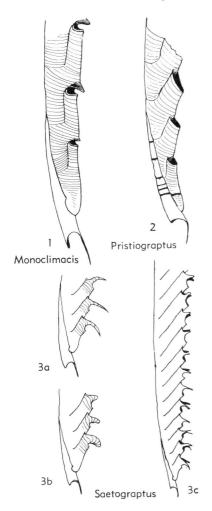

Fig. 101. Monograptidae (p.V134).

AVERIANOW, 1931, p. 11); *Lapworthograptus* BOUČEK & PRIBYL, 1952, p. 14 (type, *Cyrtograptus grayi* LAPWORTH, 1876, p. 545); *Uralograptus* KOREGN, 1962, p. 136 (type, *U. insuetus*)]. Thecae biform, hooked, or triangulate proximally with retroflexed apertures, becoming simpler distally. *M.Sil.(Wenlock),* worldwide, except S.Am.——FIG. 102,1. **C. murchisoni,* Czech.; ×2 (9).

Averianowograptus OBUT, 1949, p. 29 [**Cyrtograptus magnificus* AVERIANOW, 1931, p. 9; OD]. Like *Cyrtograptus*, but with multiple second-order cladia on 2nd thecal cladium. *M.Sil.(Wenlock),* USSR (C.Asia).——FIG. 103,1. **A. magnificus* (AVERIANOW); ×0.5 (167).

Barrandeograptus BOUČEK, 1933, p. 62 [**Cyrtograptus pulchellus* TULLBERG, 1883, p. 36; OD]. Stipes slender, thecae uniform, simple straight tubes without apertural modifications. *M.Sil.(Wenlock),* Eu.-?N.Am.——FIG. 102,2. **B. pulchellus* (TULLBERG), S.Sweden; ×2 (243).

Subfamily LINOGRAPTINAE Obut, 1957

[*nom. transl.* TELLER, 1962, p. 153 (*ex* Linograptidae OBUT, 1957, p. 18)]

Rhabdosome comprising one or more sicular cladia, with or without thecal cladia. *L.Sil.-L.Dev.*

Linograptus FRECH, 1897, p. 662 [**Dicranograptus posthumus* RICHTER, 1875, p. 267 (=*Linograptus nilssoni* FRECH, 1897, p. 662) (*non Graptolithus nilssoni* BARRANDE, 1850, p. 51; *nec Monograptus nilssoni* LAPWORTH, 1876, p. 315); OD]. Rhabdosome composed of main stipe (procladium) with at least one and generally very numerous sicular cladia; virgella with virgellarium; thecae simple, without apertural processes. *U.Sil.(low.Ludlow, Neodiversograptus nilssoni Z.)-L.Dev. (Monograptus hercynicus Z.),* Eu.-N.Am.-?Australia(NewS.Wales).——FIG. 102, 4. **L. posthumus* (RICHTER), low.Ludlow, Pol.(Silesia); ×3 (9). (Stages in development of the sicular cladia are shown in FIG. 67.)

Abiesgraptus HUNDT, 1935, p. 3 [**A. multiramosus;* SD BULMAN, 1938, p. 84] [=*Gangliograptus* HUNDT, 1939 (type, *G. hoppeianus;* SD MÜLLER, 1969)]. Rhabdosome complex, comprising procladium and 3 sicular cladia; procladium and central sicular cladium bear paired thecal cladia; thecae simple, without apertural modifications. *L.Dev.(Monograptus uniformis-M.hercynicus Z.),* C.Eu.-N.Afr.——FIG. 104,1. **A. multiramosus;* Ger.(Thuringia); ×0.7 (97).

Diversograptus MANCK, 1923, p. 283 [**D. ramosus;* SD BULMAN, 1929, p. 176]. Rhabdosome comprising one sicular cladium with or without thecal cladia; thecae hooked, with retroflexed apertures generally becoming simpler distally. *L.Sil.,* Eu.-Arctic Can.-?N.Am.——FIG. 102,3a. **D. ramosus,* Ger.(Thuringia); ×2 (140).—— FIG. 102,3b. *D. runcinatus* (LAPWORTH), up. Llandov., S.Scot.; ×4 (225).

Neodiversograptus URBANEK, 1963, p. 149 [**Monograptus nilssoni* LAPWORTH, 1876, *sensu* URBANEK, 1954, p. 300]. Rhabdosome consisting of one (? or more) sicular cladia, without thecal cladia; thecae simple, without apertural modifications. *U.Sil.(low.Ludlow, N. nilssoni Z.),* NW.Eu-N.Am.-N.Afr.-Australia. (Details of production of sicular cladium in *Neodiversograptus beklemischevi* URBANEK are shown in Fig. 66.)

Sinodiversograptus MU & CHEN, 1962, p. 152 [**S. multibrachiatus;* OD]. Like *Diversograptus*, but with numerous more or less regularly developed thecal cladia. *L.Sil.(Monograptus turriculatus Z.),* China.——FIG. 103,2. **S. multibrachiatus; 2a,* ×2; *2b,* portion enlarged to illustrate cladia production (153).

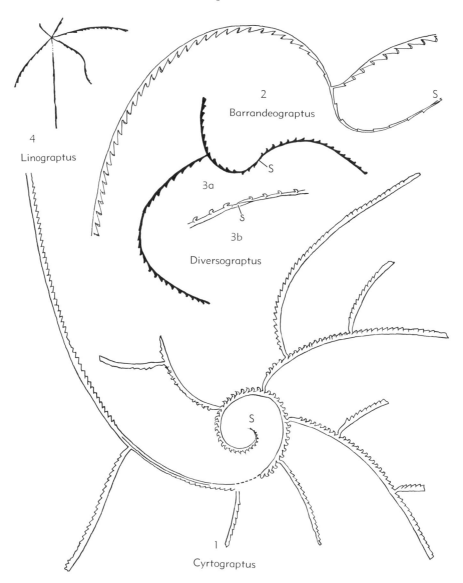

Fig. 102. Cyrtograptidae (Cyrtograptinae) *(1-2)*; (Linograptinae) *(3-4)* [*S*, sicula] (p. *V*134-*V*135).

GRAPTOLITHINA INCERTAE SEDIS

Group GRAPTOBLASTI Kozłowski, 1949

[Graptoblasti Kozłowski, 1949, p. 206]

As originally described from the Tremadoc of Poland, the graptoblasts consist of small ovoid bodies, clearly attached by their lower surface, with an upper surface which exhibits a series of ridges closely resembling the fusellar segments of Graptolithina. One end shows a rounded protuberance, the umbilicus, with a circular

Incertae Sedis—Graptoblasti

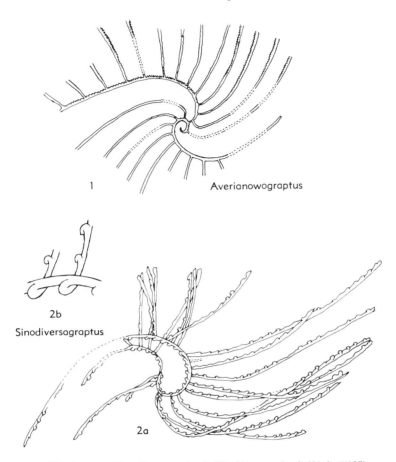

Fig. 103. Cyrtograptidae (Cyrtograptinae) *(1)*; (Linograptinae) *(2)* (p. *V*135).

opening called the cryptopyle; the other end terminates in a short spine called the filum. The vesicle itself may be undivided, or be divided into two chambers by an imperforate transverse partition, the larger of the two communicating with the exterior by the cryptopyle.

Later, Kozłowski (1962) described better-preserved material of *Graptoblastoides* from Llandeilo boulders which reveals the presence of a stolon, enclosed within a tubular stolotheca which passes into the base of the graptoblast vesicle. In the less well-preserved Tremadoc material, the filum doubtless represents a trace of the stolon, the stolotheca itself not being preserved. The upper wall in this later material is seen to be composed of two layers, a thin structureless external layer and a thick, opaque inner layer with fusellar structure. This material is intimately associated with various Crustoidea, a graptoblast occurring within the autothecal cavity of a crustoid completely filling the cavity, its walls adhering closely to those of the crustoid (see p. *V*51).

No crustoids have as yet been recorded from the Tremadoc and the relationships of the Graptoblasti remain problematic. *L.Ord.(Tremadoc-Llandeilo).*

Graptoblastus Kozłowski, 1949, p. 210 [**G. planus;* OD]. Divided by transverse partition into anterior and posterior chambers. *L.Ord.(Tremadoc),* Eu.(Pol.).——Fig. 105. **G. planus;* reconstr., ×40 (114).

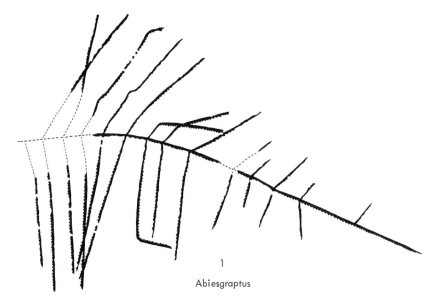

Fig. 104. Cyrtograptidae (Linograptinae) (p. V135).

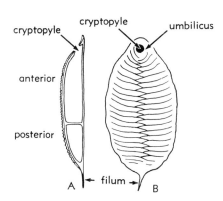

Fig. 105. Restoration of *Graptoblastus* in median section *(A)* showing anterior and posterior chambers; and in dorsal view *(B)* showing transverse ridges and median crest; approx. ×40 (114).

Graptoblastoides Kozłowski, 1949, p. 216 [**G. nowaki*; OD]. Without transverse partition. *L. Ord.(Tremadoc-Llandeilo)*, Eu.(Pol.).

Group ACANTHASTIDA
Kozłowski, 1949

[Acanthastida Kozłowski, 1949, p. 217]

Small chitinous bodies with somewhat complicated structure which appear to represent secretion of sessile colonial organism of an unknown nature. Colony discoidal, 4 to 5 mm. in diameter, attached by flattened lower surface; upper surface convex, composed of central perforated area (reticulum) bearing a few large spines surrounded by a ring of long spines; these together with the subreticular cavity constitute the spinarium. Around the spinarium lies a peripheral region with an irregularly rugose or even spinose surface, called calotte (Fig. 106). A number of radially arranged chambers underlie the calotte and spinarium; these do not communicate with one another or with the exterior but their upper portion extends into adjacent trabeculae of the reticulum. *L.Ord.*

Acanthastus Kozłowski, 1949, p. 226 [**A. luniewskii*; OD]. *L.Ord.(Tremadoc)*, Eu.(Pol.).——Fig. 106. **A. luniewskii*; reconstr., ×15 (114).

Group GRAPTOVERMIDA
Kozłowski, 1949

[Graptovermida Kozłowski, 1949, p. 204]

Small irregularly coiled chitinous tubes with fusellar structure. *L.Ord.*

Graptovermis Kozłowski, 1949, p. 206 [**G. spiralis*; OD]. Flexuous or irregularly coiled

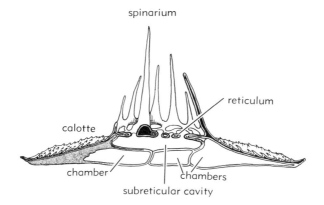

FIG. 106. Restoration of *Acanthastus*, in median section, approx. ×15 (114).

chitinous tubes with a diameter of 100 to 400 microns, attached by one surface; growth by addition of fusellar segments as in Graptolithina. L.Ord., Eu.(*Tremadoc*, Pol.; ?*up.Arenig*, Sweden).

UNRECOGNIZABLE GENERA

The following genera are not accepted as graptolites, represent unidentifiable preservational views (e.g., scalariform or subscalariform), or are too imperfectly known for description and taxonomic placement. [Most of HUNDT's genera were described in periodicals inaccessible outside Germany, but figures were published in HUNDT, 1953 and 1965.]

Birastrites GEINITZ, 1866, p. 125.
Buthograptus HALL, 1861, p. 18.
Cameragraptus HUNDT, 1951.
Cardograptus HUNDT, 1965.
Conograptus RUEDEMANN, 1947, p. 267.
Ctenograptus NICHOLSON, 1876, p. 248.
Cystoturriculograptus HUNDT, 1952.
Dawsonia NICHOLSON, 1873, p. 139 [*non* HARTT in DAWSON, 1868].
Demicystograptus HUNDT, 1942.
Dibranchiograptus HUNDT, 1949.
Didymograptoides HUNDT, 1951.
Eiseligraptus HUNDT, 1965.
Falcatograptus HUNDT, 1965.
Geminograptus HUNDT, 1951.
Labrumograptus HUNDT, 1952.
Limpidograptus KHALETSKAYA, 1962, p. 72.
Megalograptus MILLER, 1874, p. 343.
Mystiograptus HUNDT, 1965.
Nereitograptus HUNDT, 1951.
Nereograptus GEINITZ, 1852, p. 27.
Nodosograptus HUNDT, 1951.
Paradimorphograptus HUNDT, 1951.
Paragraptus HUNDT, 1965.
Phycograptus GURLEY, 1896, p. 89.
Planktograptus YAKOVLEV, 1933, p. 979.
Procrytograptus POULSEN, 1943, p. 302.
Protistograptus MCLEARN, 1915, p. 55.
Protograptus MATTHEW, 1886, p. 31.
Protovirgularia M'COY, 1850, p. 272.
Spinosidiplograptus HUNDT, 1951.
Stelechograptus RUEDEMANN, 1947, p. 279.
Strophograptus RUEDEMANN, 1904, p. 716.
Thamnograptus HALL, 1859, p. 519.
Thecocystograptus HUNDT, 1947.
Thuringiagraptus HUNDT, 1935.
Trigonograpsus NICHOLSON, 1869, p. 231 [=*Trigonograptus* LAPWORTH, 1873, ICZN Opin. 650].
Triplograptus RICHTER, 1871, p. 251.
Triplograptus HUNDT, 1965.
Undograptus HUNDT, 1949.

REFERENCES

Angelin, N. P.
(1) 1854, *Palaeontologia Scandinavica*: 92 p., 49 pl., T. O. Weigel (Lipsiae).

Barrande, Joachim
(2) 1850, *Graptolites de Bohême*: 74 p., 4 pl., the author (Prague).

Barrington, E. J. W.
(3) 1965, *The biology of Hemichordata and Protochordata*: 176 p., 82 text fig., 1 pl., Oliver & Boyd (Edinburgh).

Barrois, Charles
(4) 1893, *Sur le Ronvilligraptus Richardsoni de*

Cabrières: Soc. Géol. Nord., Ann., v. 21, p. 107-112, pl. 3,4.

Bassler, R. S.
(5) 1909, *Dendroid graptolites of the Niagaran dolomites at Hamilton, Ontario:* U.S. Natl. Museum, Bull. 65, 76 p., 91 text fig., 5 pl.

Beklemishev, V. N.
(6) 1951, *K postroeniyu sistemy zhivotnykh. Vtorichnorotye (Deuterostomia), ikh proizkhozhdenie i sostav:* Uspekhi Sovremernoi Biologii, v. 32, p. 256-270, text fig. 1-10. [*On the systematic structure of animals. Vtorichnorotye (Deuterostomia), their origin and composition.*]

Benson, W. N., Keble, R. A., King, L. C., & McKee, J. T.
(7) 1936, *Ordovician graptolites of North-West Nelson, N.Z.:* Royal Soc. New Zealand, Trans., v. 65, p. 357-382, text fig. 1-6.

Berry, W. B. N.
(8) 1960, *Graptolite faunas of the Marathon region, West Texas:* Univ. Texas Publ. 6005. p. 179, 20 pl.

Bouček, Bedřich
(9) 1933, *Monographie der obersilurischen Graptolithen aus der Familie Cyrtograptidae:* Geol.-Paleont. Ústavu Karlovy Univ. Praze, Práce, no. 1, p. 1-84, text fig. 1-19, pl. 1-7.
(10) 1941, *Ueber neue Algenreste aus dem böhmischen Silur:* K. České Společnost Nauk, Věstník, p. 1-5.
(11) 1956, *The graptolite and dendroid fauna of the Klabava Beds:* Ústřed. Ústavu Geol., Sborník, v. 22, p. 1-105, text fig. 1-9, pl. 1-6.
(12) 1957, *The dendroid graptolites of the Silurian of Bohemia:* Same, Rozpravy, v. 23, p. 1-294, text fig. 1-75, pl. 1-39.

———, **& Münch, Arthur**
(13) 1952, *The Central European Retiolites of the Upper Wenlock and Ludlow:* Ústřed. Ústavu Geol., Sborník, v. 19, p. 1-151, text fig. 1-14, pl. 1.

———, **& Přibyl, Alois**
(14) 1951, *Taxonomy and phylogeny of some Ordovician graptolites:* Acad. Tchèque Sci., Bull. Internatl., v. 52, no. 20, p. 1-17, text fig. 1-4.
(15) 1952, *Contribution to our knowledge of the cyrtograptids from the Silurian of Bohemia:* Same, v. 53, p. 1-25.

Bronn, H. G.
(16) 1834, *Lethaea geognostica:* v. 1, 768 p., E. Schweizerbart (Stuttgart).

Bulman, O. M. B.
(17) 1927, *Koremagraptus, a new dendroid graptolite:* Ann. & Mag. Nat. History, ser. 9, v. 19, p. 344-347, text fig. 1, pl. 4.
(18) 1927-67, *British dendroid graptolites:* Palaeontograph. Soc. London, Mon., lxiv+ 97 p., 43 text fig., 10 pl.
(19) 1932-36, *On the graptolites prepared by Holm, I-VII:* Arkiv f. Zoologie, v. 24A, no. 8, p. 1-46, text fig. 1-18, pl. 1-9; no. 9, p. 1-29, text fig. 1-12, pl. 1-9; v. 26A, no. 5, p. 1-52, text fig. 1-19, pl. 1-9; v. 28A, no. 17, p. 1-107, text fig. 1-30, pl. 1-4.
(20) 1936, *Rhaphidograptus, a new graptolite genus:* Geol. Mag., v. 73, p. 19-26.
(21) 1938, *Graptolithina:* In O. H. Schindewolf, Handbuch der Paläozoologie, v. 2D, p. 1-92, text fig. 1-42, Borntraeger (Berlin).
(21a) 1938, *The proximal end of Cryptograptus:* Geol. Mag., v. 75, p. 539-543.
(22) 1941, *Some dichograptids of the Tremadocian and Lower Ordovician:* Ann. & Mag. Nat. History, ser. 11, v. 7, p. 100-121, pl. 11.
(23) 1944-47, *Monograph of Caradoc (Balclatchie) graptolites from limestones in Laggan Burn, Ayrshire:* Palaeontograph. Soc., London, Mon., p. 1-78, text fig. 1-40, pl. 1-10.
(24) 1949, *A re-interpretation of the structure of Dictyonema flabelliforme Eichwald:* Geol. Fören. Stockholm, Förhandl., v. 71, p. 33-40, text fig. 1-6.
(25) 1950, *Graptolites from the Dictyonema shales of Quebec:* Geol. Soc. London, Quart. Jour., v. 106, p. 63-99, text fig. 1-4, pl. 4-8.
(26) 1950, *The structure and relations of Cyclograptus Spencer:* Jour. Paleontology, v. 24, p. 566-570, text fig. 1, pl. 73.
(27) 1953, *Some graptolites from the Ogygiocaris Series (4aα) of the Oslo District:* Arkiv Mineralogi Geologi, v. 1, p. 509-518, pl. 1,2.
(28) 1954, *The graptolite fauna of the Dictyonema shales of the Oslo Region:* Norsk Geol. Tidsskrift, v. 33, p. 1-40, text fig. 1-13, pl. 1-8.
(29) 1955, *Graptolithina:* in R. C. Moore (ed.), Treatise on invertebrate paleontology, Pt. V, xvii+101 p., 358 text fig., Geol. Soc. America and Univ. Kansas Press, (New York, N.Y.; Lawrence, Kans.).
(30) 1958, *Patterns of colonial development in graptolites:* Linnean Soc. London (Zool.), Jour., v. 44, p. 24-32, text fig. 1-7.
(31) 1958, *The sequence of graptolite faunas:* Palaeontology, v. 1, p. 159-173, text fig. 1-3.
(32) 1963, *On Glyptograptus dentatus (Brongniart) and some allied species:* Same, v. 6, p. 665-689, text fig. 1-11, pl. 96, 97.
(33) 1963, *The evolution and classification of the Graptoloidea:* Geol. Soc. London, Quart. Jour., v. 119, p. 401-418, text fig. 1-5.
(34) 1964, *Lower Palaeozoic plankton:* Same, v. 120, p. 455-476, text fig. 1-10.

(35) 1968, *The mode of development of Isograptus manubriatus:* Geol. Mag., v. 105, p. 211-215, text fig. 1,2.
(36) 1969, *"Prothecal folds" and the origin of Dicellograptus:* in K. S. Campbell (ed.), Stratigraphy and palaeontology: Essays in honour of Dorothy Hill, p. 1-16, Australian Natl. Univ. Press (Canberra).

———, & Rickards, R. B.
(37) 1966, *A revision of Wiman's dendroid and tuboid graptolites:* Univ. Uppsala, Bull. Geol. Inst., v. 43, p. 1-72, text fig. 1-46.
(38) 1968, *Some new diplograptids from the Llandovery of Britain and Scandinavia:* Palaeontology, v. 11, p. 1-15.

Carruthers, William
(39) 1858, *Dumfriesshire graptolites, with descriptions of three new species:* Royal Phys. Soc. Edinburgh, Proc., v. 1, p. 466-470.
(40) 1867, *On graptolites:* in R. I. Murchison, Siluria, 4th ed., Appendix D, p. 538-541, John Murray (London).

Castellaro, H. A.
(41) 1963-66, *Guía Paleontológica Argentina: Parte I. Faunas Cámbricas, Ordovícicas* (secc. 1-2) (1963), p. 1-165, Consejo Nac. Inves. Cient. y Tec. (Buenos Aires); Same, *Faunas Silúricas* (secc. 3) (1966), p. 1-164.

Chapman, Frederick
(42) 1919, *On some hydroid remains of Lower Palaeozoic age from Monegetta, near Lancefield:* Royal Soc. Victoria, Proc., ser. 2, v. 31, p. 388-393, pl. 19-20.

———, & Thomas, D. E.
(43) 1936, *The Cambrian Hydroida of the Heathcote and Monegetta Districts:* Royal Soc. Victoria, Proc., ser. 2, v. 48, p. 193-212, pl. 14-17.

Chen X., Sun X. R., & Han N. R.
(44) 1964, *Yushanograptus, a new graptolite genus from the Ningkuo Shale:* Acta Paleont. Sinica, v. 12, p. 236-240.

Chu M. T.
(45) 1965, *New materials of Sinograptidae:* Acta Paleont. Sinica, v. 13, p. 94-106, pl. 1.

Clark, T. H.
(46) 1924, *The paleontology of the Beekmantown Series at Levis, Quebec:* Bull. Am. Paleontology, v. 10, no. 41, p. 1-314, pl. 1-9.

Davies, K. A.
(47) 1929, *Notes on the graptolite faunas of the Upper Ordovician and Lower Silurian:* Geol. Mag., v. 66, p. 1-27.

Dawydoff, Constantin
(48) 1948, *Embranchement des Stomochordés,* in P. -P. Grassé, Traité de Zoologie, v. 11, p. 367-489, text fig. 1-113, Masson (Paris).

Decker, C. E.
(49) 1945, *The Wilberns Upper Cambrian graptolites from Mason, Texas:* Univ. Texas, Publ., 4401, p. 13-61, text fig. 1-4, pl. 1-10.

Dixon, E. E. L.
(50) 1931, *Comparison with Welsh and North American series:* in E. E. L. Dixon, T. Eastwood, S. E. Hollingworth, & B. Smith, The geology of the Whitehaven and Workington District, Geol. Survey United Kingdom, Mem., p. 31-34.

Dunham, K. C.
(51) 1961, *Black shale, oil and sulphide ore:* (British Assoc.) Adv. Sci., v. 18, p. 284-299.

Eichwald, Eduard
(52) 1955, *Beitrag zur geographischen Verbreitung der fossilen Thiere Russlands:* Soc. Impér. Naturalistes Moscou, Bull., v. 28, p. 433-466.

Eisel, Robert
(53) 1912, *Über zonenweise Entwicklung der Rastriten und Demirastriten in mittelsilurischen Graptolithenschiefern:* Gesell. Freund. Naturwiss. Gera, Jahresber., 53/54, p. 27-55, pl. 1-4.

Eisenack, Alfred
(54) 1935, *Neue Graptolithen aus Geschieben baltischen Silurs:* Paläont. Zeitschr., v. 17, p. 73-90, pl. 4-7.
(55) 1941, *Epigraptus bidens n. g., n. sp., eine neue Graptolithenart des baltischen Ordoviciums:* Zeitschr. Geschiebeforschung, v. 17, p. 24-28, pl. 1.
(56) 1942, *Über einige Funde von Graptolithen aus ostpreussischen Silurgeschieben:* Same, v. 18, p. 29-42, text fig. 1, pl. 1-2.
(57) 1951, *Retioliten aus dem Graptolithengestein:* Palaeontographica, v. 100, p. 129-163, text fig. 1-11, pl. 21-25.

Elles, G. L.
(58) 1922, *The graptolite faunas of the British Isles:* Geol. Assoc., Proc., v. 33, p. 168-200, text fig. 38-52.

———, & Wood, E. M. R.
(59) 1901-18, *Monograph of British graptolites, Pts. I-XI:* Palaeontograph. Soc., Mon., clxxi+539 p., 359 text fig., 52 pl.

Emmons, Ebenezer
(60) 1855, *American Geology:* v. 1, pt. 1, 194 p.; pt. 2, 251 p., Sprague (Albany, N.Y.).

Erdtmann, B. D.
(61) 1967, *A new fauna of early Ordovician graptolites from St. Michel, Quebec:* Canadian Jour. Earth Sci., v. 4, p. 335-355, pl. 1.

Florkin, Marcel
(62) 1965, *Paléoprotéins:* Acad. Royale Belgique (Sciences), Bull., ser. 5, v. 51, p. 156-169.

Foerste, A. F.
(63) 1923, *Notes on Medinan, Niagaran and*

Chester fossils: Denison Univ., Sci. Lab., Bull. 20, p. 37-120, pl. 4-15a.

Frech, Fritz
(64) 1897, *Lethaea geognostica;* Theil 1, *Lethaea palaeozoica,* 1.Bd., *Graptolithiden:* (Lief. 1-2 by Ferdinand Roemer; Lief.3 continued by Fritz Frech), p. 544-684, text fig. 127-226, Schweizerbart (Stuttgart).

Geinitz, H. B.
(65) 1852, *Die Versteinerungen der Grauwacken-Formation (Die Graptolithen):* 58 p., 6 pl., Verlag Wilhelm Engelmann (Leipzig).
(66) 1866, *James Hall: Graptolites of the Quebec Group (review):* Neues Jahrb. Mineralogie, Geologie, Paläontologie, Jahrg. 1866, p. 274.

Gürich, Georg
(67) 1908, *Leitfossilien:* v. 1, Kambrium und Silur, Lief 1, 95 p., 28 pl., Gebrüder Borntraeger (Berlin).

Gurley, R. R.
(68) 1896, *North American graptolites:* Jour. Geology, v. 4, p. 63-102, pl. 4-5, p. 291-311.

Haberfelner, Erich
(69) 1936, *Neue Graptoliten aus dem Gotlandium von Böhmen:* Geol. Balkanica, v. 2, p. 87-95.

Hadding, Assar
(70) 1913, *Undre Dicellograptusskiffern i Skåne:* Lunds Univ. Årsskr., new ser., pt. 2, v. 9, no. 15, p. 1-91, text fig. 1-23, pl. 1-8.
(71) 1915, *Om Glossograptus, Cryptograptus och tvenne dem närstående graptolitslägten:* Geol. Fören. Stockholm, Förhandl., v. 37, p. 303-336, pl. 5-6.
(72) 1915, *Der mittlere Dicellograptus-schiefer auf Bornholm:* Lunds Univ. Årsskr., new ser., pt. 2, v. 11, no. 4, p. 1-39, text fig. 1-3, pl. 1-4.

Hall, James
(73) 1851, *New genera of fossil corals:* Am. Jour. Sci., ser. 2, v. 11, p. 398-401.
(74) 1858, *Descriptions of Canadian graptolites:* Geol. Survey Canada, Rept. for 1857, p. 111-145, pl. 1-8 (?publ.).
(75) 1859, *Notes upon the genus Graptolithus:* Paleontology of New York, v. 3 (suppl.), p. 495-529 (Albany, N.Y.).
(76) 1861, *Report of the superintendent of the Geological Survey, exhibiting the progress of the work, January 1, 1861 (including descriptions of new species of fossils from the investigations of the survey):* 52 p., Wisconsin Geol. Survey (Madison, Wis.).
(77) 1865, *Graptolites of the Quebec group:* Geol. Survey Canada, Canad. Organic Remains, dec. 2, p. 1-151, text fig. 1-31, pl. A-B, 1-21.
(78) 1868, *Introduction to the study of the Graptolitidae:* N.Y. State Cab. Nat. History, 20th Rept., p. 169-240, 25 pl. (Albany, N.Y.).
(79) 1883, *Descriptions of new species of fossils found in the Niagaran group at Waldron, Indiana:* Geol. Survey Indiana, Rept. 11, p. 217-345, 36 pl.

Hall, T. S.
(80) 1914, *Victorian graptolites, Part IV:* Royal Soc. Victoria, Proc., v. 27, p. 104-118, text fig. 1-7, pl. 17-18.

Harris, W. J.
(81) 1926, *Victorian graptolites (new series), Part II:* Royal Soc. Victoria, Proc., v. 38, p. 55-61, pl. 1,2.
(82) 1933, *Isograptus caduceus and its allies in Victoria:* Same, v. 46, p. 79-114, pl. 1.
(83) 1916, *The palaeontological sequence of the Lower Ordovician rocks in the Castlemaine District:* Royal Soc. Victoria, Proc., v. 29, p. 50-74, pl. 1.

———, & **Keble, R. A.**
(84) 1932, *Victorian graptolite zones with correlations and description of species:* Same, v. 44, p. 25-48, pl. 3-6.

———, & **Thomas, D. E.**
(85) 1938, *Revised classification and correlation of the Ordovician graptolite beds of Victoria:* Victoria Dept. Mines, Mining & Geol. Jour., v. 1, p. 69-72, pl. 1-3.
(86) 1940, *Victorian graptolites (new series), Part VII:* Same, v. 2, p. 128-136, text fig. 1-3, pl. 1-2.
(87) 1940, *Victorian graptolites (new series), Part VIII:* Same, v. 2, p. 197-198.
(88) 1941, *Victorian graptolites (new series), Part IX:* Same, v. 2, p. 308-310, pl. 1-2.

Holm, Gerhard
(89) 1881, *Bidrag till kännedomen om Skandinaviens Graptoliter. I, II:* K. Svenska Vetenskaps. Akad., Förhandl., v. 38, no. 4, p. 71-84; no. 9, p. 45-52, pl. 12-13.
(90) 1890, *Gotlands Graptoliter:* Svenska Vetenskaps.-Akad. Handl., Bihang, v. 16, ser. 4, no. 7, p. 1-34, pl. 1-2.
(91) 1895, *Om Didymograptus, Tetragraptus och Phyllograptus:* Geol. Fören. Stockholm, Förhandl., v. 17, p. 319-359, text fig. 1-6, pl. 11-16 (Eng. transl. by G. L. Elles & E. M. R. Wood in Geol. Mag., v. 32, p. 433-492, text fig. 1-3, pl. 13-14).

Hopkinson, John
(92) 1869, *On British graptolites:* Quekett Micros. Club, Jour., v. 1, p. 151-166, pl. 8.
(93) 1871, *On Dicellograpsus a new genus of graptolites:* Geol. Mag., v. 8, p. 20-26, pl. 1.
(94) 1871, *On a specimen of Diplograpsus pristis with reproductive capsules:* Ann. & Mag. Nat. History, ser. 4, v. 7, p. 317-320, text fig. 1-2.

———, & Lapworth, Charles
(95) 1875, *Descriptions of the graptolites of the Arenig and Llandeilo rocks of St. David's:* Geol. Soc. London, Quart. Jour., v. 31, p. 631-672, pl. 33-37.

Hudson, G. H.
(96) 1915, *Ceramograptus ruedemanni, a new genus and species of Graptolitoidea:* Ottawa Naturalist, v. 28, p. 129-130, pl. 2.

Hundt, Rudolf
(97) 1935, *Die Graptolithenfauna des obersten Obersilurs Thüringens:* Zeitschr. Naturwiss. (Halle), v. 91, p. 1-31.
(98) 1942, *Der Schwebeapparat der diprionitischen Graptolithen:* Beiträge Geologie Thüringens, v. 7, p. 71-74.
(99) 1953, *Graptolithen:* p. 1-48, Die Neue Brehm-Bücherei (Leipzig).
(100) 1965, *Aus der Welt der Graptolithen:* p. 1-58, 326 text fig., Seidel & Co. (Berlin, Bonn).

Huo Shih-cheng
(100a) 1957, *Some Silurian graptolites of the Family Retiolitidae from Liangshan Hanchung:* Acta Palaeont. Sinica, v. 5, p. 513-522, pl. 1-3.

Hutt, J., & Rickards, R. B.
(101) 1967, *An improved transfer technique for the preparation and preservation of pyritized graptolites:* Geol. Mag., v. 104, p. 180-181.

Huxley, Julian
(102) 1958, *Evolutionary processes and taxonomy with special reference to grades:* Uppsala Univ., Årsskr. (Systematics of Today Symposium), v. 6, p. 21-39, text fig. 1-12.

Jaanusson, Valdar
(103) 1960, *Graptoloids from the Ontikan and Viruan (Ordov.) limestones of Estonia and Sweden:* Univ. Uppsala, Bull. Geol. Inst., v. 38, p. 289-366, text fig. 1-9, pl. 1-5.
(104) 1965, *Two multiramous graptoloids from the Lower Didymograptus shale of Scandinavia:* Geol. Fören. Förhandl., v. 86, p. 413-432, text fig. 1-10.

Jackson, D. E.
(105) 1967, *Psigraptus, a new graptolite genus from the Tremadocian of Yukon, Canada:* Geol. Mag., v. 104, p. 311-315, text fig. 1.

———, & Bulman, O. M. B.
(105a) 1970, *On the generic name Trigonograptus Nicholson, 1869:* Geol. Soc. London, Proc., no. 1662 (in press).

Jaeger, Hermann
(106) 1959, *Graptolithen und Stratigraphie des jüngsten thüringer Silurs:* Deutsch. Akad. Wiss. Berlin, Abhandl., Jahrg. 1959, no. 2, p. 1-197, text fig. 1-27, pl. 1-14.

(107) 1960, *Über Diversograptus:* Paläont. Zeitschr., v. 34, p. 13-14.

Jaekel, Otto
(108) 1889, *Ueber das Alter des sogen. Graptolithen-gesteins:* Deutsch. Geol. Gesell., Zeitschr., v. 41, p. 653-716, pl. 28-29.

Jahn, J. J.
(109) 1892, *Vorläufiger Bericht über die Dendroiden des Böhmischen Silur:* K. K. Akad. Wiss. Wien, Sitzungsber., Math-nat. Kl., p. 1-8.

Jones, W. D. V., & Rickards, R. B.
(110) 1967, *Diplograptus penna Hopkinson 1869, and its bearing on vesicular structures:* Paläont. Zeitschr., v. 41, p. 173-185.

Kaźmierczak, Józef, & Pszczółkowski, Andrzej
(110a) 1969, *Burrows of Enteropneusta in Muschelkalk (Middle Triassic) of the Holy Cross Mountains, Poland:* Acta Palaeont. Polonica, v. 14, p. 299-324, pl. 1-5 (Polish and Russian summaries).

Khaletskaya, O. N.
(111) 1962, *Stratigrafiya i paleontologiya Uzbekistana i sopredelnykh rayonu:* Glavgeol. Uzbek. SSR, Akad. Nauk Uzbek. SSR, Inst. Geol., v. 1, p. 72. [*Stratigraphy and paleontology of Uzbekistan and adjoining regions.*]

Koregn, T. N.
(112) 1962, *Novyy siluriyskiy rod Uralograptus:* Paleont. Zhurnal, v. 3, p. 137-138, text fig. 1. [*Uralograptus, a new Silurian genus.*]

Kozłowski, Roman
(113) 1938, *Informations préliminaires sur les Graptolithes du Tremadoc de la Pologne et sur leur portée théorique:* Ann. Musei Zool. Polon., v. 13, p. 183-196, text fig. 1, 2.
(114) 1949, *Les graptolithes et quelques nouveaux groupes d'animaux du Tremadoc de la Pologne:* Palaeont. Polonica, v. 3 (1948), p. 1-235, text fig. 1-66, pl. 1-42.
(115) 1952, *Sur un remarquable Graptolithe ordovicien:* Acta Geol. Polonica, v. 2 (1951), p. 86-93 (p. 291-299, Polish text), text fig. 1-5.
(116) 1954, *Sur la structure de certains Dichograptides:* Same, v. 4, p. 118-135 (p. 423-444, Polish text), text fig. 1-11, pl. 1.
(117) 1956, *Nouvelles observations sur les Corynoidideae (Graptolithina):* Acta Palaeont. Polonica, v. 1, p. 259-269, text. fig. 1-5.
(118) 1959, *Les hydroïdes ordoviciens à squelette chitineux:* Same, v. 4, p. 209-271, text fig. 1-31.
(119) 1960, *Calyxdendrum graptoloides n. gen., n. sp.—a graptolite intermediate between the Dendroidea and the Graptoloidea:* Same, v. 5, p. 107-125, text fig. 1-11.

(120) 1961, *Découverte d'un Rhabdopleuridé (Pterobranchia) ordovicien:* Same, v. 6, p. 3-16, text fig. 1-3.
(121) 1962, *Crustoidea, nouveau groupe de graptolites:* Same, v. 7, p. 3-52, pl. 1-4.
(122) 1963, *Le développement d'un graptolite tuboide:* Same, v. 8, p. 103-134, text fig. 1-16.
(123) 1966, *Graptolodendrum mutabile, n. gen., n. sp.—an aberrant dendroid graptolite:* Same, v. 11, p. 3-14, text fig. 1-13.
(124) 1966, *On the structure and relationships of graptolites:* Jour. Paleontology, v. 40, p. 489-501, text fig. 1-12.
(125) 1967, *Sur certains fossiles ordoviciens à test organique:* Acta Palaeont. Polonica, v. 12, p. 99-132, text fig. 1-13.

Kraatz, Reinhart
(125a) 1964, *Untersuchungen über die Wandstrukturen der Graptolithen (mit Hilfe des Elektronenmikroskops):* Deutsch. Geol. Gesell., Zeitschr., Jahrgang 1962, v. 114, pt. 3, p. 699-702, text fig. 1.
(125b) 1968, *Elektronenmikroskopische Beobachtungen an Monograptus-Rhabdosomen:* Der Aufschluss, no. 12, p. 357-361, text fig. 1-7.

Kraft, Paul
(126) 1926, *Ontogenetische Entwicklung und Biologie von Diplograptus und Monograptus:* Paläont. Zeitschr., v. 7, p. 207-249, text fig. 1-4, pl. 3-17.

Lapworth, Charles
(127) 1873, *On an improved classification of the Rhabdophora:* Geol. Mag., v. 10, p. 500-504, 555-560.
(128) 1876, *On Scottish Monograptidae. Part 4:* Same, v. 13, p. 544-552.
(129) 1877, *On the graptolites of County Down:* Belfast Nat. Field Club, Proc., v. 1, p. 125-144, pl. 5-7.
(130) 1879-80, *On the geological distribution of the Rhabdophora:* Ann. & Mag. Nat. History, ser. 5, v. 3, p. 245-257, 449-455; v. 4, p. 333-341, 423-431; v. 5, p. 45-62, 273-285, 359-369; v. 6, p. 16-29, 185-207 (v. 3-4, 1879; v. 5-6, 1880).
(131) 1880, *On new British graptolites:* Same, ser. 5, v. 5, p. 149-177, pl. 4, 5.
(132) 1881, *On the Cladophora (Hopk.) or dendroid graptolites . . . in the Llandovery rocks of Mid Wales:* Geol. Soc. London, Quart. Jour., v. 37, p. 171-177, pl. 7.
(133) 1897, *Die Lebensweise der Graptolithen:* in J. Walther, *Lebensweise fossiler Meeresthiere,* Deutsch. Geol. Gesell., Zeitschr., v. 49, p. 238-258.

Legrand, Philippe
(134) 1963, *Un graptolite remarquable de l'Ordovicien inférieur du Sahara algérien: Choristograptus louhai nov. gen., nov. sp.:* Soc. Géol. France, Bull., ser. 7, v. 5, p. 52-58, text fig. 1, pl. 3,4.

Levina, E. F.
(135) 1928, *Graptolites from Aq-Tenga in Turkestan Range:* Acta Univ. Asiae Mediae, ser. 7A, v. 5, p. 1-18.

McCoy, Frederick
(136) 1850, *On some new genera and species of Silurian Radiata:* Ann. & Mag. Nat. History, ser. 2, v. 6, p. 270-290.
(137) 1851, in Adam Sedgwick & Frederick McCoy, *A synopsis of the classification of the British Palaeozoic rocks by the Rev. Adam Sedgwick . . . With a detailed systematic description of the British Palaeozoic fossils by Frederick M'Coy:* pt. 1, 184 p., pl. 1A-L, Cambridge Univ. Press (London & Cambridge).
(138) 1876, *On a new Victorian graptolite:* Ann. & Mag. Nat. History, ser. 4, v. 18, p. 128-130.

McLearn, F. H.
(139) 1915, *The Lower Ordovician at St. John, New Brunswick, and the new genus Protistograptus:* Am. Jour. Sci., ser. 4, v. 40, p. 49-59.

Manck, Elfried
(140) 1923, *Untersilurische Graptolithenarten der Zone 10 des Obersilurs:* Natur, Jahrg. 14, p. 282-289 (Leipzig).

Matthew, G. F.
(141) 1886, *Illustrations of the fauna of the St. John Group, III:* Royal Soc. Canada, Trans., v. 3, p. 29-84, pl. 5-7.

Miller, S. A.
(142) 1874, *Genus Megalograptus:* Cincinnati Quart. Jour. Sci., v. 1, p. 343-346.
(143) 1889, *North American geology and palaeontology:* 664 p., Western Methodist Book Concern (Cincinnati, Ohio).

Moberg, J. C.
(144) 1892, *Om några nya graptoliter från Skånes Undre Graptolitskiffer:* Geol. Fören. Stockholm, Förhandl., v. 14, p. 339-350, pl. 8.

——, & **Törnquist, S. L.**
(145) 1909, *Retioloidea från Skånes Colonusskiffer:* Sver. Geol. Undersök, Årsbok 2 (1908), no. 5, p. 1-20, pl. 1.

Monsen, Astrid
(146) 1925, *Über eine neue ordovicische Graptolithen fauna:* Norsk Geol. Tidsskr., v. 8, p. 147-187, text fig. 1-6, pl. 1-4.
(147) 1937, *Die Graptolithenfauna im unteren Didymograptusschiefer (Phyllograptusschiefer) Norwegens:* Same, v. 16, p. 57-266, pl. 1-20.

Mu A. T.
(148) 1955, *On Spirograptus Gürich:* Acta Palaeont. Sinica, v. 3, p. 1-10.

(149) 1957, *Some new or little known graptolites from the Ningkuo Shale:* Same, v. 5, p. 369-437, pl. 1-8.

(150) 1958, *Abrograptus, a new graptolite genus from the Hulo Shale:* Same, v. 6, p. 259-265, pl. 1.

(151) 1959, in S. C. Hsü, A new graptolite fauna from the Lower Ordovician shale of Tsaidam: Same, v. 7, p. 161-191.

(152) 1963, *On the complication of graptolite rhabdosome:* Same, v. 11, p. 346-377.

———, & Chen X.

(153) 1962, *Sinodiversograptus multibrachiatus gen. et sp. nov. and its developmental stages:* Acta Palaeont. Sinica, v. 10, p. 143-154, pl. 1-2.

———, & Lee C. K.

(154) 1958, *Scandent graptolites from the Ningkuo Shale:* Acta Palaeont. Sinica, v. 6, p. 391-427, pl. 1-5.

———, ———, & Geh Mei-yü

(155) 1960, *Ordovician graptolites from Xinjiang (Sinkiang):* Acta Palaeont. Sinica, v. 8, p. 27-39, pl. 1-3.

———, & Qiao X. D.

(156) 1962, *New materials of Abrograptidae:* Acta Palaeont. Sinica, v. 10, p. 1-8, pl. 1-2.

———, & Zhan S. G.

(157) 1966, *On the probable development and systematic position of Glossograptus:* Scientia Sinica, v. 15, p. 92-98, pl. 1.

Müller, A. H.

(157a) 1969, *Monograptiden (Graptoloidea) mit Nebenzweigen 2. Ordnung aus dem Unterdevon (Lochkovium) von Thüringen:* Geologie u. Mineralogie, Monatsber., v. 11, no. 11/12, p. 913-921, pl. 1-3.

Nicholson, H. A.

(158) 1867, *On a new genus of graptolites, with notes on reproductive bodies:* Geol. Mag., v. 4, p. 256-263, pl. 11.

(159) 1867, *On some fossils from the Lower Silurian rocks of the south of Scotland:* Same, v. 4, p. 107-113, pl. 7.

(160) 1868, *Notes on Helicograpsus, a new genus of graptolites:* Ann. & Mag. Nat. History, ser. 4, v. 2, p. 23-26.

(161) 1869, *On some new species of graptolites:* Same, ser. 4, v. 4, p. 231-242, pl. 11.

(162) 1872, *Monograph of the British Graptolitidae:* x+133 p., 74 text fig., Blackwood & Sons (Edinburgh, London).

(163) 1873, *On some fossils from the Quebec Group of Point Lévis, Quebec:* Ann. & Mag. Nat. History, ser. 4, v. 11, p. 133-143, text fig. 1-3.

(164) 1875, *On a new genus and some new species of graptolites from the Skiddaw Slates:* Same, ser. 4, v. 16, p. 269-273, pl. 7.

(165) 1876, *Notes on the correlation of the graptolitic deposits of Sweden with those of Britain:* Geol. Mag., v. 13, p. 245-249, pl. 9.

———, & **Marr, J. E.**

(166) 1895, *Notes on the phylogeny of the graptolites:* Geol. Mag., v. 42, p. 529-539.

Obut, A. M.

(167) 1949, *Polevoi atlas rukovodyashchikh graptolitov verkhnego silura Kirgizskoi SSR:* Akad. Nauk SSSR, Kirgiz. Filial, Geol. Inst. Frunze, p. 1-56, pl. 1-7. [*Field atlas of leading graptolites of the Upper Silurian of Kirgiz SSR.*]

(168) 1950, *Semeistva i nekotorye rody odnoryadnykh graptolitov:* Voprosy Paleontologii (Questions of paleontology), v. 1, p. 264-272. [*Families and some genera of uniserial graptolites.*]

(169) 1957, *Klassifikatsiya i ukazatel rodov graptolitov:* Vsesoyuz. Paleont. Obshch., Ezhegodnik, v. 16, p. 11-47. [*Classification and ordering of the genera of graptolites.*]

(170) 1960, *Korrelyatsiya nekotorykh chastei razreza ordovikskikh i siluriiskikh otlozhenii Estonskoi SSR po graptolitam:* Eesti NSV Teaduste Akad., Geol. Inst., Uurimused, v. 5, p. 143-158, pl. 1-5. [*Correlation of some partial sections of Ordovician and Silurian strata of the Estonian SSR.*]

(171) See (172).

(172) 1964, *Podtip Stomochordata. Stomokhordovye:* in Yu. A. Orlov (ed.), *Osnovy paleontogii: Echinodermata, Hemichordata, Pogonophora i Chaetognatha,* p. 279-337, text fig. 1-76, pl. 1-10, Nedra Press (Moskva). [*Subphylum Stomochordata. Stomochordate*s.]

———, & **Rytsk, G. V.**

(173) 1958, *Dendroidei ordovika i silura Estonskoy SSR:* ENSV Teaduste Akad., Geol. Inst. Uurimused (Akad. Nauk Estonskoi), Inst. Geol., Trudy, v. 3, p. 125-144, pl. 1-10. [*Ordovician and Silurian Dendroidea of Estonia.*]

———, & **Sobolevskaya, R. F.**

(174) 1962, *Problemy neftegazonosnosti sovetskoy Arktiki, Paleontologiya i biostratigrafiya:* Nauchno-issledov. Inst. Geologii Arktiki, Minist. Geologii i Okhrany Nedr SSSR, Trudy, v. 127, p. 65-85, pl. 1-5. [*Problems of oil and gas occurrences in the Soviet Arctic: Paleontology and biostratigraphy.*]

(175) 1964, *Graptolity ordovika Taimyra:* Akad. Nauk SSSR, Sibir. Otdel., Inst. Geol. i Geofiz.-Nauch.-Issled., Inst. Geol. Artiki, 86 p., 16 pl. [*Graptolites from the Ordovician of Taimyr.*]

(176) 1966, *Graptolity rannego silura v Kazakh-*

stane: Akad. Nauk SSSR, Sibir. Otdel., Inst. Geol. Geofiz., Minist. Geol. SSSR, Nauchno-issledov. Inst. Geol. Arktiki, p. 1-52, 8 pl. [*Early Silurian graptolites in Kazakhstan.*]

(177) 1967, *Nekotoryye stereostolonaty pozdnego kembriya i ordovika Norilskogo rayona:* in Novyye dannye po biostratigrafii nishnego paleozoya Sibirskoy platformy [New data on biostratigraphy of Lower Paleozoic of the Siberian Platform], Akad. Nauk SSSR, Sibir. Otdel., Inst. Geol. i Geofiz., p. 45-64, pl. 6-19. [*Some stereostolonates of the late Cambrian and Ordovician of the Norilsk Region.*]

(178) 1968, in A. M. Obut, R. F. Sobolevskaya, & A. P. Merkureva, Graptolity llandoveri v kernakh burovykh skvazhin Norilskogo rayona [Graptolites of the Llandovery in core samples from boreholes in the Norilsk District]: Same, Sibir. Otdel., Inst. Geol. i Geofiz., p. 1-126, pl. 1-35, Nauka Press (Moskva).

———, ———, & Nikolaev, A. A.

(178a) 1967, *Graptolity i stratigrafiya nizhnego silura okrannykh podnyatiy Kolymskogo Massiva (Severo-Vostov SSSR):* Akad. Nauk SSSR, Sibir. Otdel., Inst. Geol. Geofiz., Minist. Geol. SSSR, Nauchno-issledov. Inst. Geol. Arktiki, p. 1-162, 20 pl. [*Graptolites and stratigraphy of the Lower Silurian of the marginal uplifts of the Kolyma massif.*]

Öpik, A. A.
(179) 1928, *Beiträge zur Kenntnis der Kukruse-(C_2-C_3-) Stufe in Eesti.* II: Univ. Tartu., Acta et Comment., v. 13, no. 11, p. 1-42, text fig. 1-7, pl. 1-4.

(180) 1930, *Beiträge zur Kenntnis der Kukruse-(C_2-C_3-) Stufe in Eesti.* IV: Same, v. 19, no. 2, p. 1-34, text. fig. 1-11, pl. 1-4.

(181) 1933, *Über einen kambrischen Graptolithen aus Norwegen:* Norsk. Geol. Tidsskr., v. 13, p. 113-115, text fig. 1,2.

Philippot, André
(182) 1950, *Les Graptolites du massif armoricain:* Soc. Géol. Univ. Bretagne, Mém., v. 7, p. 1-295.

Počta, Philippe (Fillip)
(183) 1894, *Bryozoaires, Hydrozoaires et partie des Anthozoaires:* in J. Barrande, Système Silurien du Centre de la Bohême, v. 8, pt. 1, 230 p., 21 pl., Rivnác (Praha).

Poulsen, Christian
(184) 1924, *Syrrhipidograptus Nathorsti, a new graptolite genus from the Ordovician of Bornholm:* Dansk Geol. Foren., Medd., v. 6, no. 25, p. 1-7, text fig. 1,2.

(185) 1937, *On the Lower Cambrian faunas of East Greenland:* Medd. Grønland, v. 119, no. 3, 72 p., 18 text fig., 8 pl.

(186) 1943, *Procyrtograptus garboei, a new graptolite from the Lower Silurian of Bornholm:* Dansk. Geol. Foren., Medd., v. 10, p. 301-306.

Přibyl, Alois
(187) 1941, *Pernerograptus gen. nov. und seine Vertreter aus dem böhmischen und ausländischen Silur:* K. České Společnost Nauk, Vestník., Ročník 1941, p. 1-18, pl. 1-2.

(188) 1942, *Revision der Pristiograpten aus den Untergattungen Colonograptus n. subg. und Saetograptus n. subg.:* Tschech. Akad. Wiss., Mitteil., v. 52, p. 1-22.

(189) 1947, *Classification of the genus Climacograptus Hall, 1865:* Acad. Tchèque Sci., Bull. Internatl., v. 48, no. 2, p. 1-12, pl. 1-2.

(190) 1948, *Bibliographic index of Bohemian Silurian graptolites:* Stát. Geol. Ústavu Českosl. Rep., Věstník, v. 22, p. 1-96.

(191) 1948, *Some new subgenera of graptolites from the Families Dimorphograptidae and Diplograptidae:* Same, Věstník, v. 23, p. 37-48.

(192) 1949, *Revision of the Diplograptidae and Glossograptidae of the Ordovician of Bohemia:* Acad. Tchèque Sci., Bull. Internatl., v. 50, no. 1, p. 1-51, pl. 1-5.

(193) 1967, *Monograptus (Testograptus) subgen. nov. aus dem böhmischen und europäischen Silur:* Ústřed. Ústavu Geol., Věstník, v. 42, p. 49-52.

(194) 1967, *Zur Gattung Bohemograptus gen. nov. (Graptoloidea) aus dem böhmischen und fremden Ludlovium:* Časopis Národ. Muzea, no. 3, p. 133-136, pl. 1-2.

Richter, Reinhard
(195) 1871, *Aus dem Thüringischen Schiefergebirge:* Deutsch. Geol. Gesell., Zeitschr., v. 23, p. 231-256, pl. 5.

Rickards, R. B., & Bulman, O. M. B.
(196) 1965, *The development of Lasiograptus harknessi (Nicholson 1867):* Palaeontology., v. 8, p. 272-280.

———, & Hutt, Jana

(196a) 1970, *The earliest monograptid:* Geol. Soc. London, Proc. (in press).

Roemer, Ferdinand
(197) 1897, see Frech, F.

Romariz, Carlos
(198) 1962, *Graptolitos do Silurico Portugues:* Fac. Ciencias Lisboa, Revista, v. 10, p. 115-312, 22 pl.

Ross, R. J., & Berry, W. B. N.
(199) 1963, *Ordovician graptolites of the Basin ranges in California, Nevada, Utah, and Idaho:* U.S. Geol. Survey, Bull. 1134, 177 p., 13 pl.

Ruedemann, Rudolf

(200) 1895, *Development and mode of growth of Diplograptus McCoy:* N.Y. State Geol. Survey, Ann. Rept. for 1894, p. 219-249, pl. 1-5.

(201) 1904-08, *Graptolites of New York, pts. I, II:* N.Y. State Museum, Mem. 7, p. 457-803, pl. 1-17, text fig. 105; Mem. 11, p. 457-583, pl. 1-31, text fig. 1-482.

(202) 1916, *Paleontologic contributions from the New York State Museum, I:* N.Y. State Museum, Bull. 189, p. 7-98, text fig. 1-33, pl. 1-31.

(203) 1925, *Some Silurian (Ontarian) faunas of New York:* Same, Bull. 265, p. 1-134, 41 text fig. and 24 pl.

(204) 1925, *The Utica and Lorraine Formations of New York. 1. Stratigraphy:* Same, Bull. 258, p. 1-175, text fig. 1-10, 7 pl.

(205) 1933, *The Cambrian of the Upper Mississippi Valley: Pt. 3, Graptoloidea:* Publ. Museum Milwaukee, Bull., v. 12, no. 3, p. 307-348, text fig. 1-4, pl. 46-55.

(206) 1934, *Paleozoic plankton of North America:* Geol. Soc. America, Mem. 2, p. 1-141, 6 text fig., 26 pl.

(207) 1936, *Ordovician graptolites from Quebec and Tennessee:* Jour. Paleontology, v. 10, p. 385-387, text fig. 1-13.

(208) 1937, *A new North American graptolite faunule:* Am. Jour. Sci., v. 33, p. 57-62, text fig. 1-8.

(209) 1947, *Graptolites of North America:* Geol. Soc. America, Mem. 19, p. 1-652, pl. 1-92.

Salter, J. W.

(210) 1858, *On Graptopora, a new genus of Polyzoa allied to graptolites:* Am. Assoc. Adv. Sci., v. 11, p. 63-66.

(211) 1863, *Note on the Skiddaw Slate fosssils:* Geol. Soc. London, Quart. Jour., v. 19, p. 135-140.

Schmidt, Wolfgang

(212) 1939, *Ein dendroider Graptolith aus dem Untersilur Thüringens:* Geologie von Thüringen, Beiträge, v. 5, no. 3, p. 121-126, pl. 1.

(213) See (100a).

Skevington, David

(214) 1963, *Graptolites from the Ontikan Limestones (Ordovician) of Öland, Sweden: 1. Dendroidea, Tuboidea, Camaroidea and Stolonoidea:* Univ. Uppsala, Bull. Geol. Inst., v. 42, p. 1-62, text fig. 1-81.

(215) 1965, *Graptolites from the Ontikan Limestones (Ordovician) of Öland, Sweden. II, Graptoloidea and Graptovermida:* Same, v. 43, p. 1-74, text fig. 1-72.

Skoglund, Roland

(216) 1961, *Kinnegraptus, a new graptolite genus from the Lower Didymograptus Shale of Västergötand, central Sweden:* Univ. Uppsala, Bull. Geol. Inst., v. 40, p. 389-400, text fig. 1-7, pl. 1.

Skwarko, S. K.

(217) 1967, *On the structure of Didymograptus artus Elles & Wood:* Australia Bur. Min. Res., Bull. 92, Paleont. Papers (1966), p. 171-186, text fig. 1-5, pl. 21-23.

Spencer, J. W.

(218) 1878, *Graptolites of the Niagara Formation:* Canadian Nat., ser. 2, v. 8, p. 457-463.

(219) 1883, *Occurrence of graptolites in the Niagara Formation of Canada:* Am. Assoc. Adv. Sci., Proc., v. 31, p. 363-365.

(220) 1884, *Niagara fossils: Pt. I, Graptolitidae of the Upper Silurian System:* Missouri Univ. Museum, Bull., v. 1, p. 1-43, pl. 1-6.

Spjeldnaes, Nils

(221) 1963, *Some Upper Tremadocian graptolites from Norway:* Palaeontology, v. 6, p. 121-131, text fig. 1-4, pl. 17, 18.

Størmer, Leif

(222) 1933, *A floating organ in Dictyonema:* Norsk Geol. Tidsskrift, v. 13, p. 102-112, text fig. 1-4, pl. 1.

(223) 1935, *Additional remarks on the presence of a floating organ in Dictyonema flabelliforme (Eichwald):* Same, v. 14, p. 316-318, text fig. 1.

(224) 1938, *To the problem of black graptolite shales:* Same, v. 17, p. 173-176, text fig. 1, pl. 1.

Strachan, Isles

(225) 1952, *On the development of Diversograptus Manck:* Geol. Mag., v. 89, p. 365-368, fig. 1.

(226) 1954, *The structure and development of Peiragraptus fallax, gen. et. sp. nov.:* Same, v. 91, p. 509-513, text fig. 1-2.

(227) 1959, *Graptolites from the Ludibundus Beds (Middle Ordovician) of Tvären, Sweden:* Univ. Uppsala, Bull. Geol. Inst., v. 38, p. 47-68, text fig. 1-13, pl. 1-2.

Strøm, K. M.

(228) 1936, *Land-locked waters: hydrography and bottom deposits of badly ventilated Norwegian fjords:* Norske Vidensk.-Akad. Oslo, Skrifter, v. 1, no. 7, 85 p., 43 text fig., 9 pl.

Stubblefield, C. J.

(229) 1929, *Notes on some early British graptolites:* Geol. Mag., v. 66, p. 268-285, text fig. 1-11.

Sudbury, Margaret

(230) 1958, *Triangulate monograptids from the Monograptus gregarius zone of the Rheidol Gorge:* Royal Soc. London, Philos. Trans., ser. B, v. 241, p. 485-555, text fig. 1-32, pl. 19-23.

Suess, Eduard
(231) 1851, *Ueber böhmische Graptolithen:* Haidingers Ber., v. 4, p. 89-134, pl. 7-9.

Tavener-Smith, Ronald
(231a) 1969, *Skeletal structure and growth in the Fenestellidae (Bryozoa):* Palaeontology, v. 12, p. 281-309, pl. 52-56.

Teller, Lech
(232) 1962, *Graptolite fauna and stratigraphy of the Ludlovian deposits of the Chelm borehole, eastern Poland:* Studia Geol. Polonica, v. 13, p. 1-88, text fig. 1-18, pl. 1-16.

Termier, Henri, & Termier, Geneviève
(233) 1948, *Les graptolithes dendroïdes en Africa du nord:* Soc. Géol. France, Comptes Rendus, p. 174-176.

Thomas, D. E.
(234) 1960, *The zonal distribution of Australian graptolites:* Royal Soc. New S. Wales, Jour. & Proc., v. 94, p. 1-58, 15 pl.

Thomas, H. D., & Davis, O. G.
(235) 1949, *The pterobranch Rhabdopleura in the English Eocene:* Brit. Museum (Nat. History), Bull. (Geol.), v. 1, no. 1, 19 p., 3 pl.

Thorsteinsson, R.
(236) 1955, *The mode of cladial generation in Cyrtograptus:* Geol. Mag., v. 92, p. 37-49, text fig. 1-4, pl. 3-4.

Törnquist, S. L.
(237) 1890-92, *Undersökningar öfver Siljansområdets graptoliter, pts. 1, 2:* Lunds Univ. Årsskr., v. 26, no. 4, p. 1-33, pl. 1-2; v. 28, no. 2, p. 1-47, pl. 1-3.
(238) 1893, *Observations on the structure of some Diprionidae:* Same, v. 29, no. 3, p. 1-14, pl. 1.
(239) 1897, *On the Diplograptidae and Heteroprionidae of the Scanian Rastrites beds:* K. Fysiogr. Sällsk. Lund, Handl., new ser., v. 8, p. 1-24, pl. 1-2.
(240) 1899, *Monograptidae of the Scanian Rastrites beds:* Lunds Univ. Årsskr., v. 35, pt. 2, no. 1, p. 1-25, pl. 1-4.
(241) 1901-04, *Graptolites of the lower zones of the Scanian and Vestrogothian Phyllo-Tetragraptus beds, pts. 1, 2:* Same, v. 37, pt. 2, no. 5, p. 1-26, pl. 1-3; v. 40, pt. 1, no. 2, p. 1-29, pl. 1-4.

Tullberg, S. A.
(242) 1880, *Tvenne nya graptolitslägten:* Geol. Fören. Förhandl., v. 5, 1880, p. 313-315, pl. 11.
(243) 1883, *Skånes graptoliter, I and II:* Sver. Geol. Undersök., ser. C, no. 50, p. 1-44; no. 55, p. 1-43, pl. 1-4.

Turner, J. C. M.
(244) 1960, *Faunas graptoliticas de América del Sur:* Asoc. Geol. Argentina, Revista, v. 14, p. 1-180, 9 pl.

Ubaghs, Georges
(245) 1941, *Les graptolithes dendroïdes du Marbre Noir de Denée (Viséen inférieur):* Musée Royal Histoire Nat. Belgique, Bull., v. 17, no. 2, p. 1-30, pl. 1-5.

Ulrich, E. O., & Ruedemann, Rudolf
(246) 1931, *Are the graptolites bryozoans?:* Geol. Soc. America, Bull., v. 42, p. 589-604, text fig. 1-13.

Urbanek, Adam
(247) 1953, *Sur deux espèces de Monograptidae:* Acta Palaeont. Polonica, v. 3, p. 277-297, text fig. 1-16.
(248) 1954, *Some observations on the morphology of Monograptidae:* Acta Geol. Polonica, v. 4, p. 291-306.
(249) 1958, *Monograptidae from erratic boulders of Poland:* Palaeont. Polonica, v. 9, p. 1-105, text fig. 1-68, text pl. 1-7, pl. 1-5.
(250) 1959, *On the development and structure of the graptolite genus Gymnograptus:* Acta Palaeont. Polonica, v. 4, p. 279-336, text fig. 1-18, text pl. 1-7, pl. 1-2.
(251) 1960, *An attempt at biological interpretation of evolutionary changes in graptolite colonies:* Same, v. 5, p. 127-234, text fig. 1-21, pl. 1-3.
(252) 1963, *On generation and regeneration of cladia in some Upper Silurian monograptids:* Same, v. 8, p. 135-254, text fig. 1-8, text pl. 1-17, pl. 1-3.
(253) 1966, *On the morphology and evolution of the Cucullograptinae (Monograptidae, Graptolithina):* Same, v. 11, p. 291-544, text fig. 1-27, pl. 1-47.

Waern, Bertil
(254) 1948, *The Silurian strata of the Kullatorp core:* Univ. Uppsala, Bull. Geol. Inst., v. 32, p. 433-473, text fig. 1-5, pl. 26.

Walker, Margaret
(255) 1953, *The development of a diplograptid from the Platteville Limestone:* Geol. Mag., v. 90, p. 1-16, text fig. 1-12.
(256) 1953, *The development of Monograptus dubius and Monograptus chimaera:* Same, v. 90, p. 362-373, text fig. 1-6.

Westergård, A. H.
(257) 1909, *Studier öfver Dictyograptusskiffern:* Lunds Univ., Årsskr., ser. 2, v. 5, no. 3, 79 p., 5 pl.

Wetzel, W.
(257a) 1958, *Graptolithen und ihre fraglichen Verwandten im elektronenmikroskopischen Vergleich:* Neues Jahrb. Geologie u. Paläontologie, Monatshefte, no. 7, p. 307-312, 3 text fig.

Whitfield, R. P.
(258) 1902, *Notice of a new genus of marine algae, fossil in the Niagara Shale:* Am.

Museum Nat. History, Bull., v. 16, no. 30, p. 399-400, pl. 53.

Whittington, H. B.
(259) 1954, *A new Ordovician graptolite from Oklahoma:* Jour. Paleontology, v. 28, p. 613-621, pl. 63.
(260) 1955, *Additional new Ordovician graptolites:* Same, v. 29, p. 837-851, pl. 83-84.
——, & **Rickards, R. B.**
(261) 1969, *Development of Glossograptus and Skiagraptus, Ordovician graptoloids from Newfoundland:* Same, p. 800-817, text fig. 1-11, pl. 101-102.

Wiman, Carl
(262) 1893, *Ueber Diplograptidae Lapw.:* Univ. Uppsala, Geol. Inst., Bull., v. 1, no. 2, p. 97-103, pl. 6.
(263) 1893, *Ueber Monograptus Geinitz:* Same, v. 1, no. 2, p. 113-117, pl. 7.
(264) 1895, *Über die Graptolithen:* Same, v. 2, no. 4, p. 239-316, text fig., pl. 9-15.
(265) 1897, *Über Dictyonema cavernosum, n. sp.:* Same, v. 3, no. 5, p. 1-13, pl. 1.
(266) 1897, *Über den Bau einiger gotländischer Graptoliten:* Same, v. 3, no. 6, p. 352-368, pl. 11-14.
(267) 1901, *Über die Borkholmer Schicht im mittelbaltischen Silurgebiet:* Same, v. 5, no. 10, p. 149-222, text fig. 1-11, pl. 5-8.

Yakovleva [Yakovlev], N. N.
(268) 1933, *Planktonyy graptolit iz Kazakstana:* Akad. Nauk SSSR, Izvestia, Otdel. mat. i estestven. nauk, p. 979-981, pl. 1. [*A planktonic graptolite from Kazakhstan.*]

Yin T. H.
(269) 1937, *Brief description of the Ordovician and Silurian fossils from Shihtien:* Geol. Soc. China, Bull., v. 16, p. 281-302, pl. 1-2.

Zhao Y. T.
(270) 1964, *A new multiramous graptolite from the Ningkuo Shale:* Acta Palaeont. Sinica, v. 12, p. 638-641, pl. 1-3.

ADDENDUM

CLASSIFICATION OF THE GRAPTOLITE FAMILY MONOGRAPTIDAE LAPWORTH, 1873

By O. M. B. BULMAN and R. B. RICKARDS

[Sedgwick Museum, Cambridge]

INTRODUCTION

In this second edition of Part V (Graptolithina) of the *Treatise on Invertebrate Paleontology,* the suborder Monograptina has been divided into two families, Monograptidae and Cyrtograptidae, and the latter further divided into the subfamilies Cyrtograptinae and Linograptinae. All forms which exhibit thecal or sicular cladia are there assigned to the Cyrtograptidae, and the distinction between the Linograptinae and the Cyrtograptinae rests respectively upon the presence or absence of sicular cladia. This is clearly no more than an arbitrary and provisional arrangement (a key rather than a classification), acceptable only until sufficient is known of monograptid phylogeny to attempt a more "natural" classification. On this basis, *Monograptus runcinatus* LAPWORTH is assigned to the genus *Diversograptus* (Linograptinae) although it is known in the diversograptid (bipolar) condition only by relatively few specimens; and in Britain, *Neodiversograptus nilssoni* (LAPWORTH, 1876, *sensu* URBANEK, 1954) has been recorded only recently in possession of sicular cladia. The number of such anomalies known is small, but if cladia production proves to be potentially possible in any monograptid, it is clear that this feature may cease to have much influence even on generic definitions. We do not at present know, for example, whether *Cyrtograptus* is monophyletic or whether the main lineages run through such "genera" rather than originate within them; but URBANEK (1963) has already suggested possible analogy between cladia production and the well-known developmental "stages" recognized in the Dichograptina [Didymograptina] and Diplograptina.

This note is not concerned with the issues raised above but with the attempted subdivision of *Monograptus* on the basis of thecal form and rhabdosome shape. A lengthy discussion of this is out of place in the Systematic Descriptions, but some reasoned justification is needed for the lack of recognition accorded to these genera therein and by most British and American workers. In other countries, this has been contrasted with the general recognition accorded to the genera of biserial graptolites.

Genera of the Didymograptina are based to a considerable extent on rhabdosome form. In the Diplograptina, the biserial rhabdosome is universal and genera were erected largely on thecal characters. This process began a century ago and the generic names, though they have not proved altogether satisfactory, have the sanction of long use. They are now themselves beginning to be subdivided on the basis of more subtle differences in thecal form.

The monograptids, with a comparable uniformity of rhabdosome plan, have a different history as regards taxonomy. LAPWORTH (1876) recognized a number of species groups, which he believed to be made up of closely allied species, but he erected no monograptid genera for them; and ELLES & WOOD (1901-18), after analyzing the biocharacters as then understood, modified these groups and their content, and elaborated them entirely in the manner of a key, but again refrained from designating any genera for them. This action carried the implication that further knowledge of the details of thecal structure was necessary before any satisfactory nomenclature could be achieved. A few generic names had already been proposed (*Monoclimacis* FRECH, 1897; *Pomatograptus* JAEKEL, 1889; *Pristiograptus* JAEKEL, 1889), but with the exception of *Rastrites* BARRANDE, 1850, these were not widely accepted; and in a presidential address on biological classification BATHER (1927) could write:

... it would be worth while to experiment with the Graptolites, to see whether anything would really be gained by splitting up such a genus as *Monograptus*. So long as this name is retained, at least one is told the grade of structure. A few ideal schemes might be worked out on a clean slate, and provided they were all wiped out again before publication of the selected names, no harm would be done.

This expresses the conservative attitude toward subdivision of the genus *Monograptus* by most workers, especially in Britain, until well into this century; and although the devising of an "ideal scheme" would scarcely be regarded today as a profitable exercise, the hope remained that from the portmanteau genus *Monograptus* various soundly based genera could progressively be extracted as investigation of different species provided the opportunity.

However, elsewhere the temptation to name these monograptid species groups of ELLES & WOOD has latterly proved irresistible and nearly a score of technically valid genera have been proposed since 1940. The main objection to most such genera is that their erection was not accompanied by any addition to our imperfect knowledge of their morphology and phylogeny; their content is ill-defined and their application correspondingly uncertain. The sole purpose of a key is to aid the identification of species, and the bestowal of generic names on such categories or groups inevitably tends to burden the literature with names that are at best of doubtful value. The status of these and other genera is discussed below.

In conclusion, some reference should be made to the particular difficulty introduced by the prevalence of *biform* monograptids, where proximal and distal thecae of the same rhabdosome may differ to an extent scarcely paralleled in other graptolites. ELLES & WOOD concluded that the distal (mature) thecae "have always been considered to be the more characteristic and distinctive" and should take systematic precedence over the proximal thecae; but this is an oversimplification and even in the ELLES & WOOD "groups," the treatment of such species was by no means satisfactory. With the recognition that new characters can be introduced either proximally or distally, this taxonomic problem becomes more complex and the generic naming of the ELLES & WOOD "groups" becomes still more hazardous. This complexity has so far only been surmounted satisfactorily in URBANEK'S *Lobograptus* and *Cucullograptus*, where the definition is supplemented by a convincing and comprehensive phylogeny; it has not yet been resolved in the so-called demirastritids, where several phylogenies have not yet been properly disentangled.

ACCEPTABLE GENERA

RASTRITES

Rastrites BARRANDE, 1850 (type, *R. peregrinus;* SD HOPKINSON, 1869). The distinctive appearance of *Rastrites* led to the erection of this genus as one of the first true graptolites to be distinguished from the now obsolete *Graptolithus*. Its relation to *Monograptus* was noted by LAPWORTH (1876) and it was reduced to subgeneric rank by ELLES & WOOD (1901-18) mainly with the object of emphasizing this relationship. It is now known from isolated material (HUTT, RICKARDS & SKEVINGTON, in press), and SUDBURY (1958) indicated the probable derivation of one (possibly two) species from triangulate monograptids. The genus is probably polyphyletic, but the number of species involved is relatively small and the lineages appear to be closely related.

The genus *Corymbites* OBUT & SOBOLEVSKAYA, 1967 (type, *C. sigmoidalis*; OD) appears to be a sigmoidally curved *Rastrites*. The taxonomic value of rhabdosomal curvature is discussed more fully below, under *Oktavites* (p. V152). *Stavrites* OBUT & SOBOLEVSKAYA, 1968 (*S. rossicus*; OD) appears to be a *Rastrites* in which both thecae and common canal are encased in some chloritic or other material; the pyritized thecal tubes and common canal can be seen centrally placed in several figures. No new structures were elucidated in the description and these genera are here regarded as junior synonyms of *Rastrites*.

MONOCLIMACIS

Monoclimacis FRECH, 1897 (type, *Graptolithus vomerinus* NICHOLSON, 1872; OD). The thecal structure of several species of *Monoclimacis*, including the type species, is now known from pyritized material, and the thecal hoods described by URBANEK (1958) in transparencies of the Ludlovian *M. micropoma* have now been recognized in pyritized Wenlock and Llandovery representatives (though it cannot yet be asserted that these too are composed of microfusellar tissue). The evolutionary roots of this genus are lost among the diverse lower Llandovery monograptids, but the thecal structure is reasonably well established over its long stratigraphic range and at present no indication is seen that the genus is other than monophyletic.

PRISTIOGRAPTUS

Pristiograptus JAEKEL, 1889 (type, *P. frequens*; OD). Certain species of *Pristiograptus*, though not the type species, are known in three-dimensional transparencies, and a large number of species are represented by pyritized material. The simple character of the thecae makes the interpretation even of flattened material relatively simple, though it would make any polyphyly the more difficult to detect. The genus represents a long-ranging and prolific stock (extending from lower Llandovery to upper Ludlow) and is the probable source of several genera recognized in the Ludlow, including some linograptids.

Most graptolite workers have been aware for some time of URBANEK's unpublished studies on the *Pristiograptus bohemicus* (BARRANDE) group of species, and it is regretable that PŘIBYL (1967) should at this stage have erected the genus *Bohemograptus*, with *P. bohemicus* as type species. The present definition of *Bohemograptus* differs in no significant respect from that covered by *Pristiograptus*, but presumably we can expect a redefinition by URBANEK in the near future; for the present the genus *Bohemograptus* is regarded as a junior synonym of *Pristiograptus*.

SAETOGRAPTUS

Saetograptus PŘIBYL, 1942 (type, *Graptolithus chimaera* BARRANDE; OD). The work of WALKER (1953) and URBANEK (1958) has placed *Saetograptus* on a satisfactory footing, though it depends for its recognition on structural detail not always visible in shale material. *Monograptus leintwardinensis* and similar species are probably to be included, but confirmation awaits the preparation of isolated rhabdosomes. The genus *Colonograptus* PŘIBYL, 1942 (*Graptolithus colonus* BARRANDE; OD), again elucidated by URBANEK (1958), differs only in the possession of more rounded lappets rather than spines of monofusellar tissue. Recently, isolated specimens of *M. varians* WOOD show that this species is intermediate between *Saetograptus* and *Colonograptus* (HUTT, 1969) and is in fact nearer to *Saetograptus*, as originally conceived, than to *Colonograptus*, where it was placed by PŘIBYL. The probable derivation of at least some species of *Saetograptus* appears to be through *Colono-*

graptus from a *Pristiograptus* of *P. ludensis* type. It is becoming clear that these genera were too narrowly conceived by PŘIBYL and that it is more realistic to regard *Colonograptus* as a junior synonym of *Saetograptus* rather than a subgenus.

CUCULLOGRAPTUS AND LOBOGRAPTUS

Cucullograptus URBANEK, 1954 (type, *C. pazdroi;* OD) and *Lobograptus* URBANEK, 1958 (type, *M. scanicus* TULLBERG; OD). The two closely-related genera named *Cucullograptus* and *Lobograptus* are known from magnificent three-dimensional material and form the subject of one of the most reliable and detailed investigations into graptolite phylogeny (URBANEK, 1966). *Lobograptus* is represented by five divergent lineages, two of which culminate in species of *Cucullograptus*. We have preferred to regard *Lobograptus* URBANEK, 1958 as a subgenus of the terminal member (and senior taxon) *Cucullograptus* URBANEK, 1954, although recognizing that it was in every sense precisely defined.

GENERA OF DUBIOUS VALUE

GENERA BASED ON RHABDOSOME SHAPE

The synonymy of the principal genera concerned in this category—*Spirograptus* GÜRICH, 1908 (=*Tyrsograptus* OBUT, 1949) (type, *Graptolithus turriculatus* BARRANDE; SD BULMAN, 1929) and *Oktavites* LEVINA, 1928 (=*Obutograptus* MU, 1955) (type, *Graptolithus spiralis* GEINITZ; SD OBUT, 1964) is complicated, but need not be elaborated here.

The use of rhabdosome shape together with thecal form is impossible to reconcile in a single classification. If thecal form be accepted as the ultimate basis of affinity (and hence classification), then rhabdosome shape must take second place, even if it can be shown to have any taxonomic value at all.

This inadequacy of rhabdosome shape is shown by the fact that *Monograptus communis,* one of the two groups of "*Spirograptus*" recognized by PŘIBYL in 1944, was made the type of a separate genus *Campograptus* by OBUT in 1949 largely on thecal characters (Fig. 107,*A,B*). Still more significant, *M. exiguus* (BARRANDE), a species exhibiting pronounced *ventral* curvature of the rhabdosome and a fishhook proximal end, is seen to possess thecae closely resembling those of the dorsally coiled "*Oktavites*" *spiralis* when isolated three-dimensional material is obtained (HUTT, RICKARDS & SKEVINGTON, in press).

The detailed morphology of not a single species normally included in the genus *Spirograptus* is known; even pyritized specimens of *Monograptus turriculatus* have failed to clarify the real nature of the spinose, hooked thecae and some specimens suggest the presence of more than two spines to each theca. In *Oktavites*, *M. spiralis* is the only species in which the thecal structure has yet been fully elucidated (BULMAN, 1932; SUDBURY, 1958). Such species as "*Spirograptus*" *tullbergi* (BOUČEK) could, on present evidence, be assigned to *Campograptus, Oktavites,* or *Spirograptus* (see also *Campograptus* below).

As and when full details become available, it may prove that certain groups of related species also have a tendency towards a particular rhabdosome shape, but exceptions (like *Monograptus exiguus*) appear inevitable.

Two other genera can be considered in this category because rhabdosome shape was given considerable emphasis in their diagnosis.

1) *Campograptus* OBUT, 1949 (type, *Monograptus communis* LAPWORTH; SD OBUT, 1964, p. 328) was defined as a dorsally curved monograptid with hooked thecae greatly expanded at their bases (Fig. 107,*A,B*). It was left to BULMAN (1951) and SUDBURY (1958) to illustrate the true characters of the thecae, and SUDBURY attempted to assess the phyletic relationships of this and related species. Using SUDBURY's work as a basis, it would be possible to define several "genera" more adequately. For example, it would be possible to reletter her figures 28 and 29 (p. 537, 539) showing "*Pernerograptus*" giving rise to "*Campograptus,*" restricted to the species *M. revolutus* and *M. communis* respectively. But what of the species *M. limatulus* TÖRNQUIST; and does the lineage *M. toernquisti* SUDBURY-*M. pseudoplanus* SUD-

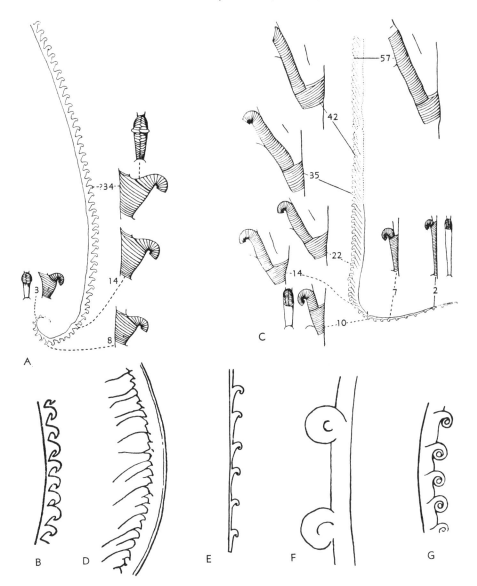

Fig. 107. Drawings of monograptids. (All figures at magnification given in original publications.)

A. *Monograptus communis* (LAPWORTH) (from Bulman, 1951).
B. *"Campograptus" communis* (LAPWORTH) (from Obut, 1949).
C. *Monograptus argenteus* (NICHOLSON) (from Bulman, 1951).
D. *Monograptus ("Testograptus") testis* (BARRANDE) (from Přibyl, 1967).
E. *"Globosograptus" wimani* (BOUČEK) (from Bouček, 1932, where it was described under the name *Monograptus wimani*).
F. *Monograptus ("Mediograptus") kolihai* BOUČEK (from Bouček & Přibyl, 1951).
G. Thecal form of *Streptograptus* YIN (from Bouček & Přibyl, 1942).

BURY-*M. planus* (BARRANDE) then require another new genus? Since new thecal types are being described in increasing numbers at this level, and the phyletic relationships appear to be complex, it is premature to propose new genera for every new variant discovered; but when the phylogenies are more completely assessed (in the manner of SUDBURY, 1958), the discrimination of useful genera containing several adequately known species may be possible.

2) *Testograptus* PŘIBYL, 1967 (type, *Graptolithus testis* BARRANDE; OD) is another genus based on general form of rhabdosome and silhouette preservation of thecae (Fig. 107,*D*); no new information was presented for the species concerned. The nature of the thecal hooks and spines in *Monograptus testis* has never been ascertained, though specimens in low relief from the Long Mountain (PALMER Coll., Trinity College, Dublin) (Fig. 108) suggest resemblance to *M. sedgwicki* (PORTLOCK). It seems probable that *M. testis* is more closely related to some of the hooked and spinous Wenlock representatives of the straighter *M. priodon*-type monograptids than to the curved, Llandovery *M. veles* (RICHTER) with which it was associated by PŘIBYL and which may well be closer to *M. turriculatus*.

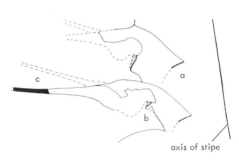

FIG. 108. *Monograptus testis* (BARRANDE), distal thecae of almost flattened specimen from the *Cyrtograptus lundgreni* Zone, Wenlock, of Long Mountain, Shropshire, ×15 (Palmer Collection no. 79 l" (i), Trinity College Dublin) [*a*, base of interthecal septum; *b*, ventral thecal wall somewhat crumpled; *c*, thecal spines (oblique shading indicates visible portions of thecal apertures)].

GENERA BASED MAINLY ON THECAL FORM

Coronograptus OBUT & SOBOLEVSKAYA, 1968 (type, *Monograptus gregarius* LAPWORTH, 1876; OD) is a name given to ELLES & WOOD's Group IA 1(a) (*M. gregarius, M. cyphus,* and *M. acinaces*) together with some new subspecies of *M. gregarius*. ELLES & WOOD's reluctance to name this group was justified by the recent discovery that JONES's specimens of *M. rheidolensis* (=*M. acinaces*) possess delicate ventral thecal processes of the kind described by HUTT (1968) in *M. tenuis*. Our examination of TÖRNQUIST's *M. acinaces* material confirm that *M. rheidolensis* and *M. acinaces* are conspecific, though TÖRNQUIST's Swedish material does not permit recognition of these delicate processes.

Recently isolated specimens of *Monograptus gregarius* show that the thecae possess a rounded geniculum and this structure deflects the apertural region of the preceding theca to give the appearance of an expanded aperture and, more important, to cause a distinct isolation of the apertural region in many specimens.

The use of *Coronograptus* to denote *Monograptus gregarius* and possibly *M. cyphus* serves little purpose and at best seems premature.

Demirastrites EISEL, 1912 (type, *Rastrites triangulatus* HARKNESS, 1851; SD BULMAN, 1929, p. 175). At first sight, there appears considerably more justification for the use of *Demirastrites* than most others in this category. SUDBURY's work (1958) has demonstrated that the type species is the precursor of one of the rastritids and she has made clear the nature of the thecal aperture characteristic of the distal portion of the rhabdosome. It would doubtless be possible to redefine *Demirastrites* so as to include precisely a small number of monograptid species similarly but less certainly related to other rastritids. Nevertheless, the phylogeny of the large group of triangulate monograptids is complex and involves more or less closely species which have been referred to *Demirastrites, Oktavites, Pernerograptus, Spirograptus,* and *Campograptus*: EISEL's six genosyntypes have already been referred by various authors to four of them. Thus the present use of any of these names would inevitably lead to repeated changes of nomenclature with increased knowledge of species morphology and phylogeny, which could only be a source of confusion to stratigraphers; and pending such further investigations we consider it preferable to retain all these species in *Monograptus*.

Globosograptus BOUČEK & PŘIBYL (in PŘIBYL, 1948) (type, *Monograptus wimani* BOUČEK, 1932; OD) was based on the silhouette appearance of a slender Llandovery species on which the thecae possess a long prothecal portion, no thecal overlap, and a seemingly enrolled metathecal portion (Fig. 107,*E*). The nature of this enrollment is indefinite in the type species or in any of the species originally assigned to the genus (1948); in *M. sartorius* TÖRNQUIST (included in the modified list of BOUČEK and PŘIBYL, 1951), it appears to comprise a sharply reflexed apertural region analogous to that figured by BULMAN (1932) in *Monograptus* sp. Very little is known of the characters of the slender monograptids as a whole, but from work in progress in this country, it appears that their thecal form is quite variable and often indicates unsuspected links with more robust species. The genus has no precise significance and its application is quite impracticable.

In *Lagarograptus* OBUT & SOBOLEVSKAYA, 1968 (type, *L. inexpeditus;* OD) the thecae are said to be hooked, but it is very difficult to determine the character of the hook from the half tone illustrations provided, the most conspicuous features in the figures being the long sicula and parallelism of the free ventral wall and dorsal wall of the stipe. Whether or not a geniculum is present and whether the hook is really in the nature of a genicular hood, as in *Monoclimacis,* is impossible to determine. On the published evidence, the erection of this genus can only be considered as highly speculative.

Mediograptus BOUČEK & PŘIBYL, in PŘIBYL, 1948 (type, *Monograptus kolihai* BOUČEK, 1931; OD). Like *Globosograptus,* the type species of *Mediograptus* is known in silhouette only, as a form with long cylindrical prothecae, no thecal overlap and vaguely "lobate" metathecae (Fig. 107,*F*). It is said to differ from *Streptograptus* and *Globosograptus* essentially by the "less coiled ends of the thecae," but the nature of the coiling is indeterminate and the figures are at best described as obscure.

In pyritized specimens of a British variety of *"Mediograptus" minimus* (BOUČEK & PŘIBYL, 1951), the dorsal wall of the metatheca consists of a reflexed shieldlike structure, transversely expanded toward its extremity, with dorsally directed winglike processes (Fig. 109,*1a-c*). To what extent this applies to other species, including the type species, is unknown; but *Monograptus antennularius* MENEGHINI, a species assigned to *"Streptograptus,"* has thecae identical with those of *"Mediograptus" minimus* (see also under *Streptograptus*) (Fig. 109,*2*).

Pernerograptus PŘIBYL, 1941 (type, *Graptolites argenteus* NICHOLSON, 1869; OD) is a name given to ELLES & WOOD's Group IB 1: monograpti in which the thecae are biform (proximally hooked and distally straight overlapping tubes) and the rhabdosome has a dorsal curvature. PŘIBYL chose *Monograptus argenteus* (NICHOLSON) as type, but made no additions to our knowledge of the group. Since then the thecae of *M. argenteus, M. revolutus,* and *M. difformis* have been more fully described by BULMAN (1951) (e.g., Fig. 107,*C*) and those of *M. revolutus* with much greater refinement by SUDBURY (1958). The value of this genus will become clearer when the structure of *M. limatulus* has been elucidated and when the relation of the group to *M. toernquisti* SUDBURY and to *"Campograptus"* can be assessed.

Pribylograptus OBUT & SOBOLEVSKAYA, 1966 (type, *Monograptus incommodus* TÖRNQUIST, 1899; OD) is a name bestowed in effect on ELLES & WOOD's Group II *(M. atavus, M. sandersoni, M. incommodus, M. tenuis* and *M. argutus*), and work by RICKARDS & RUSHTON (1968) has served to illustrate the composite nature of this group. *M. atavus* JONES and *M. tenuis* (PORTLOCK) must be excluded and *M. leptotheca* LAPWORTH should certainly be included in it. A case could be made for the erection of a genus based on *M. incommodus, M. argutus* and *M. leptotheca,* but *Pribylograptus* was prematurely erected and much remains to be done on early Llandovery monograptids before such genera can be evaluated. The situation can only become confused by present use of the name, and it may be mentioned that OBUT & SOBOLEVSKAYA (1968, pl. 16, fig. 8, pl. 17, fig. 1-5) have recently illustrated Russian specimens which they refer to *"Pribylograptus" incommodus* (TÖRNQUIST) which

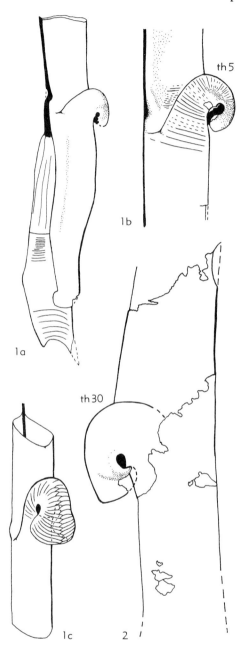

are referable neither to Törnquist's species nor to Elles & Wood's concept of it.

Streptograptus Yin, 1937 (type, *Monograptus nodifer* Törnquist, 1881; OD). An examination of some of Törnquist's material of *M. nodifer*, together with other well-preserved Scandinavian specimens, shows that this species has thecal lobes quite unlike those depicted by Elles & Wood (1901-1918) or by Bouček & Přibyl (1942) (Fig. 107,*G*). The thecae possess a conspicuous bulbous flange near the aperture and this *M. nodifer* theca is unique, though a recently isolated species shows what may be a somewhat simplified version of it (Hutt, Rickards & Skevington, in press). The other species described under the emended *Streptograptus* by Bouček & Přibyl must be accommodated elsewhere and at present can only be referred to *Monograptus sensu lato;* the example of *M. exiguus* (p. *V*152), surely considered in the past as a typical streptograptid, is a timely warning against the premature subdivision of monograptids on imperfectly understood thecal structure.

INDETERMINATE GENERA

A number of "genera" erected by Hundt (1965) are generally agreed to represent indeterminate preservational views, probably of monograptids. They are *Falcatograptus, Mystiograptus, Nodosograptus,* and *Paragraptus.*

SELECTED REFERENCES

[Following is a list of selected references used in the addendum additional to those cited in the main reference list.]

Bather, F. A.
(1) 1927, *Biological classification: past and future:* Geol. Soc. London, Proc., p. lxii-civ.

Bouček, Bedřich
(2) 1932, *Preliminary report on some new species of graptolites from the Gothlandian of Bohemia:* Stat. Geol. Ústav., Věstník, v. 8, p. 1-5.

Fig. 109. Morphological features of monograptids. *1a-c. Monograptus minimus cautleyensis* Rickards, SM A62042, from the *Cyrtograptus centrifugus* Zone, Wenlock, of the Howgill Fells, Northern England; *a,* sicula and *th¹; b, th⁵* of same specimen as in *a; c,* reconstruction of thecal hook, subventral view; ×67.

2. Monograptus antennularius (Meneghini), Wenlock, Long Mountain, Shropshire (Palmer Collection no. 60d (i), Trinity College Dublin); thecal apertures, or visible parts thereof, in black, ×67.

——, & Přibyl, Alois

(3) 1942, *Über böhmische Monograpten aus der Untergattung Streptograptus Yin*: České Akad. Rozpravy, v. 52, p. 1-23, pl. 1-3.

(4) 1951, *On some slender species of the genus Monograptus Geinitz, especially of the subgenera Mediograptus and Globosograptus*: Acad. Tchèque Sci., Bull. Internatl., v. 52, no. 13, p. 1-32, pl. 1-3.

Hutt, Jana

(5) 1968, *A redescription of the Llandoverian monograptid "Graptolithus" tenuis, Portlock, 1843*: Geol. Mag., v. 105, p. 251-255.

(6) 1969, *Development of the Ludlovian graptolite Saetograptus varians (Wood)*: Lethaia, v. 2, p. 361-368.

——, Rickards, R. B., & Skevington, D.

(7) 1970, *Isolated Silurian graptolites from the Bollerup and Klubbudden Stages of Dalarna, Sweden*: Geologica et Paleontologica (in press).

Přibyl, Alois

(8) 1944, *The Middle-European monograptids of the genus Spirograptus Gürich*: Acad. Tchèque Sci., Bull. Internatl., v. 54, no. 19, p. 1-47, pl. 1-11.

Rickards, R. B., & Rushton, A. W. A.

(9) 1968, *The thecal form of some slender Llandovery Monograptus*: Geol. Mag., v. 105, p. 264-274.

*V*158 Graptolithina

INDEX

Italicized names in the following index are considered to be invalid; those printed in roman type, including morphological terms, are accepted as valid. Suprafamilial names are distinguished by the use of full capitals and author's names are set in small capitals with an initial large capital. Page references having chief importance are in boldface type (as **V327**). Some divergences in classification reflect differences of authors concerning validity of nomenclature.

abiesgraptid stage, V89
Abiesgraptus, V88, V100, **V135**
abnormalities, V71
Abrograptidae, V8, V105, **V118**
Abrograptus, **V118**
Acanthastida, V8, V17, **V138**
Acanthastus, **V138**-V139
Acanthograpsus, V42
Acanthograptidae, V7, V26, V28-V29, V36, **V41**
Acanthograptus, V29-V30, V33-V34, V36, **V42**
Acoelothecia, **V17**
ACRANIA, V5
adapertural plate, **V8**
Adelograptus, V33-V34, **V39**, V95-V97
Aellograptus, **V55**
Agetograptus, V131
Airograptus, V32, V39
Akidograptus, V99, **V131**
"Akidograptus" acuminatus Zone, V101
Aletograptus, **V39**
ALLMAN, V7, V22
Allograptus, V105, **V118**
Alternograptus, **V38**
Amphigraptus, V106, **V121**
Amplexograptus, V78, V81-V82, V103, V107, **V125**
Amplexograptus confertus Zone, V105
anagenesis, V102
anastomosis, **V8**, V32
ancora, **V8**
ancora stage, **V8**
angular fuselli, **V8**
anisograptid fauna, V97
Anisograptidae, V7, V24, V28, **V39**, V57, V83, V104-V105
Anisograptus, V31, **V39**, V96-V97
annuli, **V8**, V59
annulus, **V8**
Anomalograptus, **V111**
Anthograptus, **V114**
ANTHOZOA, V6
apertural spine, **V8**, V66-V67
appendix, **V8**
Arachniograptus, **V128**
Archaeocryptolaria, V54-**V55**
Archaeodictyota, V43
Archaeolafoea, **V55**
Archiretiolites, V105, **V130**
Archiretiolitinae, V8, V60, V78, V108, **V130**
Ascograptus, **V55**
aseptate, **V8**

Aspidograptus, **V38**
Atopograptus, **V116**
Atubaria, **V17**
Aulograptus, V63, V75, V96, **V116**
auriculate, **V8**
auriculate group, V65
autotheca, **V8**, V27, V44, V49, V51
Averianowograptus, **V135**
axil, **V8**
AXONOLIPA, V7, V58
axonolipous, **V8**
AXONOPHORA, V7, V58
axonophorous, **V8**
Azygograptus, V106, **V116**

Balanoglossus, V5, **V13**
Balticograptus, V131
BARRANDE, V6, V18
Barrandeograptus, **V135**
BARRASS, V71
BARRINGTON, V5
basal disc, **V9**
BASSLER, V35
BATHER, V150
BECK, V18
BEKLEMISHEV, V25
BEMROSE, V20
bifidus stage, V74-V75
biform, **V9**
bilateral, **V9**
bipolar, **V9**
Birastrites, **V139**
biserial, **V9**
bitheca, **V9**, V28, V45, V50-V51
Bithecocamara, **V50**
Bithecocamaridae, V7, **V50**
"blastozooide inachevé," V14, V22
Bohemograstus, V134, V151
BOUČEK, V35
BOUČEK & PŘIBYL, V89, V156
Boučekocaulis, V42
BOURNE & HEEZEN, V13
Brachiograptus, **V111**
branch, **V9**
branching, dichotomous, **V9**
branching, lateral, **V9**
BROMELL, VON, V22
BRONGNIART, V22
BRONN, V6, V18
Bryograptus, **V39**, V96-V97, V102, V104
Bryograptus hunnebergensis Zone, V101
BRYOZOA, V25
budding individual, V9

BULMAN, V18, V20, V22, V25, V65, V73, V152, V155
Bulmanicrusta, V51-**V52**
Bulmanograptus, V131
buoyancy mechanism, V93
Buthograptus, **V139**

Cactograptus, **V55**
Calamograptus, **V113**
Callodendrograptus, V38
Callograptus, V29, V32, V36, **V38**
Calycotubus, **V48**
Calyptograpsus, V43
Calyxdendrum, V33, V36, **V39**
camara, **V9**, V49
CAMAROIDEA, V7, V17, V34, **V49**, V53
Cameragraptus, **V139**
Campograptus, V132, V152-V155
Capillograptus, V38
Cardiograptus, V69, V81, V96, V99, V102, V104, V106, **V116**
Cardograptus, **V139**
central disc, **V9**
CEPHALOCHORDATA, V5
CEPHALODISCIDA, V7, **V16**
Cephalodiscidae, V7, **V16**
Cephalodiscus, V5, V13-V14, **V17**, V21-V22, V24-V25, V91
Cephalograpsus, V125
Cephalograptus, V62, V78, V92, V99, **V125**
CEPHALOPODA, V22
Ceramograptus, **V55**
Chaunograptus, V51, V54-**V55**
CHORDATA, V5
Choristograptus, V33, V39
cladium, **V9**, V58, V85
cladogenesis, V102
Cladograpsus, V116, V121
CLADOPHORA, V7, V25
classification, V6, V36, V100, V149
clathria, **V9**, V67
Clathrograptus, V130
Clematograptus, V121
climacograptid theca, **V9**
climacograptid type, V63
Climacograptus, V6, V21, V59, V62, V64, V67, V69-V71, V78, V92, V99, V102-V103, V107, **V125**
Climacograptus baragwanathi Zone, V102
C. peltifer Zone, V101-102
Clinoclimacograptus, V63, **V126**
Clonograpsus, V39

Index

Clonograptus, V31, **V39**, V58, V94, V96-V97, V99, V102, V104
C. tenellus Zone, V101
COELENTERATA, V22
Coelograptus, **V55**
coenoecium, V6, **V9**
Coenograptus, V121
collum, **V9**, V49
Colonograptus, V134, V151
colony, **V9**
common canal, **V9**, V61
Comograptus, V125
complete septum, **V9**
Conitubus, V48
Conograptus, **V139**
conotheca, **V9**, V46
Cooper, V101
Coremagraptus, V43
corona, **V9**
corona stage, **V9**
Coronograptus, V132, V154
cortical tissue, **V9**
Corymbites, V134, V151
Corymbograptus, V116
Corynites, V60, V105, **V119**
Corynograptidae, V119
Corynograptus, V119
Corynoideae, V119
Corynoides, V59-V60, V106, **V119**
Corynoididae, V8, V105, **V119**
CORYNOIDINA, V109
Crinocaulis, **V55**
crossing canal, **V9**, V73
CRUSTOIDEA, V8, V17, V26, V34, **V51**, V137
Cryptograptidae, V8, V68, **V123**
Cryptograptus, V63, V73, V79, V85, V92, V99, V107-V108, **V123**
cryptoseptate, **V9**, V62
Ctenograptus, **V139**
Cucullograptus, V23-V24, V65, V68-V89, V100, V102, **V132**, V150, V152
Cucullograptus (Lobograptus) scanicus Zone, V101
Cyclograptus, V46, **V48**
Cymatograptus, V116
Cyrtograpsus, V134
cyrtograptid stage, V89
Cyrtograptidae, V8, V109, **V134**, V149
Cyrtograptinae, V8, V109, **V134**, V149
Cyrtograptus, V86-V88, V93, V95, V100, **V134**, V154, V156
Cyrtograptus centrifugus Zone, V101
C. ellesae Zone, V101
C. linnarssoni Zone, V101
C. lundgreni Zone, V101
C. murchisoni Zone, V101
C. rigidus Zone, V101
Cysticamara, V49-**V50**
Cysticamaridae, V8, **V50**
Cystograptus, V59, V92, V99, **V125**

Cystograptus vesiculosus Zone, V99, V101
Cystoturriculograptus, **V139**
cysts, **V9**, V51

Damesograptus, V38
Damosiograptus, V134
Dawsonia, **V139**
declined, **V9**
Decker, V36
deflexed, **V9**
Demicystograptus, **V139**
Demiothecia, V17
Demirastrites, V132, V154
Dendrograptidae, V7, **V36**
Dendrograptus, V27-29, V32-V33, V35-V36, **V38**
dendroid, **V9**
DENDROIDEA, V6-V7, V17, V21, V23-**V25**, V28, V34-V35, V44, V57, V84, V103
Dendroidea, V6, V25
Dendrotubus, V44, V46-**V48**
dentatus stage, V78
denticulate, **V9**
Denticulograptus, V41
Desmograptus, V29, V32, V35-V36, **V38**
DEUTEROSTOMIA, V6
development, V32, V46, V71
diad budding, **V9**
Dibranchiograptus, **V139**
dicalycal theca, **V9**, V61
Dicaulograptidae, V8, V107, **V128**
Dicaulograptus, V67, V70, V78, **V128**
Dicellograpsus, V121
DICELLOGRAPTA, V109
dicellograptid theca, **V9**
dicellograptid type, V24, V63
Dicellograptus, V64, V69, V76-V77, V79, V93, V95, V99, V106, **V121**
Dicellograptus anceps Zone, V101
D. complanatus Zone, V101-V102
Diceratograptus, V106, V121
Dichograpsus, V114
Dichograpti, V8, **V114**
dichograptid fauna, V97
dichograptid theca, **V9**
dichograptid type, V24, V73-V76
Dichograptidae, V7-8, V24, V69, V75, V97, V99, V103-V105, **V109**
Dichograptidi, V7
DICHOGRAPTINA, V109, V149
Dichograptus, V94, V102, V104, **V114**
dichotomous, **V9**
Dicranograptidae, V7-V8, V58, V106, **V121**
Dicranograptus, V64, V69, V77-V78, V80, V93, V95, V99, V102, V106, **V121**
Dicranograptus clingani Zone, V99, V101
D. hians Zone, V102
Dictyodendron, V38
Dictyograptus, V38

Dictyonema, V21, V28-V29, V31-V36, **V38**, V58, V94-V97, V102, V104
Dictyonema flabelliforme Zone, V101
Didymograpsus, V116
DIDYMOGRAPTA, V109
Didymograpti, V8, **V116**
DIDYMOGRAPTINA, V8, V68, V103, **V109**, V149-V150
Didymograptoides, **V139**
Didymograptus, V24, V63, V67, V72, V75-V76, V96-V97, V102, V104-V106, **V116**
Didymograptus balticus Zone, V102
D. bifidus Zone, V101, V107
D. deflexus Zone, V101
D. extensus Zone, V101, V107
D. hirundo Zone, V101, V105, V107
D. murchisoni Zone, V101
D. nitidus Zone, V101
D. protobifidus Zone, V102
Dimorphograptidae, V7-V8, V108, **V131**
Dimorphograptus, V84, **V131**
Dimykterograptus, V128
Dinemagraptus, V60, V68, **V118**
dipleural, **V9**, V58
Diplograpsis, V125
DIPLOGRAPTA, V123
diplograptid fauna, V99
diplograptid type, V73, V77, V79, V82
Diplograptidae, V7-V8, V57, V68-V69, V97, V107-V108, **V124**
DIPLOGRAPTINA, V8, V58, V103, V107-V109, **V123**, V149-V150
Diplograptus, V22, V59-V60, V71, V92, V99, V102-V103, V106-V107, **V125**
Diplograptus decoratus Zone, V102
D. magnus Zone, V101
D. multidens Zone, V101-V102
Diplospirograptus, V55-**V56**
Diprion, V6, V126
Diprionidae, V6
Discograptus, V46-**V48**
dissepiment, **V9**, V32
distal, **V9**
Dithecodendrum, V36, V54, **V56**
DITHECOIDEA, V17, V54
Dittograptus, V126
diversograptid stage, V88
Diversograptus, V86, V88, V100, **V135**, V149
Dixon, V104
dorsal, **V9**
Dunham, V91
Dyadograptus, V43

Eisel, V154
Eiseligraptus, **V139**
Eisenack, V65, V69, V73, V103

Graptolithina

ELLES, V22, V63, V66, V73, V75-V76
ELLES & WOOD, V7, V18, V20, V109, V150-V151, V154-V156
Ellesicrusta, **V52**
ENTEROPNEUSTA, V6-V7, V12-**V13**
ENTEROPNEUSTI, V13
Eocephalodiscidae, V7, **V16**
Eocephalodiscus, **V16**
Eotetragraptus, V115
Epigraptus, **V48**
Estoniocaulis, **V56**
Etagraptus, V115
everted, **V9**
Expansograptus, V116
expressivity, V66
extensiform, **V9**
Extensograptus, V104
extensus stage, V74-V75

Falcatograptus, **V139,** V156
Fasciculitubus, **V49**
flabellate, **V9**
Flexicollicamara, **V50**
FLORKIN, V21
FRECH, V7, V22, V57-V58
fusellar tissue, **V9**
fuselli, V17

Galeograptus, V45-V46, **V49**
Gangliograptus, 135
Geitonograptus, V120
Geminograptus, **V139**
genicular spine, **V9**
geniculum, **V9,** V63
geographic distribution, V35, V95
gibberulus stage, V74
Gladiograptus, V128
Gladiolites, V128
Globosograptus, V132, V153, V155
Glossograpsus, V122
Glossograptidae, V7-V8, **V122**
GLOSSOGRAPTINA, V8, V58, V103, V106, **V122**
Glossograptus, V67, V74, V79-V80, V85, V92, V99, V107-V108, **V122**
Glossograptus, V126
glyptograptid theca, **V9**
Glyptograptus, V63-V64, V70, V78, V92, V96, V103, V107, **V126**
Glyptograptus-Amplexograptus subfauna, V99
Glyptograptus austrodentatus Zone, V102
G. intersitus Zone, V102
G. persculptus Zone, V101
G. teretiusculus Zone, V99, V101-V102
gonangium, V10
Goniograpti, V8, **V111**
Goniograptus, V82, V85-V86, V94, V96, **V111**
Gothograptus, V61, V83, **V131**
Graptoblasti, V8, V17, **V136**
Graptoblastoides, **V138**

Graptoblasts, V51
Graptoblastus, **V137**-V138
Graptocamara, **V50**
graptogonophores, V59
graptolite affinities, V22
graptolite zones, V100
graptolite zooid, V22
GRAPTOLITHINA, V6-V7, V12, **V17**-V18, V34, V36, V44
Graptolithus, V6, V18, V59, V100, V151
Graptolitidae, V6-V7
Graptolodendrum, V26, V28, V33, **V39**
GRAPTOLOIDEA, V7-V8, V17-V18, V22-V23, V27, **V57,** V84
Graptoloidea, V57
Graptopora, V38
Graptovermida, V8, V17, **V138**
Graptovermis, **V138**
gymnocaulus, **V10,** V57
gymnograptid theca, **V10**
Gymnograptus, V62-V63, V78, V107, **V127**

HABERFELNER, V59
HALL, V18, V20, V22, V34, V89, V91
Hallograptus, V59, V63, V92, V99, V107, **V127**
Haplograptus, **V56**
HARRIS & THOMAS, V97, V104
Hedrograptus, V125
Helicograpsus, V120
HEMICHORDATA, V5, V7, **V12**-V13
Herrmannograptus, V40
HISINGER, V6
HOLM, V18-V20, V28
Holmicrusta, **V52**
Holmograptus, V63, **V118**
Holograptus, **V114**
Holoretiolites, **V131**
hooked type, **V63**
HOPKINSON, V7, V59
horizontal, **V10**
Hormograptidae, V8, **V52**
Hormograptus, V51, **V53**
HUNDT, V139, V156
HUTT & RICKARDS, V19
HYDROIDA, V7
hydrosome, V10
hydrotheca, V10

Idiograptus, V127
Idiothecia, **V17**
Idiotubidae, V7, V44, **V47**
Idiotubus, V46, **V48**
incomplete septum, **V10**
initial bud, **V10,** V73
Inocaulidae, V36, V41
Inocaulis, V36, **V43**
interthecal septum, **V10,** V62
introverted, **V10**
isograptid type, V74-V75, V77
Isograptidae, V105
Isograptus, V75, V80, V86, V104, **V116**

Isograptus caduceus lunata Zone, V102
I. caduceus maximodivergens Zone, V102
I. caduceus victoriae Zone, V102
I. gibberulus Zone, V101
isolate type, V64
isolation, **V10**

JAANUSSON, V85, V107
JAEGER, V89, V101
Janograptus, V85, **V116**
Jiangshanites, **V118**

KAŹMIERCZAK & PSZCZÓŁKOWSKI, V13
Kiaerograptus, V24, V28, V33, **V41,** V75, V96-V97, V104
Kinnegraptus, **V116**
Koremagraptus, V21, V30-V32, V36, **V43**
KOZŁOWSKI, V7, V18, V22, V25, V32, V44, V54, V59, V71, V90-V91, V105, V137
KRAATZ, V22
KRAFT, V18, V20, V59, V71

Labrumograptus, **V139**
lacinia, **V10,** V67-V68
lacuna stage, **V10,** V73
längsverstärkungsleisten, V59
Lagarograptus, V132, V155
languette, **V10**
LANKESTER, V15
lappet, **V10**
LAPWORTH, V6-V7, V18, V20, V22, V35, V69, V91, V149-V151
Lapworthicrusta, **V52**
Lapworthograptus, V135
lasiograptid theca, **V10**
Lasiograptidae, V8, V68, V107, **V126**
Lasiograptus, V63-V64, V69, V90, V107, **V126**
LEGRAND, V33
leptograptid theca, **V10**
leptograptid type, V63, V73, **V75,** V77
Leptograptidae, V7, V119
LEPTOGRAPTINA, V109
Leptograptus, V64, V69, V76, V99, V106, **V121**
Leveillites, V55-**V56**
Licnograptus, **V39**
ligne helicoïdale, V32
Limpidograptus, **V139**
LINNARSSON, V18
LINNÉ, V6, V17, V22
linograptid stage, V89
Linograptinae, V8, V109, **V135,** 149
Linograptus, V59-V60, V86, V88-V89, **V135**
list, **V10,** V67
lobate type, V63
Lobograptus, V65, V68, V89, V102, **V133,** V150, V152
Loganograptus, V94, V104, **V111**

Index

Lomatoceras, V132
Lonchograptus, V107, **V122**
lophophore, **V10**

M'Coy, V18
Maeandrograptus, **V116**
Magdefrau, V13
Marsipograptus, V47
Mastigograptus, **V56**
median septum, **V10**, V62
Mediograptus, V132, V153, V155
Medusaegraptus, V36, **V56**
Megalograptus, **V139**
Melanostrophus, **V53**
mesial, **V10**
Mesograptus, V125
metacladium, **V10**, V88
Metaclimacograptus, **V126**
Metadimorphograptus, V131
metasicula, **V10**, V59
metatheca, **V10**, V61
metatubus, V61
microfusellar tissue, **V10**
microtheca, **V10**, V45
Mimograptus, V104, **V114**
minutus stage, V74-V76
Monoclimacis, V35, V63, V109, **V134**, V150-V151, V155
Monoclimacis crenulata Zone, V101
monofusellar tissue, **V10**
MONOGRAPTA, V132
monograptid fauna, V99
monograptid stage, V88
monograptid type, V66, V73, V79, V84, V204
Monograptidae, V7-V8, V57, V63, V108-V109, V123, V132, **V149**
MONOGRAPTINA, V8, V58, V103, V108-V109, **V132**, V149
Monograptus, V6, V22, V24, V35, V58-V60, V63-V67, V73, V92-V93, V95, V99-V100, V102-V103, V109, **V132**, V149-V151, V153-V156
Monograptus acinaces Zone, V101
M. angustidens Zone, V101
M. atavus Zone, V101
M. bouceki Zone, V101
M. convolutus Zone, V101
M. crispus Zone, V101
M. cyphus Zone, V101
M. gregarius Zone, V101
M. griestoniensis Zone, V101
M. hercynicus Zone, V101
M. leptotheca Zone, V101
M. perneri Zone, V101
M. praehercynicus Zone, V101
M. riccartonensis Zone, V101
M. sedgwicki Zone, V101
M. triangulatus Zone, V101
M. turriculatus Zone, V101
M. uniformis Zone, V12, V101
M. vulgaris Zone, V101
monopleural, **V10**, V58
monopodial growth, **V10**
Monoprion, V6, V122
Monoprionidae, V6
Monsen, V97

morphological terms, V8
morphology, V26, V44, V49, V51, V53, V57
Münch, V61, V65
multiramous, **V10**
multiramous forms, **V8**, V111
Multitubus, V47
Murchison, V6
muscle scars, V59
Mystiograptus, **V139**, V156

Nanograptus, V107, **V122**
nema, **V10**, V57, V69
Nemagrapsus, V120
Nemagraptidae, V7-V8, V63, V106, **V119**
Nemagraptus, V82, V95, **V120**
Nemagraptus-Dicellograptus subfauna, V99
Nemagraptus gracilis Zone, V12, V99, V101-V102
nemata, **V10**
Neodiversograptus, V86, V88, V100, V109, **V135**, V149
Neodiversograptus nilssoni Zone, V101
Nephelograptus, V39
Nereitograptus, **V139**
Nereograptus, **V139**
Neurograptus, **V127**
Neurograptus, V127
Nicholson, V6, V22, V59
Nicholson & Marr, V83, V104
Nicholsonograptus, V105, **V118**
Nodosograptus, **V139**, V156
Nymphograptus, V67, V69, **V128**

Obut, V17, V36, V54, V152
Obut & Sobolevskaya, V155
Obutograptus, V132, V152
obverse, **V10**
occlusion, **V10**
Odontocaulis, V38
Öpik, V35
Oktavites, V132, V151-V152, V154
Oncograptus, V69, V81, V96, V99, V102, V104, V106, **V117**
Ophiograptus, V38
orders, **V10**
Orthoecus, **V17**
orthograptid theca, **V10**
Orthograptus, V59, V67, V70, V72, V78, V90, V99, V103, **V126**
Orthograptus-Climacograptus subfauna, V99
Orthograptus-Dicellograptus subfauna, V99
Orthoretiolites, **V130**
Oslograptus, **V112**

Palaeodictyota, V36, **V43**
paleoecology, V34, V91
Palmotophycus, V36, **V56**
Paracardiograptus, V116
Paraclimacograptus, V125
Paradimorphograptus, **V139**
Paradoxides davidis Zone, V35

Paraglossograptus, V106, **V122**
Paragraptus, **V139**, V156
Paraplectograptus, **V131**
Paratetragraptus, V115
Parazygograptus, V75, **V117**
Pardidymograptus, V118
partial septum, **V10**, V62
Parvitubus, V28, V44, **V49**
pauciramous, **V10**
pauciramous forms, V8, **V115**
pectocaulus, **V10**, V14
peduncle, V13
Peiragraptidae, V8, V107, **V128**
Peiragraptus, V108, **V128**
pendent, **V10**
Pendeograptus, V104, V115
penetrance, V66
pericalycal, **V10**, V74
pericalycal type, V79
periderm, **V10**, V21, V66
Perner, V22
Pernerograptus, V132, V152, V154-V155
Petalograptus, V92, **V126**
Petalolithus, 126
Phormograptus, **V130**
Phycograptus, **V139**
Phyllograpta, V38
Phyllograptidae, V7
Phyllograptus, V69, V82, V93, V102, V104, **V116**
phylogeny, V103
Pipiograptus, **V130**
Planctosphaera, V12
PLANCTOSPHAEROIDEA, V6-V7, **V17**
Planktograptus, **V139**
platycalycal, **V11**, V73
Plectograptinae, V8, V61, V78, V108, **V130**
Plectograptus, **V130**
Plegmatograptus, V69, **V130**
Pleurograpsus, V121
Pleurograptus, V82, V96, V106, **V121**
Pleurograptus linearis Zone, V101-V102
Počta, V35
Polygonograptus, **V56**
polymorphic, **V11**
POLYZOA, V22, V25
Pomatograptus, V132, V150
porus, **V11**, V72
Prantl, V35
preoral lobe, **V11**
Přibyl, V154-V155
Pribylograptus, V132, V155
Pristiograptus, V21, V72, V109, **V134**, V150-V151
Pristiograptus fecundus Zone, V101
P. ludensis Zone, V101
P. transgrediens Zone, V101
P. tumescens Zone, V101
P. ultimus Zone, V101
procladium, **V11**, V88
Procyrtograptus, **V139**
Prolasiograptus, V126
prosicula, **V11**, V59

prosoblastic, **V11**, V78
protheca, **V11**, V61
prothecal fold, **V11**, V61
Protistograptus, **V139**
Protograptus, **V139**
Protohalecium, **V57**
Protovirgularia, **V139**
proximal, **V11**
Pseudazygograptus, V116
Pseudobryograptus, **V112**
Pseudocallograptus, V29, **V38**
pseudocladium, **V11**, V89
Pseudoclimacograptus, V63-V64, V69, V78, V92, V98-V99, V103, V107, **V126**
Pseudodichograptus, V105, **V118**
Pseudodictyonema, V29, V35, **V39**
Pseudoglyptograptus, **V126**
Pseudoplegmatograptus, **V129**
Pseudoretiolites, V128
Pseudotrigonograptus, V116
pseudovirgula, **V11**
Pseudozygograptus, V106, V116
Psigraptus, **V41**
PTEROBRANCHIA, V6-V7, V12-**V13**, V22
Pterobranchites, **V17**
Pterograptus, **V112**
Ptilograptidae, V7, **V41**
Ptilograptus, **V41**
Ptiograptus, V32, **V39**
Ptychodera, **V13**

quadriserial, **V11**

Radiograptus, **V41**
Ramulograptus, V115
Rastrites, V64-V65, V67, V96, V102, **V134**, V150
Rastrites maximus Zone, V101
Rastrograptus, V134
Raymond, V93
reclined, **V11**
Rectograptus, V126
reflexed, **V11**
regeneration, V70
Reteograptus, **V130**
Reticulograptus, V36, V44, V46-**V47**
reticulum, **V11**
Retiograptus, V90, V130
Retiolites, V69, V83, V102, **V128**
retiolitid type, V83
Retiolitidae, V7-V8, V68, V107-V108, **V128**
Retiolitinae, V8, V78, V108, **V128**
Retioloidea, V7
retroverted, **V11**
reverse, **V11**
Rhabdinopora, V38
RHABDOPHORA, V7, V57
Rhabdopleura, V5, V12-**V15**, V21-V22, V24-V27, V45, V57
RHABDOPLEURIDA, V7, **V14**, V23, V51
Rhabdopleuridae, V7, **V15**
Rhabdopleurites, **V16**
Rhabdopleuroides, **V16**
rhabdosome, **V11**, V17, V46, V82

Rhadinograptus, **V57**
Rhaphidograptus, **V131**
Rhipidodendrum, V33-V34, **V39**
Rhizograpsus, V38
Rhizograptus, V38
Rhodonograptus, V48
Rickards, V100
Rickards & Rushton, V155
root, **V11**
Rouvilligraptus, V114
Ruedemann, V7, V18, V22, V32, V34-V36, V58-V59, V67, V90, V93-V94, V106
Ruedemannicrusta, **V52**
Ruedemannograptus, **V57**

Saccoglossus, **V13**
Saetograptus, V67, V71, V84, **V134**, V151
Saetograptus fritschi linearis Zone, V101
S. leintwardinensis Zone, V101
S. lochkovensis Zone, V101
Sagenograptus, **V39**
Salter, V18, V22
Sargassum, V35, V93-V94
scalariform, **V11**
scandent, **V11**
Schepotieff, V14, V22
Schizograpti, V8, **V114**
Schizograptus, V82, V104, **V114**
Schlotheim, von, V22
Schmidt, V93
Schraubenlinie, V32, V59, V71
sclerotized, **V11**
scopulae, V11
selvage, **V11**
semitubus, V61
septal, **V11**
septum, **V11**
Siberiodendrum, V36, **V57**
Siberiograptus, **V57**
sicula, **V11**, V57, V59
Sigmagraptus, **V112**
Sinodiversograptus, V88, **V135**
Sinograptidae, V8, V105, **V118**
Sinograptus, V63, V105, **V118**
Sinostomatograptus, **V130**
sinus, V73
sinus stage, **V11**
Skevington, V32, V62, V85
Skiagraptus, V80, V85, V104, **V117**
Skoglund, V19
Soergel, V13
solid axis, V11
Spencer, V35
Sphenoecium, V55, **V57**
Sphenophycus, V35
Sphenothallus, V57
Spinograptus, **V131**
Spinosidiplograptus, **V139**
Spirograptus, V132, V152, V154
Staurograpsus, V41
Staurograptus, V35, **V41**, V95, V102
Stavrites, V134, V151
Stelechocladia, V29, V38
Stelechograptus, **V139**

Stellatograptus, **V112**
Stephanograptus, V120
stipe, **V11**, V17
Størmer, V35
stolon, **V11**
stolon system, V50
Stolonodendridae, V8, **V53**
Stolonodendrum, **V53**
STOLONOIDEA, V8, V17, V34, **V53**
stolotheca, **V11**, V26, V44, V51
Stomatograptus, **V130**
STOMOCHORDA, V12
Strachan, V86, V109
stratigraphic distribution, V35, V96
streptoblastic, **V11**, V78
Streptograptus, V57, V132, V153, V155-156
Strøm, V93
Strophograptus, **V139**
Stubblefield, V33
Sudbury, V66, V151-V152, V154-V155
sympodial growth, **V11**
Syndyograptus, V106, **V121**
synrhabdosome, **V11**, V89
Syrrhipidograptus, V34, V38

Tangyagraptus, V121
Taphrhelminthopsis, V13
Tavener-Smith, V25
Teller, V86
Temnograpti, V8, **V113**
Temnograptus, **V113**
Testograptus, V132, V153-V154
Tetragrapsus, V115
Tetragrapti, V8, **V115**
Tetragraptus, V75, V77, V84, V86, V94-V95, V99, V102, V104-V105, **V115**
Tetragraptus approximatus Zone, V102
T. fruticosus Zone, V102
Tetraprionidae, V6
Thallograptus, V36, **V43**
Thallograptus, V53
Thamnograptus, **V139**
theca, **V11**, V26, V44, V60
thecal grouping, **V12**, V28
thecal segment, V61
thecatubus, V11
Thecocystograptus, **V139**
thecorhiza, **V12**
Thorsteinsson, V62, V86
Thuringiagraptus, **V139**
Thysanograptus, V126
Törnquist, V20, V61, V154, V156
Törnquist & Hadding, V18
Tremadoc Series, V12
triad budding, **V12**
Triaenograptus, **V112**
triangulate theca, **V12**
triangulate type, V64
Trichograptus, V82, **V112**
Tridensigraptus, V112
Trigonograpsus, V116, V139
Trigonograptus, V139

Index

Trimerohydra, V43
Triograptus, **V41,** V96-V97
Triplograptus, **V139**
Tristichograptus, V98-V99, **V116**
Trochograptus, V104, **V114**
Tubicamara, **V50**
Tubidendridae, V7, V44, **V47**
Tubidendrum, V44-V45, **V47**
TUBOIDEA, V7, V17, V34, **V44**
tunicates, V5
twig, **V12**
Tylograptus, V105, **V118**
Tyrsograptus, V132, V152

ULRICH & RUEDEMANN, V22, V59
umbellate theca, **V12**, V45

Undograptus, **V139**
uniserial, **V12**
Uralograptus, V135
URBANEK, V20, V22, V25, V60, V62, V65-V67, V71, V86, V88-V89, V100, V109, V149-V151
UROCHORDATA, V5

ventral, **V12**
vesicular diaphragm, **V12**
virgella, **V12**, V59
virgellarium, **V12**, V59
virgula, **V12**, V57

WAERN, V62

WAHLENBERG, V6, V18, V22
WALKER, V20, V60, V62, V151
WETZEL, V22
WHITTINGTON & RICKARDS, V67
WHITTINGTON & WILLIAMS, V12
WIMAN, V7, V18, V20, V22, V104
Wiman rule, **V12**, V26, V57
Wimanicrusta, **V52**
Wimanicrustidae, V8, **V52**

Yushanograptus, **V112**

zooid, **V12**, V22
Zygograptus, **V112**